MODERN TRENDS IN MACROMOLECULAR CHEMISTRY

Modern Trends in Macromolecular Chemistry

Jon N. Lee
Editor

Nova Science Publishers, Inc.
New York

NOTICE TO THE READER

The Publisher has taken reasonable care in the preparation of this book, but makes no expressed or implied warranty of any kind and assumes no responsibility for any errors or omissions. No liability is assumed for incidental or consequential damages in connection with or arising out of information contained in this book. The Publisher shall not be liable for any special, consequential, or exemplary damages resulting, in whole or in part, from the readers' use of, or reliance upon, this material.

Independent verification should be sought for any data, advice or recommendations contained in this book. In addition, no responsibility is assumed by the publisher for any injury and/or damage to persons or property arising from any methods, products, instructions, ideas or otherwise contained in this publication.

This publication is designed to provide accurate and authoritative information with regard to the subject matter covered herein. It is sold with the clear understanding that the Publisher is not engaged in rendering legal or any other professional services. If legal or any other expert assistance is required, the services of a competent person should be sought. FROM A DECLARATION OF PARTICIPANTS JOINTLY ADOPTED BY A COMMITTEE OF THE AMERICAN BAR ASSOCIATION AND A COMMITTEE OF PUBLISHERS.

LIBRARY OF CONGRESS CATALOGING-IN-PUBLICATION DATA

Modern trends in macromolecular chemistry / [edited by] Jon N. Lee.
  p. ; cm.
Includes bibliographical references and index.
ISBN 978-1-60741-252-6 (hardcover)
1. Macromolecules. I. Lee, Jon N.
[DNLM: 1. Polymers--chemistry. 2. Chemical Phenomena. QD 381 M6895 2009]
QD381.M58 2009
547'.7--dc22
          2009008028

Published by Nova Science Publishers, Inc. ✦ *New York*

CONTENTS

PREFACE

Macromolecular chemistry is the study of the physical, biological and chemical structure, properties, composition, and reaction mechanisms of macromolecules. A macromolecule is a molecule that consists of one or more types of repeated 'building blocks'. The building blocks are called monomeric units (monomers).Macromolecules (also known as polymer molecules) appear in daily life in the form of plastic, styrofoam, nylon, etc. These polymers, i.e., substances consisting of polymer molecules, are of great technological importance and are used in the manufacturing of all sorts of goods, from automobile parts to household appliances. The artificial polymer molecules usually exist of long repetitions of identical monomers, either in chains or networks.

Chapter 1 - The review provides an insight into developments concerning the various applications of phosphorus containing polymers. Organophosphorus polymers have been widely studied as such kind of polymers exhibited very interesting properties. They can improve adhesive properties *via* phosphonic groups which provided, for instance, multiple attachment sites on metallic surfaces leading to monomolecular adlayers. They also found industrial applications for anticorrosion and were used in biomedical field as drug carriers or in dental composites and as dentin adhesives. Phosphoric acid, phosphonic acid and phosphonic acid ester groups increased biocompatibility and adhesion to tooth by chelating with calcium in the tooth. They can also be used because of their fire proofing resistance. Indeed, phosphorus formed a protective layer of char and poly(phosphoric acid) derivatives during the combustion which was swollen by the products of degradation. This layer inhibited heat and oxygen transfer into the polymer bulk, diminishing the diffusion of combustible gases. Phosphorus polymers were also involved in proton-conducting fuel-cell membrane as the enhancement of the acidity of the phosphonic acid units proved to be beneficial for solubility, water uptake, and proton conductivity. As a consequence, phosphonated polymers were anticipated as potential proton exchange membranes for possible high-temperature fuel-cell applications, which enhanced the kinetics of the electrode reactions and reduced the risk of catalyst poisoning.

To conclude, organophosphorus polymers demonstrated to be involved in several industrial applications in various field making them particularly interesting.

Chapter 2 - The results obtained by investigation of the influence of synthesis procedure on the molecular structure of aliphatic hyperbranched polyesters (AHBP) are presented in this work. Different pseudo generations of these three-dimensional polymers were synthesized via an acid-catalyzed polyesterification of 2,2-bis(hydroxymethyl)propionic acid (bis-MPA) and di-trimethylolpropane (DiTMP) using two different procedures: pseudo one-step (samples of the series I) and one-step (samples of the series II). In pseudo one-step procedure the sample of certain pseudo generation was obtained starting from the sample of previous pseudo generation and adequate amount of monomer, while in simple one-step procedure polymer of the wanted pseudo generation was synthesized by mixing the stoichometric amount of all necessary reactants at once. Beside these samples, three commercial Boltorn® AHBP synthesized by pseudo one-step procedure were also investigated. Results obtained by NMR, MALDI-TOF MS and VPO show that formation of poly(bis-MPA) and cyclization through the ester and ether bonds (intramolecular) occurred as side reactions during the both procedures for the synthesis. With increasing degree of polymerization, the extent of these unwanted reactions increased and it was slightly higher when one-step procedure was applied for the preparation of AHBP samples. During the synthesis of AHBP from second till fifth pseudo generation, cyclization occurred through ester and ether bonds (intramolecular), while in the case of the higher generation samples, cycles were formed due to the intramolecular esterification. Furthermore, –OH groups belonging to the terminal units had better reactivity than linear ones. Consequently, the degree of branching for all investigated samples is lower than 0.50. Due to the side reactions occurrence, molar mass increases only up to the sixth pseudo generation and values of the molar masses are much lower than theoretical. From the obtained results it was also possible to calculate the percentage of deviation from the perfect dendrimer structure for AHBP samples of the series I and II.

Chapter 3 - This chapter discusses recent progress in the development of metallopolyyne polymers of platinum(II) with a vast range of triplet and bandgap energies. Considerable focus is placed on the design concepts toward the structural tuning of the optical bandgap and triplet state of this family of functional metallopolymers. Various applications of such class of metal-containing macromolecules in optoelectronic and photonic devices are critically discussed. The future challenges and perspectives of this research area are also highlighted.

Chapter 5 - Organic-inorganic hybrid gels have been synthesized by means of a hydrosilylation reaction of multiple Si-H functional crosslinking monomers with bi-functional vinyl spacer monomers using a Pt catalyst. We select 1,3,5,7-tetramethylcyclotetrasiloxane (TMCTS), tetrakis(dimethylsilyloxy)silane (TDSS), or silsesquioxane, 1,3,5,7,9,11,13,15-octakis(dimethylsilyloxy)pentacyclo-[9,5,1,1,1,1]octasilsesquioxane (POSS) as the crosslinking monomers. Two series of the bi-functional spacer monomers are applied in the organic-inorganic hybrid gels. One is divinyl or diallyl monomers containing Si, divinyldimethylsilane (VMS), diallyldimethylsilane (AMS), diallydiphenylsilane (APS), 1,3-divinyltetramethyldisiloxane (VMSO), or 1,4-bis(dimethylvinylsilyl)benznene (VMSB), and the other is bi-functional polar monomers, 1,6-hexanediol divinylether (HDVE), 1,6-hexanediol diacrylate (HDA), and 1,6-hexanediol dimethacrylate (HDMA). Network structures of the resulting gels have been quantitatively characterized by means of a novel scanning microscopic light scattering system. Both the structures of the crosslinking monomers and spacer monomers affect the critical gelation concentration (minimum monomers concentration which attains formation of the gel), reaction time for formation of the gels, and network structure of the gels.

Chapter 6 - Enantioselective catalytic reactions in which the chirality of an asymmetric catalyst induces the preferred formation of a given product enantiomer have been one of the most important achievements in chemistry in the 20th century. Numerous enantioselective catalytic reactions like C-H, C-C, C-O, C-N, and other bonds have been discovered. In recent years, much research has been devoted on catalytic asymmetric C-C bond formations using various chiral catalysts. Chiral products obtained by asymmetric C-C bond formation reactions are important intermediates for pharmaceuticals, vitamins, agrochemicals, flavors, fragrances and functional materials. Quite often, the cost of sophisticated chiral ligands exceeds the cost of noble metals employed in chiral catalyst. Therefore, catalyst recovery is of paramount importance for the application of enantioselective metal catalysis to large scale processes. This has led researches looking for recoverable and recyclable chiral catalysts. In recent years, inorganic materials have been extensively used as alternative supports for the heterogenization of chiral catalysts due to their excellent thermal and chemical stability, ease of handling and diversity of chemical modifications. Moreover, a large variety of preparative procedures are available leading to materials with diverse morphological properties. They can also be used at both high and low temperatures and at high pressure. Their inflexibility and noncompressibility make silica-tethered catalysts suitable for continuous-flow reactors. In this chapter, we have focused our attention on asymmetric C-C bond formations using chiral metal complexes of 'privileged chiral ligands' as catalysts tethered to inorganic materials by covalent and non-covalent interactions. Our aim is to highlight the great potential and diversity of applications, of immobilized chiral catalysts on inorganic supports in asymmetric C-C bond formations and to discuss the factors of heterogeneous chiral catalysts that influence the asymmetric catalytic performance under heterogeneous reaction conditions.

Chapter 7 - We described here the preparation, characterization, and application of the smart hybrid nanoparticles based on the stimuli-responsive PEGylated nanogels containing metal nanoparticles such as gold nanoparticles (AuNPs) and platinum nanoparticles (PtNPs). The stimuli-responsive PEGylated nanogels composed of cross-linked poly[2-(N,N-diethylamino)ethyl methacrylate] (PEAMA) gel core and poly(ethylene glycol) (PEG) tethered chains bearing a functional group (acetal group or carboxylic acid group) as a platform moiety for the installation of bio-tag were synthesized by emulsion polymerization of vinylbenzyl group-ended heterotelechelic PEG, amine monomer, and cross-linker. The resulting PEGylated nanogels showed unique reversible volume phase transition of the polyamine gel core in response to various stimuli, such as pH, ionic strength, and temperature. Note that cross-linked PEAMA gel core of the PEGylated nanogels acts as not only nanoreactor but also stimuli-responsive nanomatrix to produce and immobilize AuNPs and PtNPs. For instance, one-pot synthesis of pH-responsive PEGylated nanogels containing AuNPs was successfully carried out through the self-reduction of $HAuCl_4$ without any reducing agents. The PEGylated nanogel containing AuNPs (fluorescence quencher) in the core and fluorescence dye-labeled DEVD peptide at the tethered PEG chain end showed the recovery of pronounced fluorescence signals in response to apoptotic cells. Additionally, the PEGylated nanogels containing PtNPs showed the on-off regulation of the catalytic activity for reactive oxygen species in response to pH. Thus, smart hybrid nanoparticles based on stimuli-responsive PEGylated nanogels containing metal nanoparticles described here can be utilized to novel functional biomedical nanomaterials.

Chapter 8 - In this chapter, we review the preparation methods of the polymers with well-defined nanostructures in the polymerization field. Generally, nanostructures of the synthetic polymers have been controlled by various external forces and intermolecular interactions using previously prepared polymeric materials. On the other hand, the methods to control nanostructures of the polymeric materials during the polymerization reaction have been limited.

As the examples of these methods, we first describe the preparation of amylose-synthetic polymer inclusion complexes by the polymerization of α-D-glucose 1-phosphate catalyzed by phosphorylase in the presence of synthetic polymers, i.e., the formation of inclusion complexes was achieved during amylose-forming polymerization. As the second topic, we describe the investigation of a parallel enzymatic polymerization system to form inclusion complexes. When two enzymatic polymerizations, i.e., the aforementioned amylose-forming polymerization and the lipase-catalyzed polycondensation of dicarboxylic acids and diols leading to the aliphatic polyesters, were simultaneously performed, the inclusion complexes composed of amylose and strongly hydrophobic polyesters were produced. We also describe the preparation of synthetic cellulose with controlled crystalline structure as the third topic, which was achieved by enzymatic polymerization of β-cellobiosyl fluoride monomer using cellulase as a catalyst. Finally, we describe the preparation of crystalline nanofiber of linear polyethylene by the polymerization of ethylene using mesoporous silica fiber-supported titanocene and methylalumoxane as a cocatalyst. These methods allow the control of nanostructures of the polymeric materials during the polymerization reaction.

Chapter 9 - In this study, a detailed investigation on the application of the Differential Scanning Calorimetry (DSC) in measuring polymerization kinetics is presented. The advantages of this technique are highlighted and the limitations are explained. Use of this method in a wide diversity of polymerization reactions resulting in linear (i.e. poly(methyl methacrylate), polystyrene, etc.), branched (poly(vinyl acetate), poly(butyl acrylate)) and crosslinked (formed from bis-phenol A glycidyl dimethacrylate, urethane dimethacrylate and triethylene glycol dimethacrylate) macromolecules is explored. Polymerization techniques such as in bulk or solution are also included.

The main advantages of the DSC include the use of a small sample mass, which assures that the reaction will be carried out isothermally during the whole conversion range and especially in the autoacceleration region. In advance, the measurements are continuous and not in discrete time intervals and reactions difficult-to-study with other techniques, such as those leading to crosslinked macromolecules can be equally well investigated. Finally, one of the greatest advantages of DSC is that it provides a direct measure of instantaneous rate of polymerization rather than conversion. Conversion histories are readily obtained from integration of the raw data. This process is inherently more accurate than evaluating rates from the slope of the conversion curve. Among the limitations of the method could be mentioned the non-adequate removal of oxygen in some cases and the possible loss of monomer during reaction due to evaporation at relatively high temperatures.

Chapter 10 - In recent years the interest to the selective trimerization and tetramerization of ethylene has grown significantly. In this review paper we discuss the current state-of-the-art of the experimental mechanistic studies of the catalytic systems for trimerization and tetramerization of ethylene based on chromium and titanium compounds.

Chapter 11 - Polyoxometalates (POMs) are a class of unique inorganic oxide macromolecules with nanosized geometry. Recently, organoimido derivatives of POMs have received increasing interest in the supramolecular chemistry and chemistry of materials, since they are valuable building blocks for constructing novel nanostructured organic-inorganic hybrid molecular materials with the so-called ''value-adding properties'' and possible synergistic effects, including unique catalytic, photo and electronic properties dramatically different from the corresponding parent materials, which results from the π electrons in the organic component of such molecules may extend their conjugation to the inorganic framework and dramatically modify the electronic structure and redox properties of the corresponding parent POMs. The surface modification of POMs, that is, the substitution of one or more terminal or bridged oxo groups with organoimido ligand, is the fundamental method to synthesize organoimido derivatives of POMs via common organic chemistry reactions. In this chapter of the book, the synthetic chemistry of organoimido derivatives of POMs using different organic reagents will be reviewed, of which the novel DCC-dehydrating protocol to prepare organoimido derivatives of POMs recently developed by us and co-authors will be concentrated. As it can be expected, the reaction chemistry of organoimido derivatives of POMs stands for the fascinating future of the chemistry of organoimido derivatives of POMs since it opens not only a new road to the chemical modification of POMs, but also an exciting research arena where a variety of hybrid materials containing covalently bonded POM clusters and organic conjugated segments can be prepared in a more controllable and rational manner. Hence, in the rest of this chapter, Pd-catalyzed coupling reactions of the functionalized organoimido derivatives of POMs will be introduced in detail for this purpose. Especially, different rationally assembled hybrid macromolecules and polymers where the organic molecules and inorganic clusters are bonded together by an imido nitrogen atom will then be highlighted.

In: Modern Trends in Macromolecular Chemistry
Editor: Jon N. Lee

ISBN: 978-1-60741-252-6
© 2009 Nova Science Publishers, Inc.

Chapter 1

ORGANOPHOSPHORUS POLYMERS: APPLICATIONS AND NEW TRENDS

S. Monge and J.J. Robin*

Institut Charles Gerhardt Montpellier, Equipe Ingénierie et Architectures
Macromoléculaires, Université Montpellier II, Montpellier, France

ABSTRACT

The review provides an insight into developments concerning the various applications of phosphorus containing polymers. Organophosphorus polymers have been widely studied as such kind of polymers exhibited very interesting properties. They can improve adhesive properties *via* phosphonic groups which provided, for instance, multiple attachment sites on metallic surfaces leading to monomolecular adlayers. They also found industrial applications for anticorrosion and were used in biomedical field as drug carriers or in dental composites and as dentin adhesives. Phosphoric acid, phosphonic acid and phosphonic acid ester groups increased biocompatibility and adhesion to tooth by chelating with calcium in the tooth. They can also be used because of their fire proofing resistance. Indeed, phosphorus formed a protective layer of char and poly(phosphoric acid) derivatives during the combustion which was swollen by the products of degradation. This layer inhibited heat and oxygen transfer into the polymer bulk, diminishing the diffusion of combustible gases. Phosphorus polymers were also involved in proton-conducting fuel-cell membrane as the enhancement of the acidity of the phosphonic acid units proved to be beneficial for solubility, water uptake, and proton conductivity. As a consequence, phosphonated polymers were anticipated as potential proton exchange membranes for possible high-temperature fuel-cell applications, which enhanced the kinetics of the electrode reactions and reduced the risk of catalyst poisoning.

To conclude, organophosphorus polymers demonstrated to be involved in several industrial applications in various field making them particularly interesting.

---

* Correspondence to: S. Monge - e-mail: Sophie.Monge-Darcos@univ-montp2.fr

1. INTRODUCTION

Phosphorus is an easily available element coming from chemistry of natural phosphate and beside the industry of fertilizers, the obtention of organophosphorus compounds such as monomers is an attractive challenge for chemists. As a result, the phosphorus chemistry has been investigated for a long time but many advances in organophosphorus polymers only appeared in last decades. These polymers have been widely studied as they exhibited very peculiar and interesting properties. The different chemical environments of phosphorus made possible the synthesis of various products in the range of phosphate, phosphonate and phosphinate based compounds. Different main strategies have been followed by researchers to introduce phosphorinated moities in polymers: either the (co)polymerization of monomers bearing phosphorus atom or the grafting of phosphorus-based group onto a polymer. Beside the ester forms which are the most available compounds, monoacids and diacids can be reached and open the way to a wide range of polymers and properties. It must be noticed that the conversion from ester to acid form is sometime difficult to achieve since treatment with strong acids (which is the most convenient method) may degrade other functions of the molecules. In this case, mild conditions are required and the use of trimethyl bromosilane is preferred.

The ionization potential value of phosphonic acids R-PO(OH)$_2$ is intermediate between that of sulfonic and carboxylic acid ones due to their intermediate pKa and these acids are known for their complexing properties. They can improve adhesive properties *via* phosphonic groups which provided, for instance, multiple attachment sites on metallic surfaces leading to monomolecular adlayers. This metal attachment ability allowed industrial applications (anticorrosion, adhesion, biomedical). Phosphoric acid, phosphonic acid and phosphonic acid ester groups increased biocompatibility and adhesion to teeth by chelating with calcium onto enamel. Recently, the development of free halogen materials due to novel legislation gave them new opportunities in fire proofing retardancy. At last, the growing interest in alternative energy production gave a new challenge to phosphorus-based polymers and they have been involved in proton-conducting fuel-cell membranes as the enhancement of the acidity of the phosphonic acid units proved to be beneficial for solubility, water uptake, and proton conductivity. As a consequence, phosphonated polymers were anticipated as potential proton exchange membranes for possible high-temperature fuel-cell applications, which enhanced the kinetics of the electrode reactions and reduced the risk of catalyst poisoning.

This article provides an insight into recent developments concerning the large variety of applications of phosphorus containing polymers. The reader will find here the main applications of these peculiar polymers and mainly academic results are reported.

2. INTERACTIONS OF ORGANOPHOSPHORUS POLYMERS WITH METALS

Synthetic chelating agents are used in industrial applications because of their aptitude to bind metal [1-3]. Among all possible structures, organophosphonates exhibit interesting complexing properties [4,5] and have already been used as dispersants and corrosion inhibiting agent, for preventing deposit formation [6], or to bond metal ions, for instance.

2.1. Phosphorus-Based Materials as Complexing Agents

The presence of heavy metal ions in the environment is a major concern due to their toxicity to many life forms. Generally, removal of metal traces from water is achieved using synthetic chelating agents. The latter have already been used for many industrial applications as they can bind numerous metal ions. The use of polymers in metal ion separations has been developed over the past thirty years [7,8] and a wide range of polymers have been studied for metal ion complexations.

Among all possible chemical structures, organophosphonate group presented complexing characteristics and was mainly used for processes requiring chelation or for preventing scale formation [4]. Affinity of phosphonates moieties for metal ions in aqueous solutions have been well described in the literature [2,9] and it was shown that phosphonate groups exhibited variable affinity for various metals. The polarity of phosphonyl bond confers Lewis base properties upon the oxygen, resulting in the ability to chelate inorganic metal salts. Heavy metal ions can be removed by adsorption on solid carriers (polymer microbeads) bringing specific phosphonated sorbent (chelating agent) which interacts with the metal ions specifically. For example, phosphonate esters and phosphonates/phosphonic acid functionalized on styrene-divinylbenzene copolymers (with 7% of divinylbenzene) were prepared and evaluated for the sorption capacity of Cu^{2+}, Ca^{2+}, and Ni^{2+} from aqueous solutions [10]. Furthermore, it was shown that such chelating resins could retain ions depending on the nature of the functionalization (phosphonate esters or phosphonic acid groups). Adsorption performance depended on the number of phosphonate groups introduced in the polymeric resins. It is also interesting to notice that quaternary phosphonium salts grafted on styrene-divinylbenzene copolymers exhibited anti-bacterial activity [11].

Metal ion selectivity studies were performed with styrene-based polymer beads prepared *via* suspension polymerization bringing phosphinic acid, phosphonic acid, dimethylamine/phosphonic acid, phosphonate monoethyl ester, and phosphonate diethyl ester groups [12]. Among all results, the phosphinic acid polymer had the highest level of recognition for the mercuric ion across a wide range of conditions, and the phosphonate diester/monoester polymers interacted with silver ions. They also showed complexation with other metal ion solutions ($Fe(NO_3)_3$, $Mn(NO_3)_2$, $Zn(NO_3)_2$). Absorption of uranium was evaluated using phosphorylated resin diethyl polystyrene-methylenephosphonate [13] or poly(phosphonosiloxane) [14]. The latter dissolved and chelated metal salts like uranium nitrate ($UO_2(NO_3)_2$) forming a metal salt-polymer complex where two phosphoryl groups were used for chelation. The salt crosslinked the poly(diethylphosphonobenzyl-α,β-ethylmethylsiloxane) and resulting complex was found to be a homogeneous solution at room temperature. Other siloxane-based monomer (3-diethylphosphonatepropylmethyldimethoxy-silane) was polymerized and preliminar results showed that resulting phosphonated polysiloxane readily chelated metal salts, including uranyl nitrate, nickel chloride and copper acetate [15]. Finally, it was shown that poly(acrylaminophosphonic-carbonyl-hydrazide) fiber could be used as chelating agent. This developed fibrous sorbent presented high adsorption capacities for many metal salts such as Cu^{2+}, Pb^{2+}, Zn^{2+}, Co^{2+}, Ni^{2+}, Hg^{2+}, Cd^{2+}, Mn^{2+}, Cr^{3+}, and Ag^+. This adsorption was more or less influenced by the pH value.

The capability of phosphorus compounds to bind metal ions was also used to produce ionomers that are polymers which contain ionic groups able to lead to crosslinked networks [16] and used as viscosity modifiers [17]. The synthesis of telechelic phosphonic ionomers based on polyether or aromatic polyethylene terephthalate oligomers was achieved using a telomerization reaction [18]. Phosphonic acid groups were incorporated at the end of the polymer chain (halato ionomers). The ionic crosslinking was produced with different cations (Zn^{2+}, Ca^{2+}, Na^+) and the characterization of the network proved that an efficacious crosslinking was reached depending on the nature of the cations. Copolymerization of styrene and diethylvinylphosphonate followed by hydrolysis of the phosphonate ester provided ionomers by neutralization with sodium and zinc salts [19]. It was shown that the crosslinks were multifunctional and/or that ionic crosslinking led to significant trapped chain entanglements. The Zn-salts were more resistant to flow in the low strain than Na-ones.

It is also interesting to notice that phosphonated polymers were directly used in the preparation of ionomers (without metal salts). For example, the synthesis and characterization of isobutylene-based ammonium and phosphonium bromide ionomers was accomplished, leading to network resulting from ion-pair aggregation [20]. The dynamic mechanical analysis, solution viscosity and ^{1}H NMR relaxation studies showed that a thermoplastic network was established whose strength depended on temperature and the presence of selective plasticizers. Another work described in the literature dealt with the preparation of liquid-cristalline ionomers, synthesized by polycondensation of malonates substituted by bromide or phosphonates ester groups with mesogenic diols. Phosphonic acids were used as negative groups, and phosphonium and ammonium ions as positive groups [21]. The presence of the ionic groups led to physical crosslinking (gelation) of the polymers which displayed a rubbery consistency.

2.2. Phosphorus-Based Materials as Adhesion Promoters

Among the different ways to protect metals against corrosion, organic coatings appear to be the most important one. The main property of a corrosion protective coating is its adhesion in the presence of water or high humidity. Organophosphorus polymers have been involved in corrosion protective coatings which require maintenance of adhesion under environmental exposure. Adhesion between galvanized steel plates and polymer depends on the chemical structures of both substrates and coating and reduction of adhesion causes water penetrating at the coating/metal interface, leading to a significant reduction of adhesion. Incorporation of phosphonic groups, known as adhesion promoters, into polymer structures allowed the improvement of adhesion properties of polymers on metallic surfaces [22]. Such properties are of great interest for automotive paint.

Free-radical graft copolymerization of phosphonated (meth)acrylates onto polymers was completed using dimethyl-2-(meth)acryloxyethylphosphonate or corresponding mono or diacid derivatives (Scheme 1). Grafting of phosphonated acrylates onto PVDF powders [23] was achieved using ozone treatment to oxidize the polymer surface, and the resulting polymer was then hydrolyzed. The adhesion of the graft copolymers applied to galvanized steel substrates was studied. Results showed that the presence of phosphonic acid groups led to an significant improvement in the adhesive bond to metal compared to both dialkylphosphonate and carboxylic acid groups. Same kind of experiments were realized with low-density polyethylene powders [24] using phosphonated methacrylates following two distinct processes: grafting in the presence of free-radical initiators, and thermally induced graft copolymerization onto ozone-pretreated low density polyethylene.

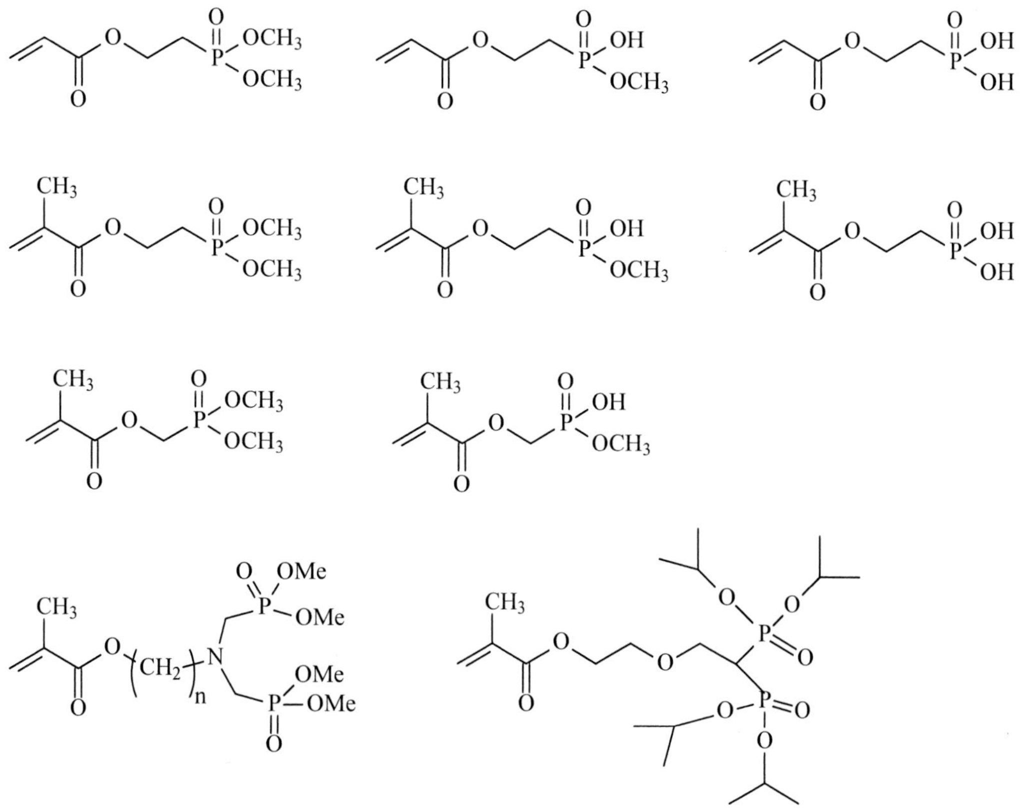

Scheme 1. Organophosphorus monomers used for anticorrosive and adhesion properties.

Emulsion polymerization was also used to produce a series of methyl methacrylate, butyl acrylate, and phosphonated methacrylates copolymers varying the nature of phosphonated methacrylates (diester, monoacid, diacid). Such kind of polymers should be used to improve the characteristics of anticorrosive formulations [25]. Other phosphonated additives were prepared by a reversible addition fragmentation chain transfer (RAFT). The living radical terpolymerization of vinylidene chloride, methyl acrylate and dimethyl-2-(meth)acryloxyethylphosphonate yielded a gradient terpolymer which was then hydrolyzed. The latter was used as polymeric additive in coating formulations based on a poly(vinylidene chloride-*co*-methyl acrylate) copolymer matrix [26-29]. Different formulations were spun cast on stainless steel surfaces and the coatings were observed by electron microscopy coupled with X-ray analyses. Phosphonic acid groups induced preferential organization of the coating as the additive segregated and migrated toward the metal inferface. As a consequence, the matrix at the air interface acted as a barrier to gas while the additive ensured adhesion at the polymer/metal interface.

Blends of organophosphorus copolymers with PVDF powders were also investigated to enhance adhesive and anticorrosive properties of fluoropolymers coatings [30]. Statistical copolymers were first synthesized using methyl methacrylate, and dimethyl-(2-methacryloyloxyethyl)phophonate, (2-methacryloyloxyethyl)phosphonic diacid or methyl(2-methacryloyloxyethyl)phosphonic hemiacid. Resulting copolymers were introduced into PVDF as adhesion promoters and anticorrosion inhibitors. Good dry and wet adhesion properties onto galvanized steel plates were obtained with blends containing high contents of phosphonic acid groups. A salt spray exposure test also showed that the phosphonic acid groups prevented the spread of corrosion at the PVDF coating-galvanized steel interface. More recently, a new phosphonated methacrylate, namely dimethyl(methacryloyloxy)methyl phosphonates, with only one methylene spacer, was prepared according to Pudovik reaction [31]. It was then copolymerized with methyl methacrylate and hydrolyzed. When blended with PVDF and coated onto a galvanized steel plate, the copolymer led to very good adhesion in both wet and dry state and provided good anticorrosion properties. At last, innovative multifunctional monomers where synthesized by David et al [32]. These molecules were incorporated into photopolymerizable mixtures of (meth)acrylates monomers and showed after UV hardening a good adhesion onto steel and an excellent resistance to corrosion under salt spray.

3. BIOMEDICAL APPLICATIONS

Synthesis of functional and reactive polymers is of great interest as polymers with highly reactive groups are attractive materials for biomedical applications. Among all possible structures, polyphosphate derivatives appeared to be very interesting as they are biodegradable through hydrolysis and can possibly be degradable by enzymatic digestion of phosphate linkages under physiological conditions [33]. They are biocompatible and present structural similarities to the naturally occurring nucleic and teichoic acids. Poly(alkyl-phosphonate)s are of great interest for biomedical applications too, notably with the synthesis of block copolymers to achieve poly(alkyl-phosphonate)-based adlayers which proved to give excellent results in terms of stability over a wide range of pH and an excellent resistance to protein adsorption when exposed to full human serum [34]. Phosphonated (meth)acrylic monomers were used in dental applications due to their high reactivity under UV radiations and also found applications in regenerative medicine which requires scaffolds of divergent physicochemical properties for different tissue-engineering applications.

3.1. Phosphorus-Based Materials for Drug Delivery

Polymers with phosphoester (P-O-C) repeating linkages in the backbone were used in drug delivery systems (Table 1). Poly(oxyethylene H-phosphonates) [35], used as precursor polymers, are particularly attractive members of the polyphosphoester family due to their biocompatibility, controlled biodegradability and the presence of a highly reactive P-H group in the building units of the polymer chain. They are water soluble polymers with low toxicity which can be functionalized to be used as drug carriers [36-38].

Immobilization of aminothiols on poly(oxyalkylene phosphates) was achieved. It has been shown that the incorporation of cysteamine, a well-known clinically tested chemical radioprotector, into poly(oxyethylene phosphate) led to a significant reduction of cysteamine toxicity and that the drug dose needed for efficient radio protection was also reduced [39]. Poly(oxyethylene H-phosphonate)s and poly(oxyethylene phosphate)s polymeric protective agents were also evaluated with other radioprotectors such as amifostine or 1-(3-aminopropyl)aminoethanethiol [40]. Fixation of other pharmacologically active amines on polyphosphonates was completed [41]. Hydrosoluble poly(3,6,9-trioxaundecamethylene phosphonates) with attached cytotoxic bis(2-chloroethyl)amine in the side chain was synthesized by chemical modification of the corresponding polyphosphonate *via* the Atherton-Todd reaction. Preliminary *in vitro* cytotoxic studies revealed that polymeric derivative exhibited an eight fold more cytotoxic potency against human hepatocellular carcinoma cell line HepG2 than against murine leukemia cell line L1210.

Poly(oxyethylene H-phosphonates) also allowed the synthesis of polyphosphoesters bearing P-H or P-OCH₃ groups in the main chain and 1,3-dioxolan-2-one rings or hydroxyurethane fragments attached to the polymer backbone through a P-C bond [42]. Resulting polymers presented characteristics such as biodegradability and versatile reactivity that enabled attachment of bioactive compounds. This was achieved with 2-phenylethylamine as model compound. Biodegradable copolymers containing both phosphonates and lactide ester linkages in the polymer backbone were investigated too. Paclitaxel delivery from novel polyphosphoesters were evaluated [43] and proved to be continuous in vitro and in vivo over at least 60 days. Paclitaxel released from microspheres had significant antitumor activity against various human cancer cells in murine models.

The potential of biodegradable polyphosphoester microspheres from poly(bis(hydroxyethyl)terephthalate-ethyl ortho-phosphorylate/terephthaloyl chloride (P(BHET-EOP/TC) notably was also evaluated for controlled delivery of neurotrophic proteins to a target tissue in order to treat various deseases of the nervous system [44]. It was proved that sustained release of nerve growth factor for a prolonged period from microspheres loaded into synthetic nerve guide conduits might improve peripheral nerve regeneration [45]. Poly(phosphoester)s were as well involved in the fabrication of nerve guide conduits due to their high biocompatibility, adjustable biodegradability, flexibility in coupling fragile biomolecules under physiological conditions, and a wide variety of physicochemical properties [46,47].

Table 1. Chemical structure of organophosphorus polymers used as drug carriers.

Entry	Chemical structure of polyphosphoesters used as drug carriers	Ref
1		[36]
2		[38]
3	$R = OH, OCH_3$	[39, 40]

4

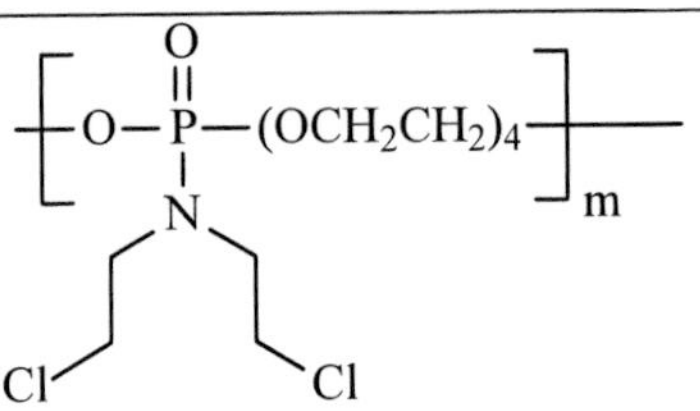

$$\left[O-\overset{\overset{\displaystyle O}{\|}}{\underset{\underset{\displaystyle N}{|}}{P}}-(OCH_2CH_2)_4 \right]_m$$

with the N bearing two CH_2CH_2Cl groups

[41]

5

$$CH_3O-\overset{\overset{\displaystyle O}{\|}}{\underset{\underset{\displaystyle (CH_2CH)_xH}{|}}{P}}-O-R\left[O-\overset{\overset{\displaystyle O}{\|}}{\underset{\underset{\displaystyle (CH_2CH)_xH}{|}}{P}}-O-R\right]_m\left[O-\overset{\overset{\displaystyle O}{\|}}{P}-O-R\right]_n-O-\overset{\overset{\displaystyle O}{\|}}{\underset{\underset{\displaystyle R'}{|}}{P}}-OH$$

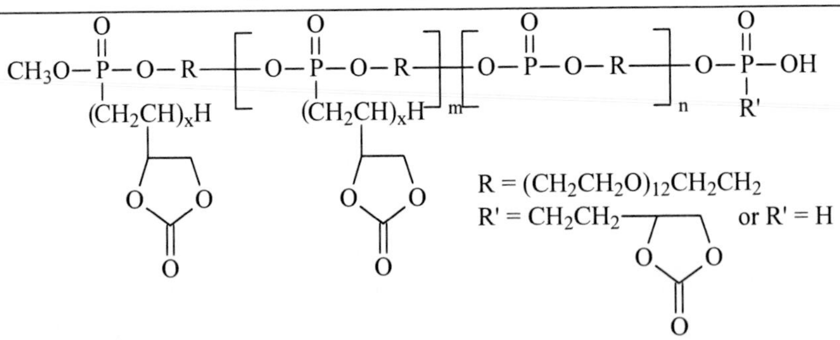

$R = (CH_2CH_2O)_{12}CH_2CH_2$

$R' = CH_2CH_2$—(cyclic carbonate)   or R' = H

[42]

6

$$-\left[OCH_2CH_2O-\overset{O}{\overset{\|}{C}}-\text{(benzene)}-\overset{O}{\overset{\|}{C}}-OCH_2CH_2O-\underset{\underset{\displaystyle OEt}{|}}{P}\right]_x\left[OCH_2CH_2O-\overset{O}{\overset{\|}{C}}-\text{(benzene)}-\overset{O}{\overset{\|}{C}}-OCH_2CH_2O-\overset{O}{\overset{\|}{C}}-\text{(benzene)}-\overset{O}{\overset{\|}{C}}\right]_y-$$

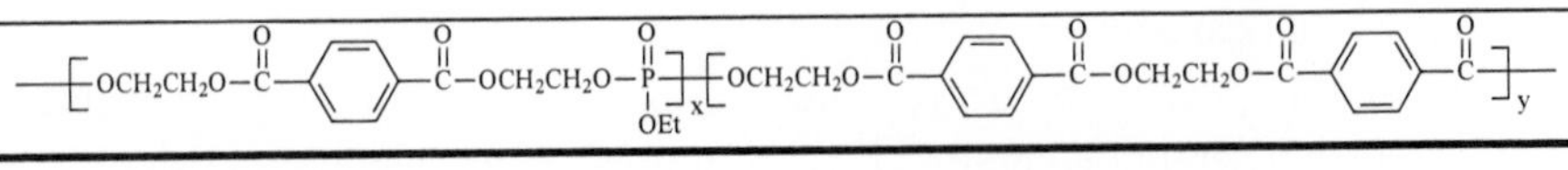

[44]

3.2. Phosphorus-Based Materials from 2-Methacryloyloxyalkyl Phosphorylcholine for Blood Compatility

When polymeric materials are in contact with living organisms, serious biological reactions such as thrombus formation and unfavorable immunoresponses are observed, inducing many problems in the treatment of patients using such biomedical devices. In such case, infusion of an anticoagulant during clinical treatments is necessary to avoid clot formation. Therefore, blood compatibility appears to be an very important parameter that must be taken into consideration. In that context, polymers having a phospholipid polar group (Scheme 2) were expected to have a strong affinity for phospholipid molecules, leading to the obtaining of a biomembrane-like structure.

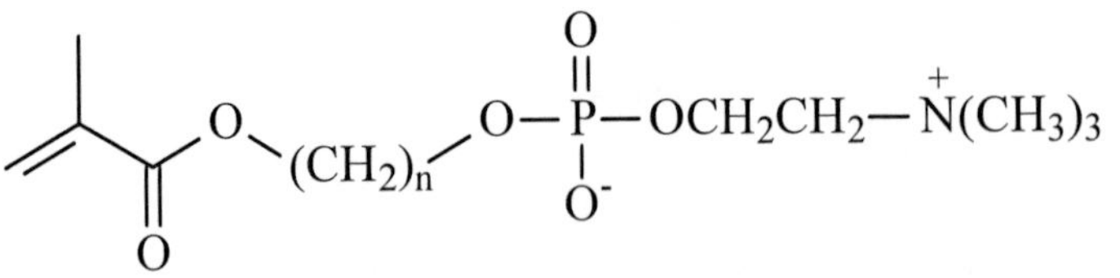

$$\text{(methacrylate)}-O-(CH_2)_n-O-\overset{\overset{\displaystyle O}{\|}}{\underset{\underset{\displaystyle O^-}{|}}{P}}-OCH_2CH_2-\overset{+}{N}(CH_3)_3$$

$n = 2$ MPC
$n = 6$ MHPC
$n = 10$ MDPC

Scheme 2. Chemical structure of monomers involved in materials for blood compatibility.

Important researches were focused on the use of a methacrylate with a phospholipid polar group in the side chain named 2-methacryloyloxyethyl phosphorylcholine (MPC) to achieve materials with blood compatibility and anti-thrombogenicity properties as polymers derived from MPC have an artificial surface which is similar to a biomembrane one. When the MPC polymers come into contact with blood, phospholipids are preferentially adsorbed on the surface rather than proteins and cells. The phospholipids could form a self-assembled biomimetic membrane structure on the surface. Statistical copolymers with both MPC and alkyl methacrylates (*n*-butyl, *tert*-butyl, *n*-hexyl, *n*-dodecyl, and *n*-stearyl groups) were synthesized [48]. Adhesion of blood cells on the MPC copolymers after contact to the whole blood was strongly influenced by MPC composition and chemical structure of the copolymer. Among all copolymers, poly(MPC-*co*-*n*-butyl methacrylate) exhibited excellent blood compatibility as shown by reduction of platelet aggregation, and suppression of protein adsorption [49,50]. Such copolymers could be used for the modification of cellulose hemodialysis membrane, polyolefin membrane for oxygenator, and covering membrane for implantable biosensor [51]. Graft polymerization of 2-methacryloyloxyethyl phosphorylcholine on polyethylene was achieved and proved to be effective for preventing platelet adhesion [52]. Other monomers quite similar to MPC such as 6-methacryloyloxyethyl phosphorylcholine (MHPC) and 10-methacryloyloxydecyl phosphorylcholine (MDPC) (Scheme 2) were also grafted to polyethylene and plasma protein adsorption and fibroblast cell adhesion were evaluated taking into account the chemical structure, surface density, mobility, and orientation of the grafted poly(ω-methacryloyloxyalkyl phosphorylcholine) [53]. Grafting of MPC onto polyvinylidene difluoride and microfiltration membranes after plasma etching was also reported [54] in the literature. Improvement in membrane performance was observed, concomitant with a reduction in protein fouling on the surface and within the matrix of the coated membranes. Finally, the last architecture prepared was amphiphilic block copolymers composed of poly(butyl acrylate) and poly(2-methacryloyloxy phosphorylcholine) reached by reversible addition fragmentation transfer polymerization (RAFT) [55]. Light scattering studies showed the self-assembly into micelles which could be suitable nanocontainers for biomedical applications such as controlled drug delivery if RAFT agent proved to be non toxic.

3.3. Phosphorus-Based Materials in Dental Applications

Dental materials based on (meth)acrylic monomers are favored polymers for dentists due to their high reactivity under UV radiations. Their high polymerization rate is greatly modified in the presence of oxygen but also leads to some structure faults in the polymerized materials. The shrinkage occurring all along the polymerization process of monomers under thermal or UV activation conducts to the formation of cracks in the materials and also at the interface between dentine (or enamel) and the polymeric composite. This phenomenon allows the passage of liquids, bacteria... in the gaps leading to some ageing (discolorations, hydrolysis...) of the dental restoration. In some cases, bacteria included in the cracks generated at the joint rupture can initiate new caries. To avoid these damages, researchers tried to develop new polymeric compositions with no or low shrinkages during polymerization and with better adhesion onto enamel or dentin. The first technique was based on acid-etching of enamel using strong acids like phosphoric acid [56]. The treatment results in the formation of microporosities making the surface rough and a micromechanical based adhesion was obtained [57,58].

Since the developments of new techniques using polymers in dental restoration, phosphorus containing polymers have been largely used. One of the problems to be solved is the adhesion of polymeric matrixes onto hydroxyapatite. The most often used compounds are phosphoric acid esters, phosphonic esters or phosphonic acids. Fu et al [59] showed using FTIR analyses that phosphoric acid esters were able to decalcify hydroxyapatite (HA) and to form two different complexes with calcium ions [60]. This phenomenon occurred simultaneously with the formation of chemical bonding with HA. As a result, etching and chemisorbing led to a better adhesion in respect to that of obtained when using carboxylic groups. The most used and studied simple monomer in the area of adhesive dental resins is represented in scheme 3 [61,62].

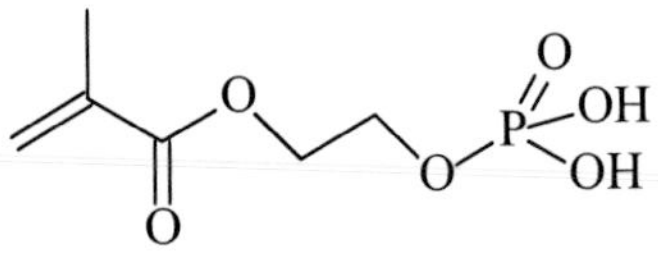

Scheme 3. Most popular monomer used as additive for adhesion.

The principle based on the coupling of a chemical function leading to adhesion onto minerals like HA with a polymerizable monomer has been largely used in dental adhesives mainly with (meth)acrylic monomers [63,64] or styrenic ones [65]. In these cases, phosphorus atom is directly linked to a hydrocarbonated chain (phosphonate ester) making this monomer more resistant to hydrolysis contrary to phosphoric acid esters (Scheme 4).

with R = H or alkyl group

Scheme 4. Example of phosphonic acid esters used as adhesion promoters.

These monomers often called coupling agents include acidic functions (carboxylic, dicarboxylic, carboxylic anhydride, phosphonic …). They are copolymerized with usual components like 2,2'-bis[4-(2-hydroxy-3-methacrylyl-oxypropoxy)phenyl]propane (Bis GMA), triethylene glycol dimethacrylate (TEGDMA) or 1,6-bis(methacryloxy-2-ethoxycarbonylamino)-2,4,4-trimethylhexane (UEDMA) [66].

More recently, some other monomers have been designed and present more complex chemical structures [67,68] (Scheme 5).

The hydrolysis of the phosphonic esters in phosphonic acids resulted in an increased reactivity of these monomers in some cases. This result was not clearly explained but supposed to be attributed to higher hydrogen-bonding in the case of acidic groups in respect to ester groups. Chains termination could decrease as a result of viscosity rise due to these bonds [67,68] and also to a pre-organization of molecules leading to double bonds close to each other, increasing polymerization rates. This observation based on kinetical data was in accordance with results published by Robin et al [69] concerning photopolymerized varnishes. This sight has been largely studied by Avci et al. who synthesized different phosphonic acid based monomers [70-74]. The authors studied the effect of the position of the $PO(OH)_2$ in the molecule, and the presence of steric hindrance effect on the polymerization rate. Nevertheless, some contradictory results were published [75] with very closed structures [76] (Scheme 6) where the authors found lower conversion. In these cases, no kinetical data (like polymerization rates) were given and only final conversion yields were cited.

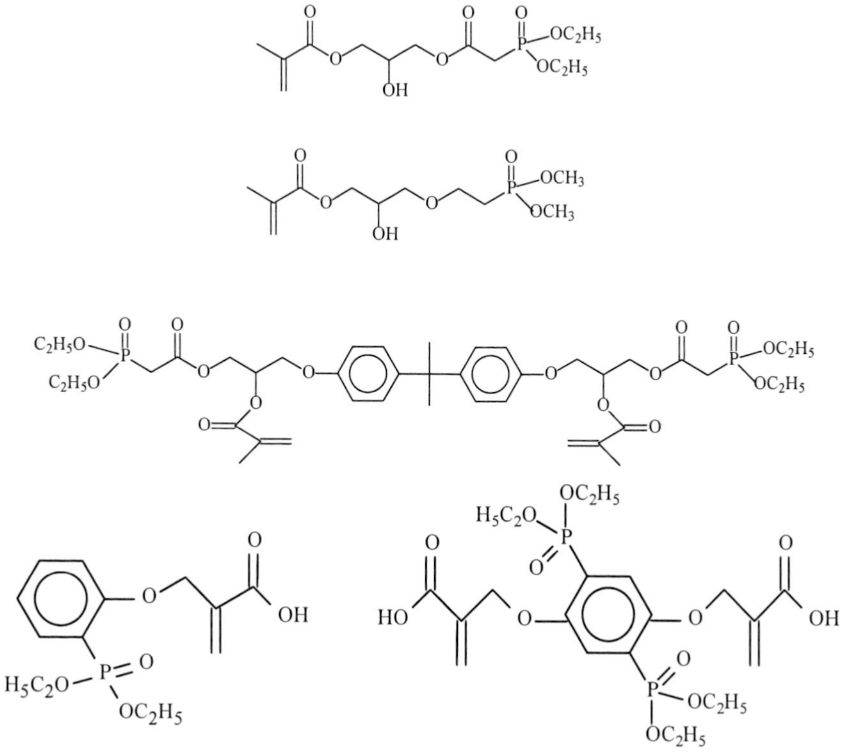

Scheme 5. Some recent chemical structures proposed as adhesion promoters.

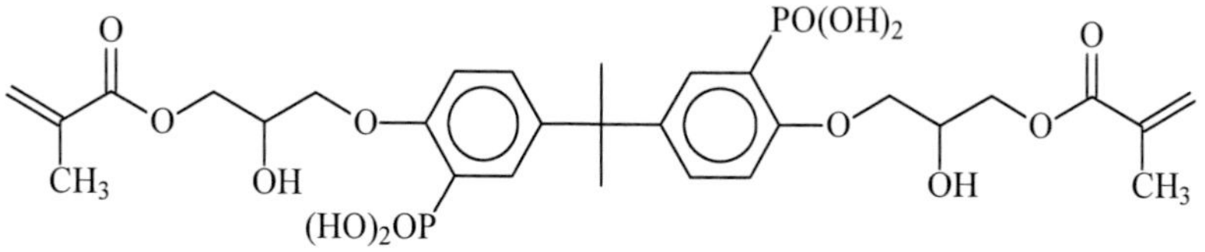

Scheme 6. Example of polymeric matrix including functions favoring adhesion.

More recently, Madec et al [77] proposed new dimethacrylic monomers including sulfur atoms but to date, no kinetical data were provided for these original structures (Scheme 7).

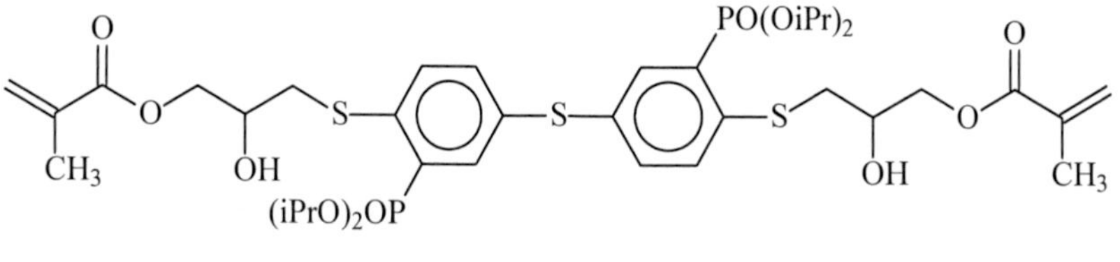

Scheme 7. New dimethacrylic monomers including sulfur atoms and phosphonic acid esters.

The same authors proposed a new class of monomers (Scheme 8) stable to hydrolysis and reactive under radical way. These compounds have been introduced in self-etching primers and showed good adhesion between composites and dentin [78].

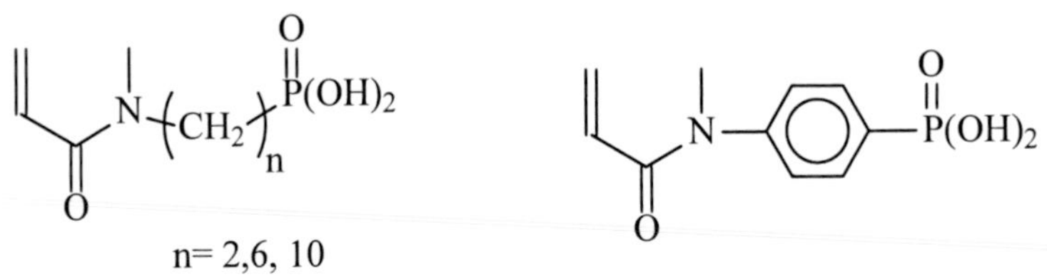

n= 2,6, 10

Scheme 8. Acrylamides used for dental applications.

Usually, phosphonic acid monomers are obtained from the conversion of the corresponding phosphonic esters. Vinyl phosphonic acid (VPA) is a very interesting monomer since it is directly commercialized in its acidic form and does not require the hydrolysis of phosphonic esters. Whereas its homopolymerization is difficult to achieve, its copolymerization with other monomers is suitable to reach high molecular weight polymers. Some authors tried to incorporate VPA in usual methacrylic systems. They observed that only low quantities (inferior to 0.25 w. %) could be added in formulations since high quantities resulted in low monomer conversion and non optimized mechanical properties [79].

Dental professionals have been trying for a long time to substitute metallic amalgams used in the old techniques of dental restoration. One of the different ways dealt with the use of glass-ionomer based materials. These ones resulted from the reaction of water soluble polymers (mainly water-soluble polyalkenoic acids) with minerals, usually containing calcium, silicium, aluminium elements plus fluorine, generally in the form of an acid decomposable glass. This acid-base neutralization reaction was most often achieved in water. (Meth)acrylic or itaconic acids were preferred acids as well as maleic anhydride. These materials involved numerous inter and intramolecular salts brigdes making these materials very rigid albeit they are plastified by water. To be exhaustive on the subject, it must be noticed that some authors developed crosslinked glass ionomers using unsaturations. This technique has been recently reviewed by Culbertsson [80] who pointed out the new developments in the area. Other acids such as vinyl phosphonic acid (VPA) have been tested. Phosphonic acid was stronger than carboxylic ones toward metal oxides and the obtained cements were more hydrolytically stable due to the very strong cation-anion bond [79,81-86]. At last, phosphonate containing polymers have been used in the formulation of whitening dentifrice or in the goal of inhibiting chemical staining of teeth. In that case, the phosphonate groups acted as calcium sequestrant.

3.4. Phosphorus-Based Materials for Tissue Engineering

Tissue engineering typically involves the seeding of biodegradable polymeric scaffolds with differentiated or pluripotent cells *in vitro*, followed by implantation of the cell-scaffold construct into the region of tissue loss or damage. Polymeric materials, especially phosphonated polymers, are of great interest in the field of tissue engineering and scaffolding.

As a matter of fact, researchers tried to promote interactions of biomaterials with bone cells and the protein interaction was favored when using phosphonated grafted polymeric surfaces. So, vinyl phosphonic acid with comonomers such as acrylamide were grafted by photopolymerization onto surfaces and showed osteoblast-like cell adhesion as well as proliferation. Methacrylate-based crosslinked hydrogels were synthesized and showed very interesting properties in water transport mechanism for medical applications [87-93]. Poly(bis(hydroxyethyl)terephthalate-ethyl   ortho-phosphorylate/terephthaloyl   chloride (P(BHET-EOT/TC)) were found to possess good physicochemical and film-forming properties. This poly(phosphoester) was ionically crosslinked with calcium ions and ionomer derivatives proved to be useful as an intermediate for the conjugation of peptides or other amino-containing entities [94].

4. PHOSPHORUS-BASED MATERIALS IN FLAME RETARDANCY

One of the main drawbacks of the usual polymeric materials is their flammability and the emission of very hazardous gases and smokes during combustion. Only few polymers are able to self-estinguish. Until the last years, the most often used additives were halogen-based organic chemicals. Recent accidents due to fire in buildings revealed that many people died of suffocation with toxic acidic gases emitted by the combustion of the formulated materials. So, recently, many researches have been conducted in the development of new halogen-free flame retardants. Phosphorus is well known to have a positive action in the flame retardance in combination or not with other elements like nitrogen, sulphur, metallic particles... Phosphorinated compounds can be added either as additives in the material or directly grafted onto the polymer backbone. In the first case, these added compounds can drastically modify the mechanical properties of the final material. These additives can act in the vapour phase or in the condensed phase, leading sometimes to a char or a protective coating (also named intumescent char) avoiding any oxygen transport toward the burning area and giving fire extinguishing.

However, the design of new phosphorinated halogen-free retardant additive is not an easy task. Many parameters have to be taken into account. First of all, no universal compound exists and each additive is efficacious with one polymer type. This observation comes from the degradation mechanisms differing from one polymer to another. Furthermore, some polymers are capable to self-generate chars when burning while others do not have this property. In the case of char formation, some phosphorus-based compounds can be easily integrated in this char and have a positive action in the formation of this porous and protective foamed material.

The characterization of the behaviour of burning materials has been extensively improved by the use of techniques like thermogravimetric analysis, cone calorimetry, pyrolysis gas chromatography/mass spectrometry, gas chromatography/Fourier Transform Infrared Spectroscopy, Nuclear Magnetic Resonance, etc... in addition to the most usual method based on the Limit Oxygen Index calculation. Nevertheless, in the most often studies, only macroscopic effects are described like ignition time since the fine determination of the char composition or of the degradation mechanism is difficult to reach. Fortunately, final users only need to have data on the efficiency of the additives.

In the following part, the use of different phosphorinated polymers as additives is described in relation with the structure of the (co)polymers in which they are incorporated. It must be pointed out that most of phosphorous based additives are molecular compounds and not polymers or oligomers. These molecular compounds are not detailed in this review focused only on macromolecular compounds.

4.1. Polystyrene and Related

Styrene based homopolymers or copolymers such as HIPS (High Impact modified Polystyrene), ABS (Acrylonitrile-Butadiene-Styrene based blends), SAN (Styrene-Acrylonitrile copolymers) are versatile materials making them favourite candidates for applications like construction materials (foams…), insulators for wires and cables, injection molded articles for automotive. Many uses require flame retardance properties that can be reached either by copolymerization of phosphorinated monomers with styrene or by adding oligomers or polymers as additives.

Brossas and Camino studied different approaches in the flame retardance. They showed that Polystyrene (PS) terminated with chloromethoxyphosphonate groups exhibited flame retardance effect ascribed to the formation of P-O-P linkages and to an increase in molecular weights during degradation, limiting the volatilization of short molecules [95-99]. Some other authors tried to copolymerize (meth)acrylic monomers bearing phosphonate or phosphate groups [100,101] (Scheme 9).

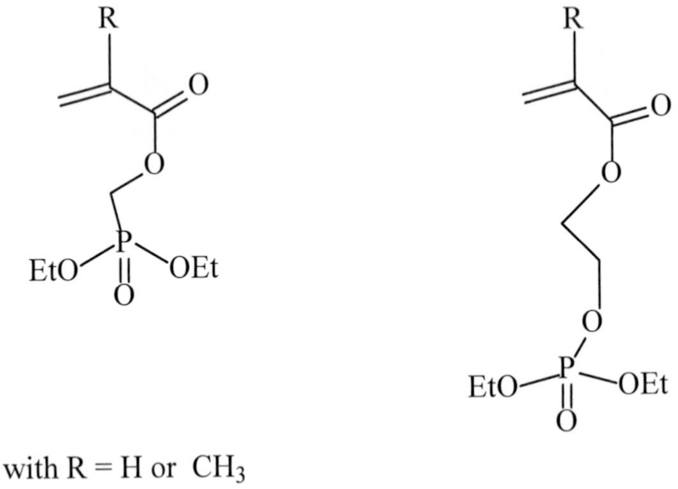

with R = H or CH$_3$

Scheme 9. Examples of simple molecules used as comonomers for flame retardancy.

These studies clearly showed that the chemical structure of phosphorus atom plays a major role in the flame retardance and that phosphonate groups are more active than the corresponding phosphate groups. Furthermore, phosphorinated groups linked to the polymer backbone and obtained by copolymerization have a low effect on mechanical and physical properties of the final material in comparison with phosphorinated compounds added as additives to PS where a decrease of these properties is obtained. The additives act in the vapor phase whereas phosphorinated copolymers are active both in the vapor and condensed phase. Some other authors developed new monomers permitting subsequent crosslinking of polystyrene during the degradation step [102] (Scheme 10).

Ebdon et al [103] synthesized a series of new monomers (vinylics, allylics, methacrylics) and evaluated their behaviour in styrene-based copolymers. All copolymers produced char and showed condensed-phase mechanism.

Polyphosphazene are well known for their high thermal stability. Some new monomers bearing phosphate groups were tested as additives in polystyrene (Scheme 11). The cyclic tetramer exhibited good effects on char formation underscoring the action in the condensed phase [104,105].

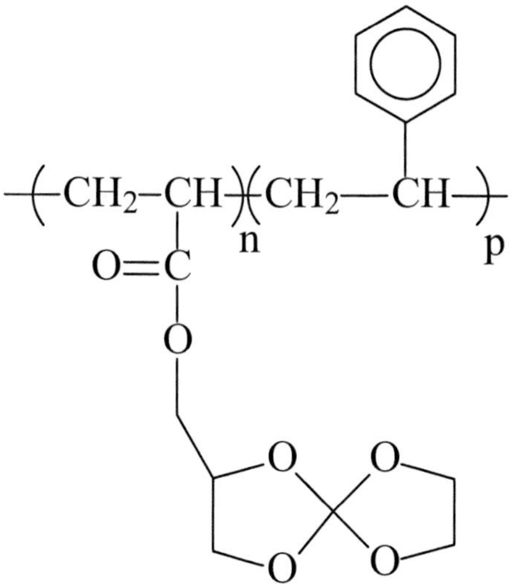

Scheme 10. Example of copolymer leading to cross-linking during thermal degradation.

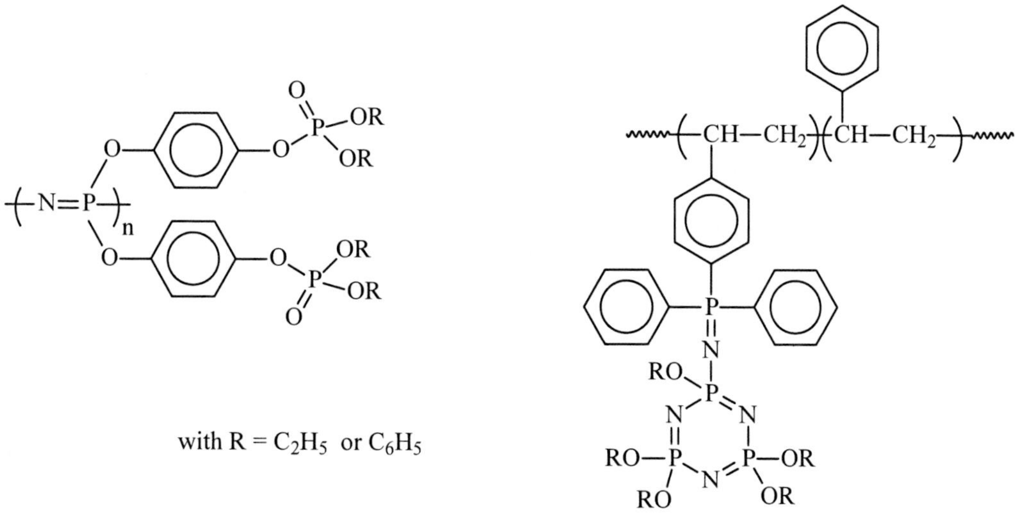

Scheme 11. Phosphazene-based polymers used as char promoters.

Other stable monomers were tested as comonomer of styrene [106]. Maleimide are well known to easily copolymerize with styrene and their thermal stability has been extensively demonstrated in the past. So, phosphate based-maleimide were synthesized (Scheme 12) and their fire behavior revealed a high char formation.

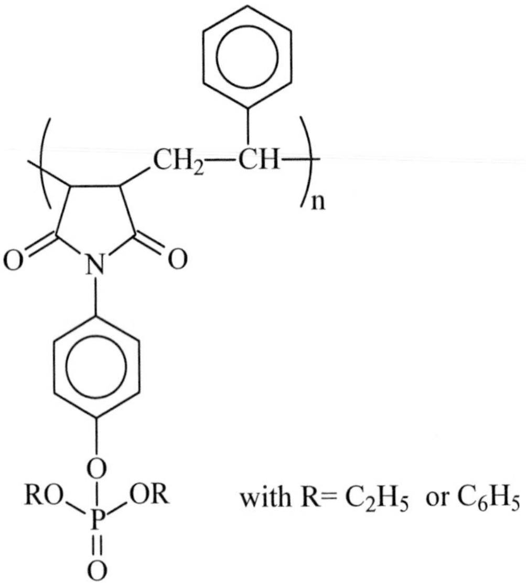

with R= C_2H_5 or C_6H_5

Scheme 12. Maleimide based polymers used as char promoters.

It must be mentioned that some authors combined phosphorinated polymers and minerals in the form of nanocomposites (most often clays). Nowadays, nanocomposites greatly expanded due to their very interesting properties such as flame retardancy which is one of the most famous and versatile one [107]. The difficulty stays in the linking of clays with polymers and researchers developed copolymers bearing anchoring functions and phosphate groups [108]. Copolymers bearing both quaternary and phosphate groups were designed and revealed reduction of the peak heat release making them good candidates for flame retardant polymers. Finally, the reader may find more information on the entire techniques devoted to flame retardance of polystyrene in the review of Weil and Levchik [109].

4.2. Polymethyl methacrylate and Related

Nowadays, meth(acrylic)-based polymers are of great interest and the improvement of their behaviour under heating or when burning is a great challenge. Most of phosphorus containing polymers used in the case of flame retardancy of poly(methyl methacrylate) (PMMA) are phosphonate groups containing methacrylate. Price and Ebdon [110] studied extensively the properties and efficiency of monomers represented on scheme 13.

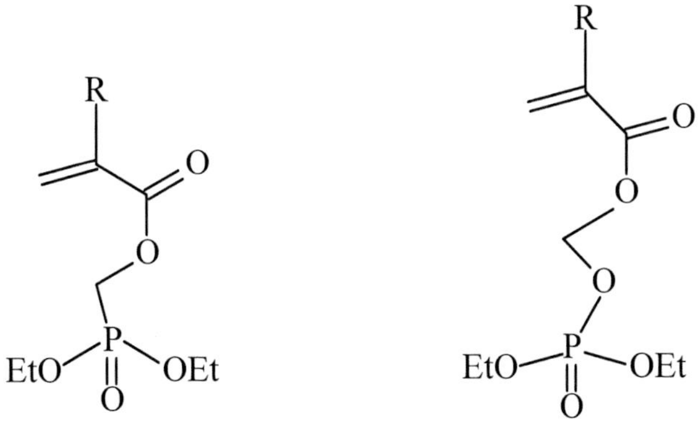

R = CH₃
Diethyl(methacryloyloxymethyl)-phosphonate
or DEMMP

R = H
Diethyl(acryloyloxymethyl)-phosphonate
or DEAMP

R = CH₃
Diethyl(methacryloyloxymethyl)-phosphate
or DEMEP

R = H
Diethyl(acryloyloxymethyl)-phosphate
or DAEP

Scheme 13. Some phosphate ester-based polymers used as flame retardance monomers.

It was shown that DEMMP acted in the condensed phase but also in the vapor phase and char formation as well as autoignition temperature were greatly increased for 10 mol. % content. Some mechanisms operating during the formation of char were proposed in a complementary study [111]. The well-known mechanism of depolymerization of PMMA was disturbed [110] in the presence of DEMMP. Free additives containing PMMA afforded lower flame retardancy properties than that additives incorporated in PMMA chains [112]. Furthermore, the authors showed that phosphonate monomers were more efficacious than the corresponding phosphate monomer [113]. More recently, the same authors studied in details the mechanisms occurring during thermal degradation of PMMA and showed the effect of the chemical structure of the phosphonated comonomer (acrylate or methacrylate) on the PMMA backbone decomposition [101].

Thermosetting resins like UV curable resins have also been studied and the effect of phosphonated monomers was evaluated. The phosphonated monomer (Scheme 14) showed a tidy effect on the flame retardancy of the cured resins [70,114-118]:

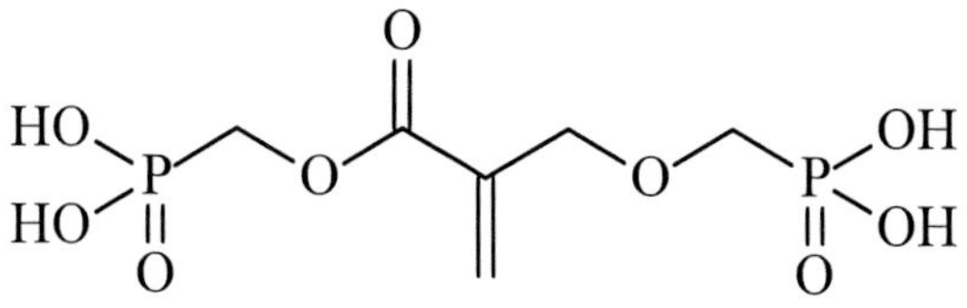

Scheme 14. Example of difunctional monomers used in thermosetting resins.

At last, like in other areas, surface modified nanoparticles were tested in nanocomposite-based PMMA. The treatment of alumina particles led to a better dispersion of the mineral in PMMA matrix and to an improvement in fire behavior, mainly when phosphonic acid-based silicone was used as surface treatment agent. The modification of the pathways occurring during PMMA decomposition was assumed [119].

4.3. Thermoplastic Polyesters

Polyethylene terephthalate (PET) and polybutylene terephthalate (PBT) are commodity plastics used as fibers, injection molded articles, bottles and tanks... These semi-crystalline polymers behave as rigid materials with high deflexion temperature and good electrical insulation properties. Most of injection molded articles are used in electrical or electronic applications where high temperatures, electrical short circuits or electric sparks may cause damages and burning. So, flame retardancy of polyesters is a great challenge since these polymers are very sensible to hydrolysis and require high temperature to be processed in extruders or injection molding devices. These drastic conditions may degrade the chemical structure of additives or of the modified (co)polymers. The hydrolysis of PET in the presence of phosphorus-based flame retardant has been studied and Sato et al [120] concluded that PET with retardant inserted in the main chain were hydrolysed two times faster than that with the same additive as side chain.

The efficiency in flame retardancy was assumed to be directly proportional to the phosphorus content in the polymer and related to the environment of phosphorus atom in the following order: phosphine oxide > phosphonate > phosphate [121].

Many additives have been proposed either as additive or as reactive co-monomer with PET or PBT monomers. Cycloaliphatic phosphine oxide [122] (Scheme 15) and aromatic phosphine with different structures have been studied and showed a tidy influence on the limit oxygen index (LOI) values with a phosphorus content ranging from 5 to 10 % (LOI values from 17 for the original polymer to 25-32 for the modified polymers) [123,124].

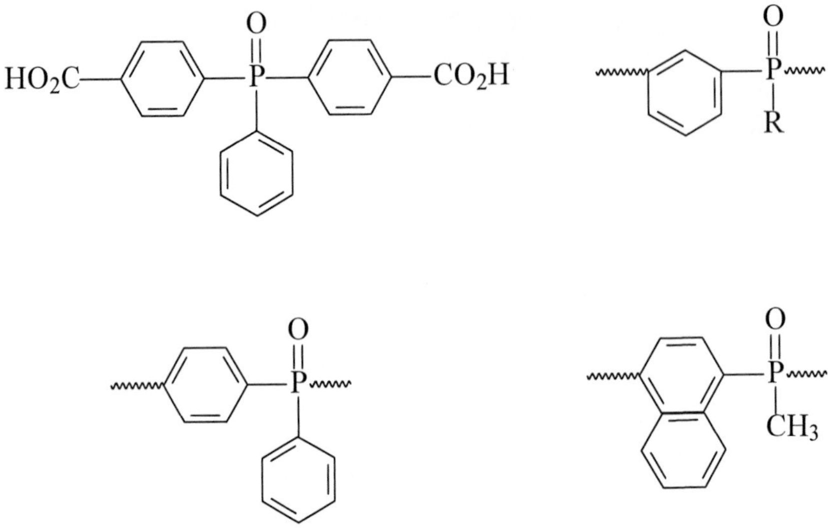

Scheme 15. Some phosphinic oxide structures used as flame retardants.

Phosphinic acid were deeply investigated, mainly 2-carboxyethyl(methyl)phosphinic acid and 2-carboxyethyl(phenyl)phosphinic acid [125,126]. The latter could be incorporated in high content in PET leading to polyester with very interesting properties [125]. The former has been largely used in fibers for fabrics and the action of the phosphinate was demonstrated to be simultaneously active in the solid and vapor phases. The use of nanocomposites has also been investigated by Wang et al [127]. Polyester/Montmorillonite nanocomposites were prepared by intercalation method. The presence of clays impacted the crystallization and significantly enhanced the flame retardancy property. The combination of phosphorus moieties and clays permitted to reduce the co-monomer content in the PET backbone.

Cyclic phosphonates are of great importance since they can react with preformed PET or can be condensed with PET or PBT monomers. Wang et al has successfully synthesized a series of copolyesters including spirocyclic pentaerithrytol-based phosphonates (Scheme 16) leading to good properties like anti-dripping, char formation or flame retardancy. The action of these compounds in the char formation was detailed and chemical reaction were proposed [128,129] [130]]. Due to their efficiency, these cyclic phosphonates have been largely patented in the past.

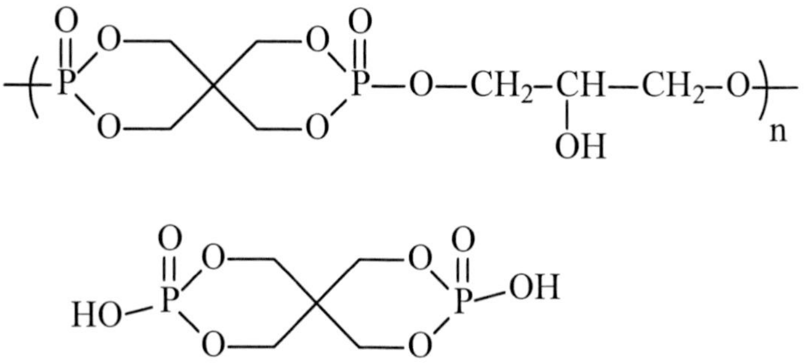

Scheme 16. Example of spirocyclic pentaerithrytol based phosphonates used in flame retardancy of PET.

In a similar way, PET was copolymerized with cyclic phosphoric acid esters moieties to give cellular char used in intumescent coatings [131]. One of the most active molecules recently discovered is 9,10-dihydro-9-oxa-10-phosphaphenanthrere-10-oxide (DOPO) (Scheme 17). This compound has been chemically modified to be copolymerized with other monomers or to be grafted onto polymer backbones. The reaction of DOPO with itaconic acid gave a diacid that could be inserted in polyester chains by copolymerization with PET monomers. Some terpolymers PET/PEN/itaconic modified DOPO were also prepared. Due to its excellent properties, this comonomer has been extensively studied in terms of chemical reactivity, physical and mechanical properties by the academic community [132-137].

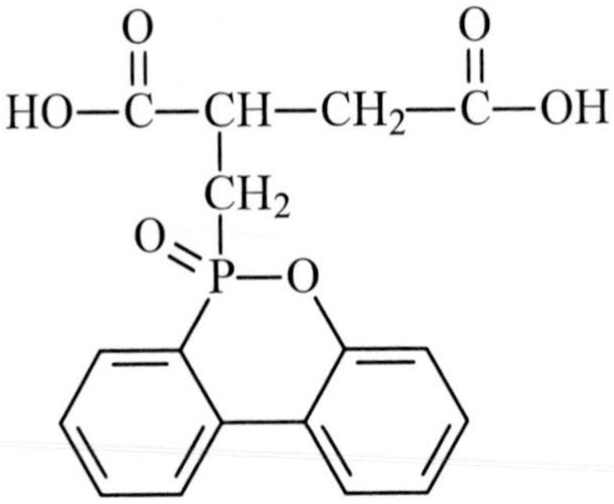

Scheme 17. Diacid based modified DOPO used in flame retardancy.

More recently, a new modified DOPO molecule (Scheme 18) was described and polymerized to give polymers that were added in PET matrix [136]. Interesting flame retardant properties were obtained while mechanical properties of modified PET stayed closed to that of virgin PET.

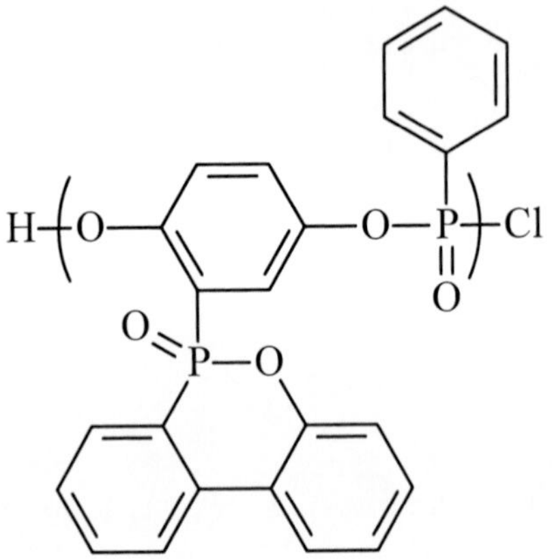

Scheme 18. Example of polyaromatic moiety used in PET flame retardancy.

At last, researchers will find more detail on all specific techniques devoted to flame retardance of polyesters in the review of Weil and Levchik [138].

4.4. Epoxy Resins

These thermosetting resins are most often used in harsh environment and flame retardancy is required for electric and electronic applications based on these polymers. As examples, spirocyclic pentaerythritol bisphosphonate were tested and led to char formation and good LOI indexes [139]. Phosphoryl containing aryl novolacs based on DOPO were synthesized and used as hardener in o-cresol formaldehyde epoxy resin and showed a sharp effect on LOI and char formation [140]. Other authors studied the use of organo-phosphorus modified clays. The effect of the minerals was clearly demonstrated and flame retardancy was observed [141,142]. At last, dialkyl and diarylphosphate [143], DOPO modified diacids

[144,145] and polyphosphazene [146] were also evaluated and showed flame retardancy effect.

4.5. Polyurethanes (PU)

This range of materials is very versatile since they can be processed as thermoplastics or thermosetting polymers and can be soft or hard materials. Flame retardancy is very often required for various applications (insulation, hot atmosphere…). Youssef et al synthesized phosphonated diols that were incorporated in a further step in the reactive two components blends. A sharp effect on the LOI values and char formation was observed when phosphorus was introduced in PU [147-150]. The authors grafted phosphonate groups onto hydroxytelechelic polybutadiene [148] that was crosslinked with a diisocyanate in a second step. The char formation was enhanced by the presence of the grafted phosphorus. The originality stays in the development of chain extender bearing phosphorus. Celebi et al. [151] synthesized a new phosphine oxide amine that was used in the chain extension process. Mechanical properties were not modified in respect to conventional similar PU whereas flame retardancy was reached. In a similar way, diols with phosphorus containing groups like phenylbis(hydroxyalkyl)phosphonate was used as extender and gave similar effect in term of benefit in fire behavior [152-154]. At last, some authors developed new diisocyanate like Bis(3-isocyanatophenyl)phenyl Phosphine Oxide (BIPPPO) (Scheme 19) and observed tidy effects in burning conditions [155].

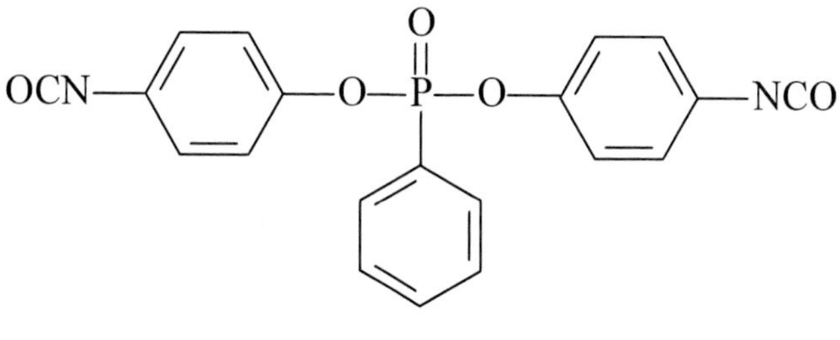

Scheme 19. A new diisocyanate used in polyurethane flame retardancy: bis(3-isocyanatophenyl)phenyl phosphine oxide.

5. ORGANOPHOSPHORUS POLYMERS FOR FUEL CELLS APPLICATIONS

Fuel cells with proton-exchange polymer electrolyte membranes (PEM) are currently being developed as they emerge as an environment friendly and efficient power source while fossil fuel reserves are diminishing. Indeed, they are an attractive energy conversion system that can be used in many industrial applications such as electric vehicles, mobile phones for instance [156]. They provide an electric current *via* redox reactions (oxidation/reduction) using an ion-conductive polymer electrolyte membrane (Figure 1). To be efficient, the latter must present numerous characteristics [157] such as high protonic conductivity, low fuel and water permeability, good mechanical properties in both dry and hydrated states, low price, etc…

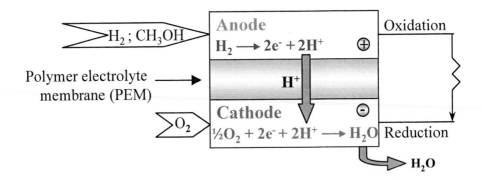

Figure 1. Schematic representation of a fuel cell.

Traditionally, polymer electrolyte membrane fuel cell technology is based on sulfonic acid functionalized polymers such as Nafion®, which showed required high proton conductivity and longevity [158]. Unfortunately, sulfonic acid-functionalized polyelectrolytes also present drawbacks [159]. Among the most important ones, excessive water swelling can lead to poor mechanical properties [160] and high fuel permeability of swollen sulfonated polyelectrolytes reduces membrane lifetime [161]. Sulfonated polyelectrolytes are also sensitive to dehydration at high temperature. This problem was solved using more sophisticated water management system leading in return to higher costs [162]. In this context, new polymeric materials bringing phosphonic site were developed as it was known that phosphonic acid moieties had higher chemical and thermal stability than sulfonic acid ones. They also presented greater ability to retain water than sulfonic acid derivatives which was important to maintain high conductivity at elevated temperatures [163,164]. Kotov et al. [165] synthesized the first proton conductive membrane carrying phosphonic acid units. Synthesis of novel fluoropolymers based on the co- and ter-polymers of tetrafluoroethylene and fluorinated vinylethers were prepared and the resulting films showed promising electrochemical properties, such as good proton conductivity for instance. The phosphonic acid containing materials are considered to be most promising proton-conducting polyelectrolytes.

In a more general manner, phosphonated polyelectrolyte can be classified in different groups depending of the nature of the polymer backbone.

5.1. Poly(arylene ether sulfone)s with Pendant Phosphonic Acid Units

Poly(arylene ether sulfone)s are constituted by phenylene units linked by ether or sulfone groups. Aromatic-based polymers exhibited interesting properties as they showed greater thermal stability and lower flammability than aliphatic polymers [166]. Different kinds of phosphonated poly(arylene ether sulfone)s were developed (Table 2) *via* the modification of poly(arylene ether sulfone) main chains as polymerization of phosphonic acid units led to poor results due to the strong aggregation and condensation of phosphonic acid units [163,167].

Phosphonic acid groups can be directly linked to aryl groups (Table 2, Entry 1-3). Jacoby et al [168] reported the bromination of the polymer and the subsequent bromine-phosphorus exchange using a Pd(0)-catalyzed P-C coupling reaction. Nickel-catalyzed arylphosphonylation with tris(trimethylsilyl)phosphite was also applied to brominated poly(arylene ether sulfone)s leading to polymeric phosphonate and phosphonic acid functional materials [169]. Jannasch et al [163,170] established a non catalytic route for the phosphonation of polysulfones in which lithiated sites on polysulfones reacted with an excess of chlorophosphonic acid ester through a nucleophilic substitution at phosphorus. In contrast to sulfonated poly(arylene ether sulfone)s, phosphonic poly(arylene ether sulfone)s exhibited improved thermal and chemical stabilities combined with far lower water swelling in accordance with the required characteristics for polyelectrolyte membrane fuel cell applications [169].

Poly(arylene ether sulfone)s carrying benzoyl(difluoromethylenephosphonic acid) side chains were also prepared *via* chemical grafting technique [171]. Phosphonated polysulfones were prepared using a multistep pathway: lithiation, reaction with methyl iodobenzoate, CuBr-mediated cross-coupling reaction involving ((diethoxyphosphinyl)difluoromethyl)zinc bromide and finally, dealkylation with bromotrimethylsilane. Resulting phosphonated poly(arylene ether sulfone)s (Table 2, Entry 4) were evaluated for use as proton-conducting fuel cell membranes. Conductivities up to 5 mS.cm^{-1} at 100 °C were measured. On the reverse, the thermal and hydrothermal stability of aryl-CF$_2$-P was found to be limited and was lower than those found for sulfonated derivatives. Nevertheless, the improvement of the acidity of the phosphoric acid units proved to be beneficial for many properties such as solubility, water uptake, and proton conductivity.

Copolymers with phosphonic acid moieties were also evaluated for fuel cell applications [172-174]. Meng et al [172,173] developed copolymers containing phosphonic acid groups (Table 2, Entry 5), as side chains to aromatic poly(ether sulfone), prepared by solution nucleophilic polycondensation. Synthesized copolymers dissolved in common solvents and led to tough and smooth films. The proton conductivities of these polymers were measured and showed good results. Strong and flexible membranes with proton-conducting phosphonic acid were also achieved by grafting poly(vinylphosphonic acid) side chains onto poly(arylene ether sulfone)s [174]. The graft copolymers (Table 2, Entry 6) were prepared by anionic polymerization of diethyl vinylphosphonate initiated by polymers first lithiated followed by hydrolysis in acidic medium to reach multifunctional membranes. The latter notably showed high proton conductivity, under both humidified and nominally dry conditions and thermal stability. Results could be ascribed to the large concentration of phosphonic acid groups which was necessary to form bulky, hydrogen bonded aggregates which promoted a high degree of self-dissociation [175].

It was established that phosphonated membranes with a proper macromolecular design might show advantages for fuel cell applications. The copolymers provided interesting results, providing a way to obtain lower cost proton-exchange polymer electrolyte membranes.

Table 2. Poly(arylene ether sulfone)s with phosphonated groups.

Entry	Chemical structure of poly(arylene ether sulfone) phosphonated polymers	Ref
1		[163,168]
2		[169]
3		[163,170]
4		[171]
5		[172,173]
6		[174]

5.2. Poly(arylene ether, aryl or alkyl)s with Pendant Phosphonic Acid Units

Polymers with arylene ether, aryl and alkyl backbones (Table 3) with pendant phosphonic acid units were also evaluated for fuel cell applications. For example, poly(4-phenoxybenzoyl-1,4-phenylene) was phosphonated by a three step reaction (Table 3, Entry 1) to afford poly(arylene ether) with phosphonic acid units in the side chain [176] which gave interesting results for conductivity or thermal and electrical stabilities. Other arylene ether polymer structures were prepared *via* direct polycondensation in ionic liquid: phosphonated poly(1,3,4-oxadiazole)s [177] led to transparent flexible films. Membranes with pendant phosphonic acid groups were thermally and chemically stable and possessed proton conductivity in the range of 4.10^{-7} to 5.10^{-6} S.cm^{-1} at 100% relative humidity which remained lower than that of Nafion$^\circledR$.

Table 3. Poly(aryl ether or alkyl)s phosphonated polymers.

Entry	Chemical structure of poly(aryl ether or alkyl) phosphonated polymers	Ref
1		[176]
2		[178]
3		[179]
4		[180,181]

Other kind of structures were also reported in the literature. Phosphonic acid-bearing styrene-ethylene/butylene-styrene block copolymer were synthesized (Table 3, Entry 2) and characterized by spectroscopic, thermal, and conductivity measurements [178]. Such copolymer showed good ion-exchange capacity and proton conductivity similar to what has been already observed for other phosphonated polymers. These results were obtained despite a relatively low degree of phosphonation, demonstrating the ability of the phase separated nature of block copolymers to promote proton conductivity.

Furthermore, phosphonated copolymers were synthesized using atom transfer radical polymerization [182,183]. Copolymerization of diisopropyl-p-vinylbenzyl phosphonates and 4-vinylpyridine was studied in the presence of different catalytic systems using DMF as solvent [179]. A proton-conducting statistical copolymer was then obtained by complete hydrolysis of the phosphonates groups (Table 3, Entry 3). A series of vinylbenzylphosphonic acid-*stat*-4-vinylpyridine copolymers with various compositions were obtained. For 90 % of 4-vinylpyridine in molar ratio the proton conductivity rose by several order of magnitude with increasing water content. At low hydration level, this copolymer had a high conductivity of 10^{-2} S.cm^{-1} at 25 °C.

Last described example concerns proton conducting polymer electrolytes based on poly(vinylphosphonic acid)-heterocycle composite material. Honma et al [180,181] prepared the acid-base composite materials by mixing of the strong phosphonic acid polymer with a high proton-exchange capacity and an organic base heterocycle, such as imidazole, pyrazole or 1-methylimidazole (Table 3, Entry 4). Such kind of composite material exhibited a large proton conductivity (7.10^{-3} S.cm^{-1} at 150 °C under anhydrous conditions). Additionally, the thermal stability increased by the mixing of a heterocycle molecule. All results obtained showed that the acid-base composite material might present potential for the polymer electrolyte membrane fuel cells operated at intermediate temperature under anhydrous or extremely low humidity conditions.

5.3. Poly(fluoroalkyl)s with Pendant Phosphonic Acid Units

Different proton conducting perfluorinated membranes carrying phosphonic acid were reported in the literature with polymers and copolymers containing fluoroaromatic or fluoroalkyl groups. Synthesis of several co- and ter-fluoropolymers based on tetrafluoroethylene and phosphonated perfluorovinylethers by redox-initiated emulsion polymerization was reported [165]. Films were achieved from both the polymeric phosphonates and phosphonic acids by compression molding. Ion-exchange capacity within the range 2.5-3.5 milliequivalents per gram and thermal stability of 300-350 °C were measured with perfluorinated phosphonic acid membranes.

Fluorinated aromatic polymers were also studied (Table 4, Entry 1). Fluorinated poly(aryl ether) containing a phosphonic acid pendant group [159,184] possessed excellent thermal stability, oxidative resistance, low methanol permeability (1.07×10^{-8} cm^2.s^{-1}), and reasonable proton conductivity (up to 2.6×10^{-3} S.cm^{-1} in water at room temperature and 6.0×10^{-3} S.cm^{-1} at 95 % relative humidity at 80 °C), allowing its use as polymeric electrolyte membrane for fuel cell applications. Other phosphonated structure containing tetraphenylphenylene ether and octafluorobiphenylene units was also achieved [159,185] (Table 4, Entry 2) and showed hydrophilic properties as well as hydrolytic and thermal stability, indicating potential applications as polymer electrolyte membranes.

Phosphonated perfluorocarbon polymers (Table 4, Entries 3 and 4) with different pendant groups, phosphonated perfluoroalkyl groups [186] or phosphonated phenyl groups [187], were prepared and evaluated as membrane for fuel cells. Both derivatives showed promising results.

Table 4. Poly(fluoroalkyl)s phosphonated polymers.

Entry	Chemical structure of fluoro alkyl phosphonated polymers	Ref
1		[159,184]
2	X = H or P(O)(OH)$_2$	[159,185]
3		[186]
4		[187]

5.4. Poly(aryloxyphosphazene)s with Pendant Phosphonic Acid Units

Polyphosphazene-based cation-exchange membranes were attractive materials for proton exchange membrane mainly due to their reported chemical and thermal stability [188]. As sulfonated polyphosphazene membranes led to promising results, phosphonated polyphosphazene were also evaluated as polymer electrolyte membrane in the development of direct-methanol fuel cells. It has been shown that under ambient temperature conditions, phenyl phosphonic acid functionalized poly(aryloxyphosphazene) membranes had a low methanol permeability, low water swelling ratios, satisfactory mechanical properties, and a conductivity comparable to that of Nafion 117[®]. The synthesis of the first poly(aryloxyphosphazenes) (Table 5, Entry 1) bearing phosphonic acid units was reported by Allcock et al. [189]. Polyphosphazenes bearing bromophenoxy side groups were treated with *n*-butyllithium followed by diethyl chlorophosphate. The resultant phosphonates ester groups were then converted to phosphonic acid ones through basic hydrolysis and subsequent acidification. Membranes of phosphonated poly(aryloxyphosphazenes) had proton conductivity between 10^{-2} and 10^{-1} S.cm^{-1}. They appeared to have lower methanol diffusion characteristics than Nafion 117[®] and to be superior to the sulfonated polyphosphazene derivatives [190]. Analogous structures were prepared [191] (Table 5, Entry 2) and are currently under evaluation as proton conduction membranes for use in fuel cells. Some published results [192] showed that the methanol diffusion coefficients of phenyl phosphonic acid functionalized poly(arylphosphazene) membranes in an aqueous methanol solution (50 % v/v) were about 40 times lower than for Nafion 117[®] and about 10 to 20 times lower than for sulfonated polyphosphazene membranes. These results were of prime importance as the high resistance of phosphonated membranes to methanol permeation should allow for an increase in the methanol concentration of the fuel feed.

More simple phosphonated poly(aryloxyphosphazene) structures were also prepared (Table 5, Entry 3), once again bringing aromatic ring with phosphonic acid functions [193]. The proton conductivity and methanol permeability of resulting membranes were determined at temperatures up to 120 °C. Among all results, it was demonstrated that the permeability of a phosphonated polyphosphazene derivative was about 9 times lower at 120 °C than that of the Nafion membrane which was an interesting improvement over the behavior of Nafion 117[®].

5.5. Poly(siloxane)s with Phosphonic Acid Units

Some examples involving phosphonated polysiloxane structures were developed in the literature with phosphonic acid functionalized oligosiloxanes [194] or organic-inorganic hybrid proton exchange membranes [195,196] (Table 6).

In a first example, phosphonic acid groups were tethered to cyclic siloxanes *via* flexible spacers [194]. Contrary to conventional systems, where proton conductance takes place within an aqueous environment, the proton conductivity of the present material occurred within a dynamical hydrogen bond network formed by the phosphonic acid groups, because of their very high density and the presence of an appropriated spacer. Analysis proved that such kind of materials presented insufficient properties for fuel cell applications as they were too soft and because their proton conductivity remained about one order of magnitude too low.

Table 5. Poly(aryloxyphosphazene)s phosphonated polymers.

Entry	Chemical structure of poly(aryloxyphosphazenes) phosphonated polymers	Ref
1		[189,190]
2		[191]
3		[192,193]

Organic-inorganic membranes were also evaluated as they combine in a single solid both the attractive properties of a mechanically and thermally stable inorganic backbone and the specific chemical reactivity and flexibility of the organo functional groups, which is of great interest (Table 6, entries 2 and 3). Moreover, such kind of materials are generally synthesized through easy procedures with low cost and few environmental impact [197]. Phosphonic acid-grafted hybrid inorganic-organic polymer membranes were prepared using a sol-gel process in acidic conditions. Poly(vinyl alcohol)-silica composite acted as an excellent methanol barrier possessing good hydrophilicity and proton conductivity [195]. Other organic-inorganic polysiloxane membrane [196] had high proton conductivity at low relative humidity and thus could be used as electrolytes for high temperature, low humidity proton exchange membrane fuel cells and other electrochemical applications.

The development of membranes *via* sol-gel procedures in aqueous media is probably a more environment-friendly and cost effective process as the use of harmful organic solvent is limited. The proton displacement mechanism of these organic-inorganic derivatives is not yet known and has to be determined for future potential practical applications.

Table 6. Poly(siloxane)s phosphonated polymers.

Entry	Chemical structure of poly(siloxanes) phosphonated polymers	Ref
1		[194]
2		[195]
3		[196]

6. CONCLUSION

Organophosphorus polymers are of great interest as they can be used for various applications due to their excellent properties as a function of the nature of the phosphorinated group. It was shown that polymers with phosphonic acid (in the pendant group) was mainly used for adhesion or complexing capabilities with metal ions, leading to applications as anticorrosive materials or in the removal of heavy metals in the environment which is a big challenge since these last years with more and more important environmental concerns. They were also developed as fuel cell proton-exchange polymer electrolyte membrane as they present high protonic conductivities, good mechanical properties and low fuel and water permeability, and can be considered as serious competitive of the sulfonic acid functionalized polymers usually employed for such purpose. Polyphosphoesters (phosphoester group in the main chain) are mainly used for their low toxicity, their biocompatibility and biodegradability specificities in an other important domain which is the biomedical field. For instance, these organophosphorus polymers were successfully used as drug carriers or for blood compatibility, as they possess similar structures than natural nucleic acids. Other chemical structures based on phosphorus atom led to other goals: polyphosphazenes found applications as additive in flame retardancy where new developments are due to the ban of halogen-based organic chemical which proved to be very toxic when burning.

All examples developed in this overview have shown that organophosphorus polymers are currently employed in important fields with possible industrial developments which make them very attractive for academic research. The only limitation concerning such polymers could be the synthesis of functionalized phosphorinated monomers as it was shown that some reactions on the phosphorinated groups could take place with other functions brought by the monomers, making the production of such molecules more difficult. Nevertheless, as imagination of scientists for the elaboration of new strategies of synthesis will never be limited, we can assume that chemistry of organophosphorus polymers will be successfully further developed in the next years.

REFERENCES

[1] Clearfield, A. *Curr. Opin. Solid State Mat. Sci.* 1996, *1*, 268-278.
[2] Clearfield, A. *Curr. Opin. Solid State Mat. Sci.* 2002, *6*, 495-506.
[3] Clearfield, A. *J. Alloy. Compd.* 2006, *418*, 128-138.
[4] Knepper, T. P. *Trac-Trends Anal. Chem.* 2003, *22*, 708-724.
[5] Matczak-Jon, E.; Videnova-Adrabinska, V. *Coord. Chem. Rev.* 2005, *249*, 2458-2488.
[6] Herrera-Taboada, L.; Guzmann, M.; Neubecker, K.; Goethlich, A. *PCT Int. Appl.* 2008, 28.
[7] Sahni, S. K.; Cabasso, I. *J. Polym. Sci. Part A: Polym. Chem.* 1988, *26*, 3251-3268.
[8] Kantipuly, C.; Katragadda, S.; Chow, A.; Gesser, H. D. *Talanta* 1990, *37*, 491-517.
[9] Demadis, K. D. In *Solid State Chemistry Research Trends*; Buckley, R. W., Ed.; Nova Science Publishers, Inc.: Hauppauge, NY, 2007, p 109-172.
[10] Popa, A.; Davidescu, C. M.; Negrea, P.; Ilia, G.; Katsaros, A.; Demadis, K. D. *Indust. Eng. Chem. Res.* 2008, *47*, 2010-2017.
[11] Popa, A.; Davidescu, C. M.; Trif, R.; Ilia, G.; Iliescu, S.; Dehelean, G. *React. Funct. Polym.* 2003, *55*, 151-158.
[12] Alexandratos, S. D.; Crick, D. W.; Quillen, D. R. *Ind. Eng. Chem. Res.* 1991, *30*, 772-778.
[13] Kennedy, J.; Burford, F. A.; Sammes, P. G. *J. Inorg. Nucl. Chem.* 1960, *14*, 114-122.
[14] Lin, S.; Cabasso, I. *J. Polym. Sci. Part A: Polym. Chem.* 1999, *37*, 4043-4053.
[15] Gallagher, S. *J. Polym. Sci. Part A: Polym. Chem.* 2002, *41*, 48-59.
[16] Capek, I. *Adv. Colloid Interface Sci.* 2005, *118*, 73-112.
[17] Batra, A.; Cohen, C.; Duncan, T. M. *Macromolecules* 2006, *39*, 2398-2404.
[18] Essahli, M.; Colomines, G.; Monge, S.; Robin, J. J.; Collet, A.; Boutevin, B. *Polymer* 2008, *49*, 4510-4518.
[19] Wu, Q.; Weiss, R. A. *J. Polym. Sci. Part B: Polym. Phys.* 2004, *42*, 3628-3641.

[20] Parent, J. S.; Penciu, A.; Guillen-Castellanos, S. A.; Liskova, A.; Whitney, R. A. *Macromolecules* 2004, *37*, 7477-7483.

[21] Passmann, M.; Zentel, R. *Macromol. Chem. Phys.* 2002, *203*, 363-374.

[22] Yang, H. S.; Adam, H.; Kiplinger, J. *JCT Coatingstech* 2005, *2*, 44-52.

[23] Brondino, C.; Boutevin, B.; Parisi, J.-P.; Schrynemackers, J. *J. Appl. Polym. Sci.* 1999, *72*, 611-620.

[24] Gaboyard, M.; Robin, J.-J.; Hervaud, Y.; Boutevin, B. *J. Appl. Polym. Sci.* 2002, *86*, 2011-2020.

[25] Gaboyard, M.; Jeanmaire, T.; Pichot, C.; Hervaud, Y.; Boutevin, B. *J. Polym. Sci. Part A: Polym. Chem.* 2003, *41*, 2469-2480.

[26] Rixens, B.; Severac, R.; Boutevin, B.; Lacroix-Desmazes, P.; Hervaud, Y. 2005, *POLY-223*, U4101-U4101.

[27] Rixens, B.; Severac, R.; Boutevin, B.; Lacroix-Desmazes, P.; Hervaud, Y. *Macromol. Chem. Phys.* 2005, *206*, 1389-1398.

[28] Rixens, B.; Severac, R.; Boutevin, B.; Lacroix-Desmazes, P. *Polymer* 2005, *46*, 3579-3587.

[29] Rixens, B.; Severac, R.; Boutevin, B.; Lacroix-Desmazes, P. *J. polym. Sci. Part A: Polym. Chem.* 2006, *44*, 13-24.

[30] Bressy-Brondino, C.; Boutevin, B.; Hervaud, Y.; Gaboyard, M. *J. Appl. Polym. Sci.* 2002, *83*, 2277-2287.

[31] Zhor, Z. E.; Chougrani, K.; Negrell-Guirao, C.; David, G.; Boutevin, B.; Loubat, C. *J. Polym. Sci. Part A: Polym. Chem.* 2008, *46*, 4794-4803.

[32] Chougrani, K.; Boutevin, B.; David, G.; Seabrook, S.; Loubat, C. *J. Polym. Sci. Part A: Polym. Chem.* 2008, *46*, 7972-7984.

[33] Renier, M. L.; Kohn, D. H. *J. Biomed. Mater. Res.* 1997, *34*, 95-104.

[34] Zoulalian, V.; Monge, S.; Zurcher, S.; Textor, M.; Robin, J. J.; Tosatti, S. *J. Phys. Chem. B* 2006, *110*, 25603-25605.

[35] Bezdushna, E.; Ritter, H.; Troev, K. *Macromol. Rapid Commun.* 2005, *26*, 471-476.

[36] Kossev, K.; Vassilev, A.; Popova, Y.; Ivanov, I.; Troev, K. *Polymer* 2003, *44*, 1987-1993.

[37] Bai, L.; Chen, R. Y.; Zhu, Y. Y. *Chin. Chem. Lett.* 2002, *13*, 29-32.

[38] Tzevi, R.; Novakov, P.; Troev, K.; Roundhill, D. M. *J. Polym. Sci. Part A: Polym. Chem.* 1997, *35*, 625-630.

[39] Georgieva, R.; Tsevi, R.; Kossev, K.; Kusheva, R.; Balgjiska, M.; Petrova, R.; Tenchova, V.; Gitsov, I.; Troev, K. *J. Med. Chem.* 2002, *45*, 5797-5801.

[40] Troev, K.; Tsatcheva, I.; Koseva, N.; Georgieva, R.; Gitsov, I. *J. Pol. Sci. Part A: Polym. Chem.* 2007, *45*, 1349-1363.

[41] Fontaine, L.; Marboeuf, C.; Brosse, J. C.; Maingault, M.; Dehaut, F. *Macromol. Chem. Phys.* 1996, *197*, 3613-3621.

[42] Koseva, N.; Bogomilova, A.; Atkova, K.; Troev, K. *React. Funct. Polym.* 2008, *68*, 954-966.

[43] Dordunoo, S. K.; Vineek, W. C.; Chaubal, M.; Zhao, Z.; Lapidus, R.; Hoover, R.; Dang, W. B. In *Polymeric Drug Delivery II: Polymeric Matrices and Drug Particle Engineering*; Svenson, S., Ed.; American Chemical Society: Washington DC, 2006; Vol. 924, p 44-68.

[44] Xu, X. Y.; Yu, H.; Gao, S. J.; Mao, H. Q.; Leong, K. W.; Wang, S. *Biomaterials* 2002, *23*, 3765-3772.

[45] Xu, X. Y.; Yee, W. C.; Hwang, P. Y. K.; Yu, H.; Wan, A. C. A.; Gao, S. J.; Boon, K. L.; Mao, H. Q.; Leong, K. W.; Wang, S. *Biomaterials* 2003, *24*, 2405-2412.

[46] Wang, S.; Wan, A. C. A.; Xu, X. Y.; Gao, S. J.; Mao, H. Q.; Leong, K. W.; Yu, H. *Biomaterials* 2001, *22*, 1157-1169.

[47] Wan, A. C. A.; Mao, H. Q.; Wang, S.; Leong, K. W.; Ong, L.; Yu, H. *Biomaterials* 2001, *22*, 1147-1156.

[48] Ueda, T.; Oshida, H.; Kurita, K.; Ishihara, K.; Nakabayashi, N. *Polym. J.* 1992, *24*, 1259-1269.

[49] Iwasaki, Y.; Fujiike, A.; Kurita, K.; Ishihara, K.; Nakabayashi, N. *J. Biomater. Sci.-Polym. Ed.* 1996, *8*, 91-102.

[50] Iwasaki, Y.; Kurita, K.; Ishihara, K.; Nakabayashi, N. *J. Biomater. Sci.-Polym. Ed.* 1996, *8*, 151-163.

[51] Ishihara, K.; Iwasaki, Y.; Nakabayashi, N. *Mater. Sci. Eng. C-Biomimetic Supramol. Syst.* 1998, *6*, 253-259.

[52] Ishihara, K.; Iwasaki, Y.; Ebihara, S.; Shindo, Y.; Nakabayashi, N. *Colloid Surf. B-Biointerfaces* 2000, *18*, 325-335.

[53] Iwasaki, Y.; Sawada, S.; Nakabayashi, N.; Khang, G.; Lee, H. B.; Ishihara, K. *Biomaterials* 1999, *20*, 2185-2191.

[54] Akhtar, S.; Hawes, C.; Dudley, L.; Reed, I.; Stratford, P. *J. Membr. Sci.* 1995, *107*, 209-218.

[55] Stenzel, M. H.; Barner-Kowollik, C.; Davis, T. P.; Dalton, H. M. *Macromol. Biosci.* 2004, *4*, 445-453.

[56] Buonocore, M. G. *J. Dent. Res.* 1955, *34*, 849-853.

[57] Perdigao, J.; Swift, E. J.; Denehy, G. E.; Wefel, J. S.; Donly, K. J. *J. Dent. Res.* 1994, *73*, 44-55.

[58] Van Meerbeek, B.; Vanherle, G.; Lambrechts, P.; Braem, M. *Curr. Opin. Dent.* 1992, *2*, 117-127.

[59] Fu, B.; Sun, X.; Qian, W.; Shen, Y.; Chen, R.; Hannig, M. *Biomaterials* 2005, *26*, 5104-5110.

[60] Albayrak, A. Z.; Bilgici, Z. S.; Avci, D. *Macromol. React. Eng.* 2007, *1*, 537-546.

[61] Moszner, N.; Zeuner, F.; Fischer, U. K.; Rheinberger, V. *Macromol. Chem. Phys.* 1999, *200*, 1062-1067.

[62] Zeuner, F.; Moszner, N.; Volkel, T.; Vogel, K.; Rheinberger, V. *Phosphorus, Sulfur Silicon Relat. Elem.* 1999, *144-146*, 133-136.

[63] Yamanouchi, J.; Masuhara, E.; Nakabayashi, N.; Shibatani, K.; Wada, I. *Jpn. Tokkyo Koho*, 1980; JP 55005790 19800209, CAS 93:225653, 15 pp.

[64] Tanaka, H.; Koro, Y.; Torii, K.; Fujii, T. In *Jpn. Kokai Tokkyo Koho*, 2006; JP 2006045179, 22 pp.

[65] Farley, E. P.; Jones, R. L.; Anbar, M. *J. Dent. Res.* 1977, *56*, 943-952.

[66] Peutzfeldt, A. *Eur. J. Oral Sci.* 1997, *105*, 97-116.

[67] Yeniad, B.; Albayrak, A. Z.; Olcum, N. C.; Avci, D. *J. Polym. Sci. Part A: Polym. Chem.* 2008, *46*, 2290-2299.

[68] Sahin, G.; Albayrak, A. Z.; Sarayli, Z.; Avci, D. *J. Polym. Sci. Part A: Polym. Chem.* 2006, *44*, 6775-6781.

[69]   Senhaji, O.; Monge, S.; Chougrani, K.; Robin, J. J. *Macromol. Chem. Phys.* 2008, *209*, 1694-1704.

[70]   Avci, D.; Albayrak, A. Z. *J. Polym. Sci. Part A: Polym. Chem.* 2003, *41*, 2207-2217.

[71]   Avci, D.; Mathias, L. J. *Abstracts of Papers, 220th ACS National Meeting, Washington, DC, United States, August 20-24, 2000* 2000, POLY-143.

[72]   Avci, D.; Mathias, L. J. *J. Polym. Sci. Part A: Polym. Chem.* 2002, *40*, 3221-3231.

[73]   Avci, D.; Mathias, L. J. *Polym. Bull.* 2005, *54*, 11-19.

[74]   Salman, S.; Albayrak, A. Z.; Avci, D.; Aviyente, V. *J. Polym. Sci. Part A: Polym. Chem.* 2005, *43*, 2574-2583.

[75]   Adusei, G.; Deb, S.; Nicholson, J. W.; Mou, L.; Singh, G. *J. Appl. Polym. Sci.* 2003, *88*, 565-569.

[76]   Mou, L.; Singh, G.; Nicholson, J. W. *Chem. Commun.* 2000, 345-346.

[77]   Sibold, N.; Madec, P.-J.; Masson, S.; Pham, T.-N. *Polymer* 2002, *43*, 7257-7267.

[78]   Catel, Y.; Degrange, M.; Le Pluart, L.; Madec, J.-P.; Pham, T.-N.; Picton, L. *J. Polym. Sci. Part A: Polym. Chem* 2008, *46*, 7074-7090.

[79]   Adusei, G. O.; Deb, S.; Nicholson, J. W. *Dent. Mater.* 2005, *21*, 491-497.

[80]   Culbertson, B. M. *Prog. Polym. Sci.* 2001, *26*, 577-604.

[81]   Ellis, J.; Wilson, A. D. *Polym. Int.* 1991, *24*, 221-228.

[82]   Adusei, G. O.; Nicholson, J. W.; Deb, S. *J. Dent. Res.* 2003, *82*, 527-527.

[83]   Akinmade, A. O.; Braybrook, J. H.; Nicholson, J. W. *Polym. Int.* 1994, *34*, 81-88.

[84]   Akinmade, A. O.; Braybrook, J. H.; Nicholson, J. W. *J. Mater. Sci. Lett.* 1994, *13*, 91-92.

[85]   Anstice, H. M.; Nicholson, J. W. *J. Mater. Sci. Mater. Med.* 1995, *6*, 420-425.

[86]   Braybrook, J. H.; Nicholson, J. W. *J. Mater. Chem.* 1993, *3*, 361-365.

[87]   Gemeinhart, R. A.; Bare, C. M.; Haasch, R. T.; Gemeinhart, E., J. *Journal of biomedical materials research. Part A* 2006, *78*, 433-440.

[88]   George, K. A.; Wentrup-Byrne, E.; Hill, D. J. T.; Whittaker, A. K. *Biomacromolecules* 2004, *5*, 1194-1199.

[89]   Grondahl, L.; Cardona, F.; Chiem, K.; Wentrup-Byrne, E. *J. Appl. Polym. Sci.* 2002, *86*, 2550-2556.

[90]   Grondahl, L.; Cardona, F.; Chiem, K.; Wentrup-Byrne, E.; Bostrom, T. *J. Mater. Sci. Mater. Med.* 2003, *14*, 503-510.

[91]   Grondahl, L.; Suzuki, S.; Wentrup-Byrne, E. *Chem. Commun.* 2008, 3314-3316.

[92]   Suzuki, S.; Rintoul, L.; Monteiro, M. J.; Wentrup-Byrne, E.; Grondahl, L. *Polymer Preprints* 2007, *48*, 430-431.

[93]   Suzuki, S.; Whittaker Michael, R.; Grondahl, L.; Monteiro Michael, J.; Wentrup-Byrne, E. *Biomacromolecules* 2006, *7*, 3178-3187.

[94]   Wan, A. C. A.; Mao, H. Q.; Wang, S.; Phua, S. H.; Lee, G. P.; Pan, J. S.; Lu, S.; Wang, J.; Leong, K. W. *J. Biomed. Mater. Res. Part B* 2004, *70B*, 91-102.

[95]   Costa, L.; Camino, G.; Chiotis, A.; Clouet, G.; Brossas, J.; Bert, M.; Guyot, A. *Polym. Degrad. Stab.* 1984, *6*, 177-188.

[96]   Costa, L.; Camino, G.; Guyot, A.; Bert, M.; Chiotis, A. *Polym. Degrad. Stab.* 1982, *4*, 245-260.

[97]   Costa, L.; Camino, G.; Guyot, A.; Bert, M.; Clouet, G.; Brossas, J. *Polym. Degrad. Stab.* 1986, *14*, 85-93.

[98] Costa, L.; Camino, G.; Guyot, A.; Clouet, G.; Brossas, J. *Polym. Degrad. Stab.* 1986, *15*, 251-64.

[99] Camino, G.; Costa, L.; Clouet, G.; Chiotis, A.; Brossas, J.; Bert, M.; Guyot, A. *Polym. Degrad. Stab.* 1984, *6*, 105-21.

[100] Price, D.; Bullett, K. J.; Cunliffe, L. K.; Hull, T. R.; Milnes, G. J.; Ebdon, J. R.; Hunt, B. J.; Joseph, P. *Polym. Degrad. Stab.* 2005, *88*, 74-79.

[101] Price, D.; Cunliffe, L. K.; Bullet, K. J.; Hull, T. R.; Milnes, G. J.; Ebdon, J. R.; Hunt, B. J.; Joseph, P. *Polym. Adv. Technol.* 2008, *19*, 710-723.

[102] Canadell, J.; Hunt, B. J.; Cook, A. G.; Mantecon, A.; Cadiz, V. *Polym. Degrad. Stab.* 2007, *92*, 1482-1490.

[103] Ebdon, J. R.; Price, D.; Hunt, B. J.; Joseph, P.; Gao, F.; Milnes, G. J.; Cunliffe, L. K. *Polym. Degrad. Stab.* 2000, *69*, 267-277.

[104] Allcock, H. R.; Taylor, J. P. *Polym. Eng. Sci.* 2000, *40*, 1177-1189.

[105] Allcock, H. R.; Hartle, T. J.; Taylor, J. P.; Sunderland, N. J. *Macromolecules* 2001, *34*, 3896-3904.

[106] Shu, W. J.; Perng, L. H.; Chin, W. K.; Hsu, C. Y. *J. Macromol. Sci. Pure Appl. Chem.* 2003, *A40*, 897-913.

[107] Morgan, A. B.; Wilkie, C. A.; Morgan, A. B., Wilkie, C. A., Eds.; John Wiley & Sons, Inc.: Hoboken, NJ, 2007, p 421 pp.

[108] Zheng, X.; Wilkie, C. A. *Polym. Degrad. Stab.* 2003, *81*, 539-550.

[109] Levchik, S. V.; Weil, E. D. *Polym. Int.* 2008, *57*, 431-448.

[110] Price, D.; Pyrah, K.; Hull, T. R.; Milnes, G. J.; Wooley, W. D.; Ebdon, J. R.; Hunt, B. J.; Konkel, C. S. *Polym. Int.* 2000, *49*, 1164-1168.

[111] Ebdon, J. R.; Hunt, B. J.; Joseph, P.; Konkel, C. S.; Price, D.; Pyrah, K.; Hull, T. R.; Milnes, G. J.; Hill, S. B.; Lindsay, C. I.; McCluskey, J.; Robinson, I. *Polym. Degrad. Stab.* 2000, *70*, 425-436.

[112] Price, D.; Pyrah, K.; Hull, T. R.; Milnes, G. J.; Ebdon, J. R.; Hunt, B. J.; Joseph, P.; Konkel, C. S. *Polym. Degrad. Stab.* 2001, *74*, 441-447.

[113] Price, D.; Pyrah, K.; Hull, T. R.; Milnes, G. J.; Ebdon, J. R.; Hunt, B. J.; Joseph, P. *Polym. Degrad. Stab.* 2002, *77*, 227-233.

[114] Youssef, B.; Lecamp, L.; El Khatib, W.; Bunel, C.; Mortaigne, B. *Macromol. Chem. Phys.* 2003, *204*, 1842-1850.

[115] Asif, A.; Liang, H.; Zhu, S.; Shi, W. In *Photochemistry and UV Curing*; Fouassier, J. P., Ed.; Research Signpost: Trivandrum, India, 2006, p 355-371.

[116] Zhu, S.; Shi, W. *Polym. Degrad. Stab.* 2003, *82*, 435-439.

[117] Zhu, S.; Shi, W. *Polym. Degrad. Stab.* 2003, *81*, 233-237.

[118] Zhu, S.; Shi, W. *Polym. Int.* 2004, *53*, 266-271.

[119] Cinausero, N.; Azema, N.; Cochez, M.; Ferriol, M.; Essahli, M.; Ganachaud, F.; Lopez-Cuesta, J.-M. *Polym. Adv. Technol.* 2008, *19*, 701-709.

[120] Sato, M.; Endo, S.; Araki, Y.; Matsuoka, G.; Gyobu, S.; Takeuchi, H. *J. Appl. Polym. Sci.* 2000, *78*, 1134-1138.

[121] Stackman, R. W. *Ind. Eng. Chem. Prod. Res. Dev.* 1982, *21*, 328-331.

[122] Aufmuth, W.; Levchik, S. V.; Levchik, G. F.; Klatt, M. *Fire Mater.* 1999, *23*, 1-6.

[123] Wang, L.-S.; Wang, X.-L.; Yan, G.-L. *Polym. Degrad. Stab.* 2000, *69*, 127-130.

[124] Wan, I. Y.; Keifer, L. A.; McGrath, J. E.; Kashiwagi, T. *Polym. Prepr. (Am. Chem. Soc., Div. Polym. Chem.)* 1995, *36*, 491-492.

[125] Asrar, J.; Berger, P. A.; Hurlbut, J. *J. Polym. Sci. Part A: Polym. Chem.* 1999, *37*, 3119-3128.

[126] Wu, B.; Wang, Y.-Z.; Wang, X.-L.; Yang, K.-K.; Jin, Y.-D.; Zhao, H. *Polym. Degrad. Stab.* 2002, *76*, 401-409.

[127] Ge, X.-G.; Wang, D.-Y.; Wang, C.; Qu, M.-H.; Wang, J.-S.; Zhao, C.-S.; Jing, X.-K.; Wang, Y.-Z. *Eur. Polym. J.* 2007, *43*, 2882-2890.

[128] Chen, D.-Q.; Wang, Y.-Z.; Hu, X.-P.; Wang, D.-Y.; Qu, M.-H.; Yang, B. *Polym. Degrad. Stab.* 2005, *88*, 349-356.

[129] Liu, W.; Chen, D.-Q.; Wang, Y.-Z.; Wang, D.-Y.; Qu, M.-H. *Polym. Degrad. Stab.* 2007, *92*, 1046-1052.

[130] Wang, D.-Y.; Ge, X.-G.; Wang, Y.-Z.; Wang, C.; Qu, M.-H.; Zhou, Q. *Macromol. Mater. Eng.* 2006, *291*, 638-645.

[131] Ma, Z.; Zhao, W.; Liu, Y.; Shi, J. *J. Appl. Polym. Sci.* 1997, *63*, 1511-1515.

[132] Wang, C. S.; Shieh, J. Y.; Sun, Y. M. *J. Appl. Polym. Sci.* 1998, *70*, 1959-1964.

[133] Chang, S.-J.; Chang, F.-C. *J. Appl. Polym. Sci.* 1999, *72*, 109-122.

[134] Wang, D.-Y.; Wang, Y.-Z.; Wang, J.-S.; Chen, D.-Q.; Zhou, Q.; Yang, B.; Li, W.-Y. *Polym. Degrad. Stab.* 2004, *87*, 171-176.

[135] Wang, C. S.; Shieh, J. Y.; Sun, Y. M. *Eur. Polym. J.* 1999, *35*, 1465-1472.

[136] Chang, Y.-L.; Wang, Y.-Z.; Ban, D.-M.; Yang, B.; Zhao, G.-M. *Macromol. Mater. Eng.* 2004, *289*, 703-707.

[137] Chang, S.-J.; Chang, F.-C. *Polym. Eng. Sci.* 1998, *38*, 1471-1481.

[138] Levchik, S. V.; Weil, E. D. *Polym. Int.* 2005, *54*, 11-35.

[139] Chen, G.-H.; Yang, B.; Wang, Y.-Z. *J. Appl. Polym. Sci.* 2006, *102*, 4978-4982.

[140] Liu, Y. L.; Wu, C. S.; Hsu, K. Y.; Chang, T. C. *J. Polym. Sci. Part A: Polym. Chem.* 2002, *40*, 2329-2339.

[141] Camino, G.; Tartaglione, G.; Frache, A.; Manferti, C.; Finocchiaro, P.; Falqui, L. *ACS Symp. Ser.* 2006, *922*, 21-35.

[142] Hussain, M.; Varley, R. J.; Mathys, Z.; Cheng, Y. B.; Simon, G. P. *J. Appl. Polym. Sci.* 2004, *91*, 1233-1253.

[143] Derouet, D.; Morvan, F.; Brosse, J. C. *J. Appl. Polym. Sci.* 1996, *62*, 1855-1868.

[144] Wang, C.-S.; Shieh, J.-Y. *Polymer* 1998, *39*, 5819-5826.

[145] Buckingham, M. R.; Lindsay, A. J.; Stevenson, D. E.; Muller, G.; Morel, E.; Costes, B.; Henry, Y. *Polym. Degrad. Stab.* 1996, *54*, 311-315.

[146] Cho, C.-S.; Fu, S.-C.; Chen, L.-W.; Wu, T.-R. *Polym. Int.* 1998, *47*, 203-209.

[147] El Khatib, W.; Youssef, B.; Bunel, C.; Mortaigne, B. *Polym. Int.* 2003, *52*, 146-152.

[148] Schapman, F.; Youssef, B.; About-Jaudet, E.; Bunel, C. *Eur. Polym. J.* 2000, *36*, 1865-1873.

[149] Spirckel, M.; Regnier, N.; Mortaigne, B.; Youssef, B.; Bunel, C. *Polym. Degrad. Stab.* 2002, *78*, 211-218.

[150] Youssef, B.; Mortaigne, B.; Soulard, M.; Saiter, J. M. *J. Therm. Anal. Calorim.* 2007, *90*, 489-494.

[151] Celebi, F.; Polat, O.; Aras, L.; Guenduez, G.; Akhmedov, I. M. *J. Appl. Polym. Sci.* 2004, *91*, 1314-1321.

[152] Chang, T. C.; Shen, W. S.; Chiu, Y. S.; Ho, S. Y. *Polym. Degrad. Stab.* 1995, *49*, 353-360.

[153] Chang, T. C.; Shen, W. Y.; Chiu, Y. S.; Chen, H. B.; Ho, S. Y. *J. Polym. Res.* 1994, *1*, 353-9.

[154] Sivriev, C.; Zabski, L. *Eur. Polym. J.* 1994, *30*, 509-514.

[155] Liu, Y.-L.; Hsiue, G.-H.; Lan, C.-W.; Chiu, Y.-S. *J. Polym. Sci. Part A: Polym. Chem.* 1997, *35*, 1769-1780.

[156] Kerres, J. A. *J. Membrane Sci.* 2001, *185*, 3-27.

[157] Hickner, M. A.; Ghassemi, H.; Kim, Y. S.; Einsla, B. R.; McGrath, J. E. *Chem. Rev.* 2004, *104*, 4587-4611.

[158] Li, Q. F.; He, R. H.; Jensen, J. O.; Bjerrum, N. J. *Chem. Mater.* 2003, *15*, 4896-4915.

[159] Bock, T.; Mohwald, H.; Mulhaupt, R. *Macromol. Chem. Phys.* 2007, *208*, 1324-1340.

[160] Kreuer, K. D.; Paddison, S. J.; Spohr, E.; Schuster, M. *Chem. Rev.* 2004, *104*, 4637-4678.

[161] Fritts, S. D.; Gervasio, D.; Zeller, R. L.; Savinell, R. F. *J. Electrochem. Soc.* 1991, *138*, 3345-3349.

[162] Kreuer, K. D. *Solid State Ionics* 1997, *97*, 1-15.

[163] Lafitte, B.; Jannasch, P. *J. Polym. Sci. Part A: Polym. Chem.* 2005, *43*, 273-286.

[164] Lassegues, J. C.; Grondin, J.; Hernandez, M.; Maree, B. *Solid State Ionics* 2001, *145*, 37-45.

[165] Kotov, S. V.; Pedersen, S. D.; Qiu, W.; Qiu, Z.-M.; Burton, D. J. 1997, *82*, 13-19.

[166] McGrail, P. T. *Polym. Int.* 1996, *41*, 103-121.

[167] Rikukawa, M.; Sanui, K. *Prog. Polym. Sci.* 2000, *25*, 1463-1502.

[168] Jakoby, K.; Peinemann, K. V.; Nunes, S. P. *Macromol. Chem. Phys.* 2003, *204*, 61-67.

[169] Bock, T.; Muelhaupt, R.; Moehwald, H. *Macromol. Rapid Commun.* 2006, *27*, 2065-2071.

[170] Jannasch, P. *Fuel Cells* 2005, *5*, 248-260.

[171] Lafitte, B.; Jannasch, P. *J. Polym. Sci. Part A: Polym. Chem.* 2007, *45*, 269-283.

[172] Meng, Y. Z.; Tjong, S. C.; Hay, A. S.; Wang, S. J. *Eur. Polym. J.* 2003, *39*, 627-631.

[173] Meng, Y. Z.; Tjong, S. C.; Hay, A. S.; Wang, S. J. *J. Polym. Sci. Part A: Polym. Chem.* 2001, *39*, 3218-3226.

[174] Parvole, J.; Jannasch, P. *Macromolecules* 2008, *41*, 3893-3903.

[175] Steininger, H.; Schuster, M.; Kreuer, K. D.; Kaltbeitzel, A.; Bingol, B.; Meyer, W. H.; Schauff, S.; Brunklaus, G.; Maier, J.; Spiess, H. W. *Phys. Chem. Chem. Phys.* 2007, *9*, 1764-1773.

[176] Yanagimachi, S.; Kaneko, K.; Takeoka, Y.; Rikukawa, M. *Synthetic Met.* 2003, *135*, 69-70.

[177] Shaplov, A. S.; Lozinskaya, E. I.; Odinets, I. L.; Lyssenko, K. A.; Kurtova, S. A.; Timofeeva, G. I.; Iojolu, C.; Sanchez, J. Y.; Abadie, M. J. M.; Voytekunas, V. Y.; Vygodskii, Y. S. *React. Func. Polym.* 2008, *68*, 208-224.

[178] Subianto, S.; Choudhury, N.; Dutta, N. *J. Polym. Sci. Part A: Polym. Chem.* 2008, *46*, 5431-5441.

[179] Jiang, F.; Kaltbeitzel, A.; Meyer, W. H.; Pu, H.; Wegner, G. *Macromolecules* 2008, *41*, 3081-3085.

[180] Honma, I.; Yamada, M. *B. Chem. Soc. Jpn.* 2007, *80*, 2110-2123.

[181] Yamada, M.; Honma, I. *Polymer* 2005, *46*, 2986-2992.

[182] Monge, S.; Giani, O.; Ruiz, E.; Cavalier, M.; Robin, J. J. *Macromol. Rapid. Commun.* 2007, *28*, 2272-2276.

[183] Monge, S.; Darcos, V.; Haddleton, D. M. *J. Polym. Sci. Part A: Polym. Chem.* 2004, *42*, 6299-6308.

[184] Liu, B. J.; Robertson, G. P.; Guiver, M. D.; Shi, Z. Q.; Navessin, T.; Holdcroft, S. *Macromol. Rapid. Commun.* 2006, *27*, 1411-1417.

[185] Miyatake, K.; Hay, A. S. *J. Polym. Sci. Part A: Polym. Chem.* 2001, *39*, 3770-3779.

[186] Yamabe, M.; Akiyama, K.; Akatsuka, Y.; Kato, M. *Eur. Polym. J.* 2000, *36*, 1035-1041.

[187] Stone, C.; Daynard, T. S.; Hu, L. Q.; Mah, C.; Steck, A. E. *J. New Mat. Electrochem. Syst.* 2000, *3*, 43-50.

[188] Guo, Q. H.; Pintauro, P. N.; Tang, H.; O'Connor, S. *J. Membr. Sci.* 1999, *154*, 175-181.

[189] Allcock, H. R.; Hofmann, M. A.; Ambler, C. M.; Morford, R. V. *Macromolecules* 2002, *35*, 3484-3489.

[190] Allcock, H. R.; Hofmann, M. A.; Ambler, C. M.; Lvov, S. N.; Zhou, X. Y.; Chalkova, E.; Weston, J. *J. Membr. Sci.* 2002, *201*, 47-54.

[191] Allcock, H. R.; Hofmann, M. A.; Wood, R. M. *Macromolecules* 2001, *34*, 6915-6921.

[192] Fedkin, M. V.; Zhou, X. Y.; Hofmann, M. A.; Chalkova, E.; Weston, J. A.; Allcock, H. R.; Lvov, S. N. *Mater. Lett.* 2002, *52*, 192-196.

[193] Zhou, X.; Weston, J.; Chalkova, E.; Hofmann, M. A.; Ambler, C. M.; Allcock, H. R.; Lvov, S. N. *Electrochim. Acta* 2003, *48*, 2173-2180.

[194] Steininger, H.; Schuster, M.; Kreuer, K. D.; Maier, J. *Solid State Ionics* 2006, *177*, 2457-2462.

[195] Binsu, V. V.; Nagarale, R. K.; Shahi, V. K. *J. Mater. Chem.* 2005, *15*, 4823-4831.

[196] Li, S.; Zhou, Z.; Abernathy, H.; Liu, M.; Li, W.; Ukai, J.; Hase, K.; Nakanishi, M. *J. Mater. Chem.* 2006, *16*, 858-864.

[197] Matejka, L.; Dukh, O.; Meissner, B.; Hlavata, D.; Brus, J.; Strachota, A. *Macromolecules* 2003, *36*, 7977-7985.

In: Modern Trends in Macromolecular Chemistry
Editor: Jon N. Lee

ISBN: 978-1-60741-252-6
© 2009 Nova Science Publishers, Inc.

Chapter 2

SYNTHESIS AND MOLECULAR STRUCTURE INVESTIGATION OF ALIPHATIC HYPERBRANCHED POLYESTERS

J. Vukovic[1], S. Jovanovic[2], M.D. Lechner[3] and Vesna Vodnik[4]

[1]ICTM - Center for Chemistry, Belgrade, Serbia
[2]Faculty of Technology and Metallurgy, University of Belgrade, Belgrade, Serbia
[3]Institute for Chemistry, University of Osnabrück, Osnabrück, Germany
[4]Institute of Nuclear Science "Vinča", Belgrade, Serbia

ABSTRACT

The results obtained by investigation of the influence of synthesis procedure on the molecular structure of aliphatic hyperbranched polyesters (AHBP) are presented in this work. Different pseudo generations of these three-dimensional polymers were synthesized via an acid-catalyzed polyesterification of 2,2-bis(hydroxymethyl)propionic acid (bis-MPA) and di-trimethylolpropane (DiTMP) using two different procedures: pseudo one-step (samples of the series I) and one-step (samples of the series II). In pseudo one-step procedure the sample of certain pseudo generation was obtained starting from the sample of previous pseudo generation and adequate amount of monomer, while in simple one-step procedure polymer of the wanted pseudo generation was synthesized by mixing the stochometric amount of all necessary reactants at once. Beside these samples, three commercial Boltorn® AHBP synthesized by pseudo one-step procedure were also investigated. Results obtained by NMR, MALDI-TOF MS and VPO show that formation of poly(bis-MPA) and cyclization through the ester and ether bonds (intramolecular) occurred as side reactions during the both procedures for the synthesis. With increasing degree of polymerization, the extent of these unwanted reactions increased and it was slightly higher when one-step procedure was applied for the preparation of AHBP samples. During the synthesis of AHBP from second till fifth pseudo generation, cyclization occurred through ester and ether bonds (intramolecular), while in the case of the higher generation samples, cycles were formed due to the intramolecular esterification. Furthermore, –OH groups belonging to the terminal units had better reactivity than linear ones. Consequently, the degree of branching for all investigated samples is lower than 0.50. Due to the side reactions occurrence, molar mass increases only up to the sixth pseudo generation and values of the molar masses are much lower than theoretical. From the obtained results it was also possible to calculate the percentage of deviation from the perfect dendrimer structure for AHBP samples of the series I and II.

Keywords: Hyperbranched polymers; synthesis; molecular structure; MALDI-TOF MS.

INTRODUCTION

The unique, nonlinear structure and specific physical and chemical properties of dendritic polymers, i.e. dendrimers and hyperbranched polymers, were intensively studied during recent years. The synthesis of the perfect dendrimers was first introduced by Vogtle in 1978 and in 1985 developed by Tomalia and Newkome [1-3]. Dendrimers represent regularly branched, well-defined, monodisperse polymers composed of AB_x type ($x \geq 2$) monomers attached in concentric layers around a central B_y core (Figure 1a). Each layer is called a generation (G1, G2), where all B functional groups have reacted with A functional groups from the next layer, while unreacted B groups can be found only as end groups (last generation). With increasing generation number, molar mass of dendrimers considerably increases. These polymers can be obtained only in a time consuming and very expensive synthesis.

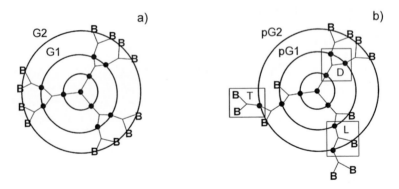

Figure 1. Schematic representation of AB_2/B_3 a) dendrimer (G1 and G2 represent different number of generation) and b) hyperbranched polymer (pG1 and pG2 represent different number of pseudo generation, while L, D and T are different units)

On the other side, hyperbranched (HB) polymers can be prepared by the one-step polycondensation of AB_x-monomers (with or without multifunctional core B_y) at a reasonable cost [4]. Although this simple procedure yields randomly branched polymers with broad molar mass distribution and without perfect globular structure (Figure 1b), HB polymers resemble dendrimers in many physical and chemical properties and therefore are usually considered as irregular analogues of dendrimers. According to the increasing number of publications, the interest in these complex, branched macromolecules is growing rapidly. Due to their compact structure and presence of a large number of functional end groups, HB polymers similar to dendrimers show high solubility and reactivity, low viscosity in solution and melt and good compatibility with other materials, which allows them to be used in numerous applications [5]. The imperfection of HB structure originates from the fact that beside terminal (T) and fully reacted dendritic (D) units, molecular structure of these polymers is also composed from the linear units (L), which have one unreacted B group (Figure 1b). As a consequence of that, the inner layers of these polymers are usually called pseudo generations.

In contrast to HB polymers obtained by self-condensation of AB_x monomers, the polydispersity and appearance of the side reactions can be significantly lowered by copolymerization of AB_x monomers with B_y core molecules [6]. According to Hult et al., by varying the core/monomer ratio it is possible to obtain HB polymers of different molar masses [6]. This reaction can be carried out using one-step or pseudo one-step procedure [7]. During the synthesis of aliphatic HB polyesters from 2,2-bis(hydroxymethyl)propionic acid (bis-MPA) and 2-ethyl-2-(hydroxymethyl)-1,3-propanediol, Hult et al. have used pseudo one-step procedure in order to minimize appearance of side reactions [6]. The same procedure was also applied for the synthesis of aliphatic HB polyethers in order to prevent cyclization [8]. Hyperbranched polymers synthesized using pseudo one-step and simple one-step procedure should generally have different properties. However, there are no literature evidences showing how big these differences are. Furthermore, the most often investigated bis-MPA HB polyesters were mostly examined only up to the fifth or sixth pseudo generation [9,10]. Therefore, the aim of this work was the synthesis of hydroxy-functional aliphatic HB polyesters (AHBP) by one-step and pseudo one-step procedure, in order to compare the structure and properties between samples obtained using these two procedures, as well as between samples of different pseudo generations (from second till tenth). At the same time, the influence of the procedure for the synthesis on the amount of the side reaction products was investigated. Obtained results of the adequate AHBP synthesized in this work were further compared with results determined for the commercial Boltorn® AHBP, synthesized by pseudo one-step procedure.

EXPERIMENTAL PART

Materials

Hydroxy-functional aliphatic hyperbranched polyesters of different pseudo generations were synthesized starting from bis-MPA (Aldrich), as an AB_2 monomer, and di-trimethylolpropane, DiTMP, (Fluka Chemika), as the tetrafunctional core molecule. Methanesulphonic acid (MSA), purchased from Aldrich, was used as a catalyst. In this work three commercially available aliphatic HB polyesters (Boltorn®) of the second (BH-2), third (BH-3) and fourth (BH-4) pseudo generation, supplied by Perstorp (Specialty Chemicals AB, Sweden), were also investigated. According to the supplier's data, commercial AHBP were synthesized via pseudo one-step procedure from the bis-MPA as monomer and a tetrafunctional ethoxylated pentaerythrytol (PP50) core. All other chemicals were obtained from Aldrich and used as received, without further purification.

Synthesis of AHBP Samples

Two series of hydroxy-functional AHBP were synthesized in an acid-catalyzed polyesterification reaction. Samples of the series I (AHBP-2I, AHBP-3I, AHBP-4I, AHBP-5I, AHBP-6I, AHBP-8I and AHBP-10I) were prepared using pseudo one-step procedure, which is in the following text described for the synthesis of the third pseudo generation AHBP (AHBP-3I). 95.00 g of the previously synthesized AHBP-2I was placed in a four necked flask and heated at ~ 110 ^{0}C for one hour with an oil bath. After that, 124.05 g (0.93 mol, calculated from the theoretical molar mass) of bis-MPA and 0.89 g MSA were added into the flask. N_2 inlet, a drying tube and a mechanical stirrer were connected to the flask, while the temperature was increased up to 140 ^{0}C. The mixture was left to react for two hours under a stream of N_2, which was used to remove the formed water. After that, a reduced pressure was applied to the flask until the reaction reached the completion. The course of the reaction was controlled by the acid number titration. The synthesis of other AHBP of the higher pseudo generations was done in the same manner, starting from the previous pseudo generation and adding the stoichometric amount of bis-MPA and an adequate amount of MSA. On the other hand, polyester AHBP-2I and samples of the series II (AHBP-4II, AHBP-6II and AHBP-8II) were synthesized using one-step procedure, by adding an adequate, stoichometric amount of DiTMP and bis-MPA at once to obtain AHBP of wanted pseudo generation number. Further reaction was done as it was previously described for the synthesis of series I AHBP. All samples were obtained as transparent, light yellow solids. After the synthesis, samples were dried in a vacuum oven until constant mass was achieved.

Characterization

The acid (N_{AN}) and hydroxyl number (N_{HN}) of investigated AHBP were determined according to the procedures described in ref [11]. The acid number (N_{AN}) of the AHBP was determined as follows. Samples of 0.2-0.3 g were taken out from the reaction mixture at different times during the synthesis. The samples were dissolved in 50 cm^3 of acetone/ethanol (1:1 by volume) and titrated with 0.50 M KOH using phenolphthalein as an indicator. Blank tests were done on the pure solvent mixture. Acid number was then calculated according to the following equation:

$$N_{AN} = \frac{56.1 \times c_{KOH} \times \left(V_{KOH} - V_{BT}\right)}{m_{sample}}$$

(1)

where c_{KOH} is the concentration of the KOH solution, V_{KOH} is the volume of KOH used for the sample titration, V_{BT} is the volume of the KOH used for the blank test and m_{sample} is the mass of the sample. Since V_{BT} was not possible to measure (less than one drop of KOH), it was taken that $V_{BT} = 0$. Acid number titrations were done during the synthesis in order to control the course of the reaction. Under the given conditions for the synthesis, the reaction was carried out up to the lowest possible acid number, i.e. until N_{AN} became constant.

The hydroxyl number (N_{HN}) was determined after the synthesis, by dissolving ~ 1 g of the sample in 15 cm^3 of a mixture acetic anhydride/pyridine (1:9 by volume) in a 250 cm^3

erlenmeyer flask. The solution was then heated for 1 h with reflux. After cooling, ~ 50 cm^3 of distilled water was slowly added through the drying tube, and the other 50 cm^3 of distilled water were added into the solution when the erlenmeyer flask was disconnected from the drying tube. The solution was then titrated with 0.5 M KOH using phenolphthalein as an indicator. Blank tests were done on the pure solvent mixture. Finally N_{HN} was calculated as follows:

$$N_{HN} = \frac{56.1 \times c_{KOH} \times (V_{BT} - V_{KOH})}{m_{sample}} \quad (2)$$

For the purpose of this measurement pyridine was dried above solid KOH for several days. Values of theoretical hydroxyl number, $(N_{HN})_{theor}$, were calculated using the equation (3) and assuming that during the synthesis all acid groups have reacted ($N_{AN} = 0$):

$$(N_{HN})_{theor} = \frac{56100 \times (n_{OH})_{theor}}{M_{theor}} \quad (3)$$

where 56100 is the molar mass of KOH (in mg · mol^{-1}).

^{1}H and ^{13}C NMR spectra of synthesized samples were recorded on Bruker (250 MHz) NMR spectrometer at room temperature. ^{13}C NMR spectra were obtained using deuterated dimethylsulfoxide (DMSO-d_6) as solvent, while for recording ^{1}H NMR spectra, samples were dissolved in DMSO-d_6 and in the mixture deuterium oxide (D$_2$O)/ DMSO-d_6 (1:4 by volume). Signal of the DMSO-d_6 resonate at around 2.5 ppm in the ^{1}H NMR spectra and at around 39.6 ppm in the ^{13}C NMR spectra, while D$_2$O resonate at around 4.0 ppm in ^{1}H NMR spectra. Tetramethylsilane was used as an internal reference. The reliability of the NMR measurements was confirmed by comparison of the results obtained in this work with results published by other authors, which have performed NMR measurements with similar AHBP, as well as with a comparable model compounds of known composition [6].

Vapour pressure osmometry (VPO) of synthesized and commercial samples was carried out using a Knauer vapour pressure osmometer. Measurements were done in N,N-dimethylformamide as a solvent at 90 ^{0}C. For the calibration benzil was used.

MALDI-TOF mass spectrometry (MS) was performed on a BIFLEX III instrument (Bruker Saxonia Analytik GmbH, Leipzig, Germany) in the reflection mode using 2,5-dihydroxybenzoic acid as matrix and tetrahydrofuran as solvent.

RESULTS AND DISCUSSION

Schematic representation of the AHBP synthesis is presented in Figure 2. All self-synthesized and commercial AHBP were investigated using the methods described in experimental part of this work. Selected calculated and experimentally determined characteristics of the examined AHBP are listed in Table 1 and presented in Figure 3.

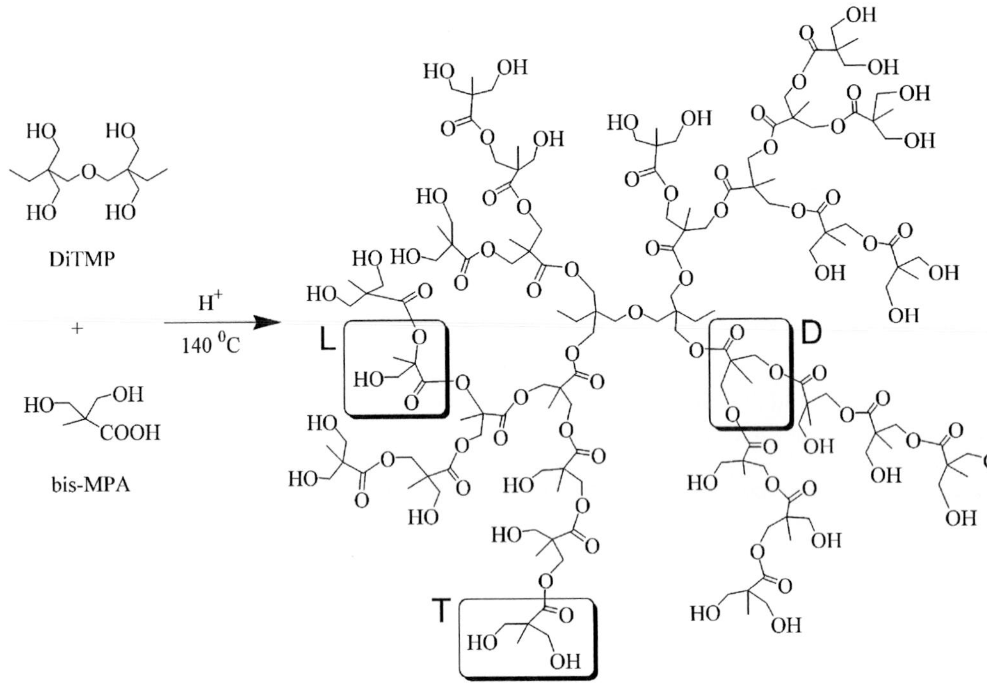

Figure 2. Schematic representation of the AHBP synthesis (L, D and T represent different units)

Table 1. Values of the molar ratio core/monomer used for the synthesis, theoretical number of pseudo generation, $(n)_{theor}$, theoretical number of end –OH groups, $(n_{OH})_{theor}$, theoretical molar mass, M_{theor}, and molar mass determined by VPO, $(M_n)_{VPO}$, of investigated AHBP

Sample	Molar ratio core/monomer	$(n)_{theor}$	$(n_{OH})_{theor}$	M_{theor} [g · mol^{-1}]	$(M_n)_{VPO}$ [g · mol^{-1}]
AHBP-2I	1/12	2	16	1642	-
AHBP-3I	1/28	3	32	3498	2027
AHBP-4I	1/60	4	64	7210	2819
AHBP-5I	1/124	5	128	14634	3044
AHBP-6I	1/252	6	256	29482	3575
AHBP-8I	1/1020	8	1024	118570	3571
AHBP-10I	1/4092	10	4096	474922	3552
AHBP-4II	1/60	4	64	7210	5415
AHBP-6II	1/252	6	256	29482	-
AHBP-8II	1/1020	8	1024	118570	3284
BH-2	1/12	2	16	1747	1343
BH-3	1/28	3	32	3604	3081
BH-4	1/60	4	64	7316	2716

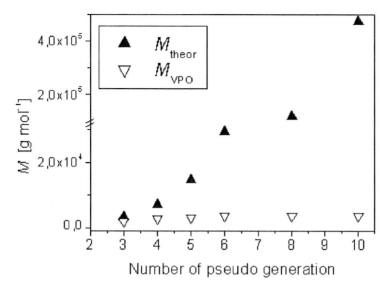

Figure 3. Calculated (theoretical) and experimentally determined molar masses of AHBP from the series I

The theoretical molar mass, M_{theor}, and the theoretical number of end –OH groups, $(n_{OH})_{theor}$ (Table 1) were calculated with the assumptions that the reaction was carried out up to the conversion of 100 % (without occurrence of side reactions) and that no linear units were formed during the synthesis (degree of branching equal to one). Values presented in Table 1 show that number average molar mass determined using VPO results, $(M_n)_{VPO}$, slightly increases only up to the sixth pseudo generation and at the same time $(M_n)_{VPO}$ values are much lower than M_{theor}, which is especially pronounced for the higher pseudo generation samples (Figure 3). The detected differences are a consequence of the side reactions occurrence during the synthesis, which have prevented formation of the high molar mass AHBP, due to the competition between unwanted reactions and polycondensation. According to the results presented in the literature, the most important side reactions which might occur during the synthesis of this type of hyperbranched polyesters are [12-19]:

1) The formation of a HB macromolecule poly(bis-MPA) by self-condensation of the monomer.
2) Intramolecular esterification (cyclization) between focal –COOH group of poly(bis-MPA) and –OH group. In this way, poly-B-functional macro-cycles are formed in the course of the reaction. The cyclization due to the intramolecular esterification can also occur as hydroxy-ester interchange, leading to the chain scission and formation of the HB molecules with one cyclic branch and without core unit.
3) Intramolecular etherification (cyclization) between two –OH groups.

According to the suggestions of some authors, occurrence of previously mentioned side reactions should depend on the applied procedure for the synthesis and may have an influence on the structure, molar mass, molar mass distribution and other properties of hyperbranched polyesters [12-19]. However, there are no experimental evidences showing the amount of these side products formed during the each procedure.

In this work, ^{13}C NMR spectroscopy was used to detect the presence of certain side reaction products in examined samples of AHBP. This technique also gives possibility to calculate degree of branching, *DB*, of HB polymers by comparing the integrals of signals corresponding to the terminal, linear and dendritic units (Figure 2) and using equations developed by Fréchet and Frey [20,21]. As an example, ^{13}C NMR spectrum and magnification of the quaternary carbon region of AHBP-8I are presented in Figure 4. In the ^{13}C NMR spectra of the samples synthesized in this work, the signals of the quaternary carbon belonging to the terminal, linear and dendritic units were found at 50.3, 48.3 and 46.3 ppm, respectively (Figure 4) [6]. Beside these signals, in ^{13}C NMR spectra of the AHBP carbonyl groups resonate between 172 and 175 ppm, methylene groups between 63 and 66 ppm and methyl groups at 17 ppm [6,15].

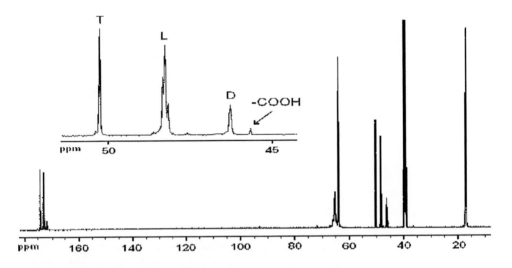

Figure 4. ^{13}C NMR spectrum and magnification of the quaternary carbon region of AHBP-8I in DMSO-d_6

Table 2. Results obtained from the ^{13}C NMR spectra of AHBP

Sample	Integrated intensities of the signals			
	$I(T)^a$	$I(L)^a$	$I(D)^a$	$I(COOH)$
AHBP-2I	1.00	1.41	0.27	/
AHBP-3I	1.00	1.71	0.40	0.02
AHBP-4I	1.00	1.92	0.49	0.03
AHBP-5I	1.00	1.92	0.56	0.05
AHBP-6I	1.00	2.04	0.54	0.04
AHBP-8I	1.00	2.09	0.56	0.08
AHBP-10I	1.00	2.13	0.54	0.08
AHBP-4II	1.00	1.94	0.43	/
AHBP-6II	1.00	1.84	0.38	0.08
AHBP-8II	1.00	2.29	0.33	0.07

[a]Results presented in ref. [22]

The integrated intensities of signals belonging to the different units are listed in Table 2, while calculated values of DB are presented in Table 3. For AHBP of series I (except for AHBP-5I), value of $DB_{Fréchet}$ decreases, while value of DB_{Frey} increases with increasing number of pseudo generation. The reason for such behaviour is the difference in definition of $DB_{Fréchet}$ and DB_{Frey}. According to Fréchet's equation, terminal groups of unreacted monomers are also used for the calculation of DB, which leads to the overestimation of the calculated DB values [20]. On the other side, for the calculation of DB_{Frey} active number of growth directions of the polymer chain is compared to the maximum number of growth directions [21]. Value of $DB_{Fréchet}$ decreases from 1 (degree of conversion of acid groups equal to zero ($p_A = 0$)) to 0.5 at complete conversion, while value of DB_{Frey} increases from zero ($p_A = 0$) to 0.5. Obtained results show that values of DB_{Frey} for the commercial AHBP are very similar to those obtained for the adequate synthesized samples. On the other side, DB values of AHBP-4II, AHBP-6II and AHBP-8II are somewhat lower than DB of AHBP-4I, AHBP-6I and AHBP-8I, indicating certain influence of the procedure for the synthesis on degree of branching. This is especially uttered for the AHBP-8II, since during the synthesis of this sample the ratio between monomer and core was the highest. From these results it was concluded that sequential addition of the stoichometric amount of a monomer to the previously synthesized pseudo generation (pseudo one-step procedure) increases degree of branching of the AHBP to a certain amount.

In ^{13}C NMR spectra (Figure 4) of most AHBP (except in spectra of samples AHBP-2I and AHBP-4II) a small peak at 45.6 ppm was also detected. This signal belongs to the quaternary carbon attached to an unreacted –COOH group and two reacted -OH groups, i.e. focal –COOH group of poly(bis-MPA) [19]. The same signal was also observed by other authors in the ^{13}C NMR spectra of the Boltorn® samples [15]. The presence of this signal indicates that during the synthesis of AHBP, hyperbranched structures without the DiTMP core molecule were also formed and that the reaction proceeded not only by addition of monomer to the growing HB molecule, but also between the monomers and oligomers. Since the relative intensity of this signal increases with increasing pseudo generation number (Table 2), it was concluded that during the synthesis of the higher pseudo generation samples, more –COOH groups were left unreacted.

The mass fractions of the end –OH groups, w_{OH}, (Table 3), were calculated using the values of experimentally determined hydroxyl number, N_{HN}, (Table 3), $(n_{OH})_{theor}$ (Table 1) and values of theoretical hydroxyl number, $(N_{HN})_{theor}$ (equation (3)). On the other side, the ratio between the –OH groups belonging to the linear and terminal units was for every sample calculated using the corresponding integrals from ^{13}C NMR spectra and knowing that terminal units have two unreacted –OH groups, while linear units have only one. Using all these parameters the percentage of terminal –OH groups per molecule was calculated (Table 3). These values show that around 50 % of all –OH groups present in the molecule of HB polyesters belongs to the terminal groups, while the rest is situated on the linear units. Beside that, the percentage of end –OH groups per molecule decreases for the samples of higher pseudo generation, which has significant influence on the behaviour of AHBP in solution [22]. Results obtained by rheological investigation of AHBP indicate the presence of higher extent of polar interactions between different molecules in concentrated solutions of lower pseudo generation AHBP in N-methyl-2-pyrrolidinon, due to the higher fraction of end –OH groups per molecule, in comparison to the samples of the higher pseudo generation [22].

Table 3. Values of $DB_{Fréchet}$, DB_{Frey}, N_{AN}, N_{HN}, mass fraction of end –OH groups, w_{OH}, and percentage of terminal –OH groups per molecule

Sample	$DB_{Fréchet}$	DB_{Frey}	N_{AN} [mg KOH · g^{-1}]	N_{HN} [mg KOH · g^{-1}]	w_{OH} [mass %]	% of term. –OH groups
AHBP-2I	0.47	0.28	5.87	541.1	16.4	58.7
AHBP-3I	0.45	0.32	8.88	507.2	15.4	53.8
AHBP-4I	0.44	0.34	10.82	492.5	14.9	51.1
AHBP-5I	0.45	0.37	11.83	487.4	14.8	51.1
AHBP-6I	0.43	0.35	12.21	480.2	14.5	49.5
AHBP-8I	0.43	0.35	14.11	476.9	14.4	48.9
AHBP-10I	0.42	0.34	14.57	473.7	14.3	48.4
AHBP-4II	0.42	0.30	9.00	488.5	14.8	50.1
AHBP-6II	0.43	0.29	10.98	467.6	14.2	52.1
AHBP-8II	0.37	0.22	13.23	464.7	14.1	46.6
BH-2	0.43[a]	0.30[a]	7.50[b]	501.1	15.2	-
BH-3	0.42[a]	0.31[a]	7.10[b]	474.1	14.4	-
BH-4	0.40[a]	0.34[a]	6.50[b]	470.5	14.3	-

[a]Data of Luciani et al. [10]
[b]Data from supplier.

Portions of different monomer units in AHBP, calculated from ^{13}C NMR results, are in Table 4 compared with the same parameters calculated using the equations developed by Frey [7c]. According to Frey's model, the equal reactivity of –OH groups belonging to the terminal and linear units and absence of the undesired side reactions are assumed. The experimentally determined portions of linear units are for all AHBP higher than values calculated according to Frey. This indicates that reactivity of different –OH groups was not the same, i.e. that terminal –OH groups had bigger reactivity than linear. As a consequence of that, experimentally determined portions of terminal units are for all AHBP higher, while the portions of dendritic units are lower than in the ideal case. Samples of the series II (synthesized by one-step procedure) have slightly lower perfection of the structure compared with the adequate samples synthesized by pseudo one-step procedure ((x_D)$_{exp}$ (II series AHBP) < (x_D)$_{exp}$ (I series AHBP)).

Table 4. Portions of terminal, x_T, linear, x_L, and dendritic, x_D, units of AHBP, experimentally determined and theoretically calculated [7c]

Sample	$(x_T)_{exp}$ [%]	$(x_L)_{exp}$ [%]	$(x_D)_{exp}$ [%]	$(x_T)_{theor}$ [%]	$(x_L)_{theor}$ [%]	$(x_D)_{theor}$ [%]
AHBP-4I	29.3	56.2	14.5	25.4	50.0	24.6
AHBP-4II	29.7	57.6	12.7	25.0	50.0	25.0
AHBP-6I	28.0	56.9	15.1	25.6	50.0	24.4
AHBP-6II	31.0	57.2	11.8	26.2	50.0	23.8
AHBP-8I	27.4	57.1	15.5	26.0	50.0	24.0
AHBP-8II	27.6	63.3	9.1	25.0	50.0	25.0

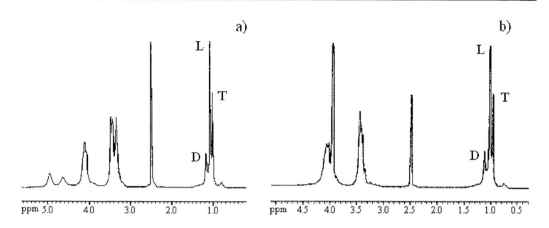

Figure 5. ^{1}H NMR spectra of the sample AHBP-4I determined a) in DMSO-d_6 and b) in mixture D$_2$O/DMSO-d_6

The presence of terminal, linear and dendritic units was also detected in ^{1}H NMR spectra of AHBP. When DMSO-d_6 is used as solvent, methyl protons belonging to the terminal, linear and dendritic units resonate at 1.01, 1.07 and 1.16 ppm, respectively (Figure 5a) [14]. In ^{1}H NMR spectra obtained using the mixture D$_2$O/DMSO-d_6 to dissolve samples (Figure 5b), these signals are slightly shifted ($\Delta\delta$ = -0.04 ppm). However, signals of different units are poorly resolved in both types of solvents and therefore they were not used to calculate DB of the AHBP. Additional signals of the methyl groups from the linear units were found in ^{1}H NMR spectra at 1.12 ppm in DMSO-d_6 and at around 1.09 ppm in the mixture D$_2$O/DMSO-d_6 [15]. On the other side, methyl protons from the DiTMP resonate at 0.80 ppm in DMSO-d_6 and at 0.75 ppm in the mixture D$_2$O/DMSO-d_6. This signal is very well detectable in the ^{1}H NMR spectra of the lower pseudo generation HB polyesters, i.e. up to the fourth pseudo generation. The intensity of this signal decreases in the ^{1}H NMR spectra of the samples of higher pseudo generations, due to the higher ratio between monomer and core molecule. The same conclusions are valid for signals of methylene protons from the core (at 1.3 and 3.2 ppm). When ^{1}H NMR measurements were performed in DMSO-d_6, two signals which correspond to the hydroxyl group protons were detected. The first signal appears at 4.64 ppm (protons of the –OH groups from the terminal units) and another at 4.95 ppm (protons of the –OH groups from the linear units) [15].

In DMSO-d_6, the methylene groups of AHBP in vicinity of reacted hydroxyl groups (-CH$_2$OR) resonate at 4.10 ppm, while methylene groups attached to unreacted hydroxyl groups (-CH$_2$OH) resonate at 3.40 ppm [13]. However, since the measurements were done at room temperature, signal of water protons coincided with signal of the –CH$_2$OH (Figure 5a) [15]. This makes it difficult to obtain the real integral of the –CH$_2$OH signal, which is necessary to have for the calculation of the number average degree of polymerization, $(P_n)_{NMR}$. On the other side, when a small portion of D$_2$O is added into the solution of the samples in DMSO-d_6, the signal of the water protons is shifted to 4.0 ppm (Figure 5b). Therefore, the real integral of the –CH$_2$OH protons signal was calculated by comparing the integrals of the signals for –CH$_2$OH and –CH$_3$ groups (belonging to the monomer units) obtained in DMSO-d_6 and in the mixture D$_2$O/DMSO-d_6 in the following way:

Table 5. Integrated intensities of different signals from the ^{1}H NMR spectra of AHBP ($I(OH) = I(OH)_L + I(OH)_T$)

Sample	Integrated intensities of the signals				
	$I(CH_3)_{HB}$	$I(CH_3)_{B4}$	$I(CH_2OH)$	$I(CH_2OR)$	$I(OH)$
AHBP-2I	5.93	1.09	4.86	3.64	2.35
AHBP-3I	5.46	0.42	4.19	3.49	2.04
AHBP-4I	5.77	0.21	4.10	3.68	2.03
AHBP-5I	5.10	-	3.74	3.30	1.84
AHBP-6I	5.20	-	3.62	3.43	1.80
AHBP-8I	5.14	-	3.55	3.33	1.78
AHBP-10I	4.96	-	3.44	3.18	1.72
AHBP-4II	5.31	0.28	3.76	3.24	1.79
AHBP-6II	5.88	-	4.08	3.93	1.95
AHBP-8II	5.00	/	3.43	3.29	1.80

$$I(CH_2OH)_{DMSO} = \frac{I(CH_3)_{HB,DMSO}\, I(CH_2OH)_{mix}}{I(CH_3)_{HB,mix}} \tag{4}$$

In equation (4), $I(CH_2OH)_{DMSO}$ and $I(CH_2OH)_{mix}$ are the integrals of the –CH$_2$OH group and $I(CH_3)_{HB,DMSO}$ and $I(CH_3)_{HB,mix}$ represent the integrals of the –CH$_3$ groups in DMSO-d_6 and in the mixture D$_2$O/DMSO-d_6, respectively. When a small portion of D$_2$O is added into the solution of AHBP in DMSO-d_6, it is no longer possible to detect signals of the –OH and –CH$_2$OR groups. Therefore, integrals of these signals were calculated from the spectra obtained in DMSO-d_6 and these values are listed in Table 5 together with integrated intensities of other signals from ^{1}H NMR spectra. Using the results presented in Table 5, $(P_n)_{NMR}$, and $(M_n)_{NMR}$ of the AHBP were calculated (Table 6 and Table 7, respectively). The presence of the newly formed B$_2$ core was also included in the calculation. This was done by combination of the results obtained from ^{13}C and ^{1}H NMR spectroscopy. A new B$_2$ core molecule actually represents AB$_2$ monomer (bis-MPA) with deactivated A group (-COOH) [15]. Beside that, the calculation of $(P_n)_{NMR}$ and $(M_n)_{NMR}$ was carried out with assumption that other side reactions (cyclization through ether or ester bonds) did not occur during the synthesis. These side reactions will be discussed later in the text.

The calculation of $(P_n)_{NMR}$ is for samples up to the fourth pseudo generation relatively simple, since the portion of the DiTMP core molecule, N_{B4}, can be calculated from the integral of its methyl protons signal, $I(CH_3)_{B4}$, in the following way:

$$N_{B4} = \frac{I(CH_3)_{B4}}{6} \tag{5}$$

On the other side, for the polyesters of higher pseudo generation (from fifth till tenth), N_{B4} was determined in another way, because the signal of the methyl protons from DiTMP core is too small to be used for calculation. Since the DiTMP core has 16 methylene protons, the portion of the B$_4$ core for the higher pseudo generation samples was calculated according to equation (6):

$$N_{B4} = \frac{I(CH_2)_{B4}}{16} \tag{6}$$

where $I(CH_2)_{B4}$ was calculated in the manner described by Žagar et al. for the commercial BH-4 sample [15].

From the integral $I(CH_3)_{HB}$, which represents the integral of the signal of methyl protons belonging to the HB structure and B_2 core molecule, i.e. bis-MPA, the portion of the monomer units plus the newly formed B_2 core, N_{M+B2}, was calculated using the equation (7):

$$N_{M+B2} = \frac{I(CH_3)_{HB}}{3} \tag{7}$$

Consequently, the ratio N_{M+B2}/N_{B4} represents the average number of monomer units plus B_2 core per B_4 core molecule:

$$\frac{N_{M+B2}}{N_{B4}} = n_T + n_L + n_D + n_{B2} \tag{8}$$

where n_T, n_L, n_D and n_{B2} are the average numbers of terminal, linear, dendritic units and B_2 core molecule, respectively. Values of the n_T, n_L, n_D and n_{B2} were calculated from the ratio N_{M+B2}/N_{B4} (results obtained from the ^1H NMR spectra, Table 5) and from the ratio of corresponding integrals of terminal, linear, dendritic units and the integral of the quaternary carbon attached to the unreacted –COOH group (results obtained from the ^{13}C NMR spectra, Table 2).

Values of $(P_n)_{NMR}$ were calculated as follows:

$$(P_n)_{NMR} = \frac{N_M}{N_{B2} + N_{B4}} \tag{9}$$

where the portion of the newly formed B_2 core, N_{B2}, is:

$$N_{B2} = \frac{n_{B2}}{n_T + n_L + n_D + n_{B2}} N_{M+B2} = n_{B2} N_{B4} \tag{10}$$

and the portion of the monomer units, N_M, is:

$$N_M = N_{M+B2} - N_{B2} \tag{11}$$

Calculated values of n_T, n_L, n_D and n_{B2} are together with $(P_n)_{NMR}$ values collected in Table 6.

Table 6. Average numbers of terminal, n_T, linear, n_L, dendritic units, n_D, B_2 core molecule, n_{B2}, and number average degree of polymerization, $(P_n)_{NMR}$ of investigated AHBP

Sample	n_T	n_L	n_D	n_{B2}	$(P_n)_{NMR}$
AHBP-2I	4.1	5.7	1.1	0	11
AHBP-3I	8.4	14.3	3.3	0.1	24
AHBP-4I	15.9	30.4	7.8	0.4	39
AHBP-5I	33.3	63.8	18.6	1.5	46
AHBP-6I	64.8	131.9	34.9	2.6	64
AHBP-8I	218.8	456.4	123.6	16.6	45
AHBP-10I	883.1	1879.4	474.9	70.6	45
AHBP-4II	11.1	21.6	4.7	0	37
AHBP-6II	56.5	103.9	21.5	4.6	32
AHBP-8II	161.3	369.2	53.3	11.9	45

Finally, values of $(M_n)_{NMR}$ were calculated using equation (12):

$$(M_n)_{NMR} = (P_n)_{NMR}\left[M_{bis-MPA} - M_{water}\right] + x_{B4}M_{B4} + x_{B2}M_{B2} \tag{12}$$

where $M_{bis-MPA}$, M_{water}, M_{B4} and M_{B2} are the molar masses of the monomer, water, DiTMP (B_4 core) and B_2 core (134 g $\cdot$ mol^{-1}), respectively, while x_{B4} and x_{B2} are the fractions of the B_4 and B_2 core:

$$x_{B4} = \frac{N_{B4}}{N_{B4} + N_{B2}} \quad \text{and} \quad x_{B2} = 1 - x_{B4} \tag{13}$$

The calculated values of x_{B4}, x_{B2} and $(M_n)_{NMR}$ are listed in Table 7.

Table 7. Fractions of the B_4, x_{B4}, and B_2 core, x_{B2}, number average molar mass, $(M_n)_{NMR}$, values of the deviation from the perfect dendrimer structure, Δ, the percentage of the formed ether groups, x_{ether}, and extent of cyclization, ξ, of investigated AHBP

Sample	x_{B4}	x_{B2}	$(M_n)_{NMR}$ [g $\cdot$ mol^{-1}]	Δ [%]	x_{ether} [%]	ξ [%]
AHBP-2I	1.00	0	1514	72.5	3.4	-
AHBP-3I	0.91	0.09	2977	72.5	2.5	31.9
AHBP-4I	0.72	0.28	4707	72.1	0.7	40.1
AHBP-5I	0.40	0.60	5540	69.0	0.1	45.0
AHBP-6I	0.28	0.72	7637	71.8	0	53.2
AHBP-8I	0.06	0.94	5396	75.7	0	33.8
AHBP-10I	0.01	0.99	5378	76.8	0	33.9
AHBP-4II	1.00	0	4588	83.2	5.0	-
AHBP-6II	0.18	0.82	3925	82.7	3.1	-
AHBP-8II	0.08	0.92	5398	89.5	0	39.2

Since the sample AHBP-2I has approximately 11 monomer units in its structure and the average number of terminal, linear and dendritic units per molecule is known (Table 6) it was possible to draw the most probable structure of this sample (Figure 6). As can be seen from Figure 6, one of the –OH groups from the DiTMP core did not react, while six linear units can be differently incorporated in the structure of the AHBP-2I sample. The drawing of the most probable structure for the higher pseudo generation samples is much complicated, due to the numerous combinations in arrangements between different monomer units plus the influence of the other side reactions on the final structure. By comparison of the obtained results from this study with the perfect structure of the dendrimer, it was calculated that the deviation from the perfect dendrimer structure, Δ, is for the samples of the series I up to the sixth pseudo generation 72 % and 76 % for the eighth and tenth pseudo generation, while for the samples of the series II the deviation is 83 % up to the sixth pseudo generation and 89 % for the eighth pseudo generation sample (Table 7). These results show that samples synthesized by pseudo one-step procedure have Δ which is approximately 10 % lower than for the samples synthesized by one-step procedure.

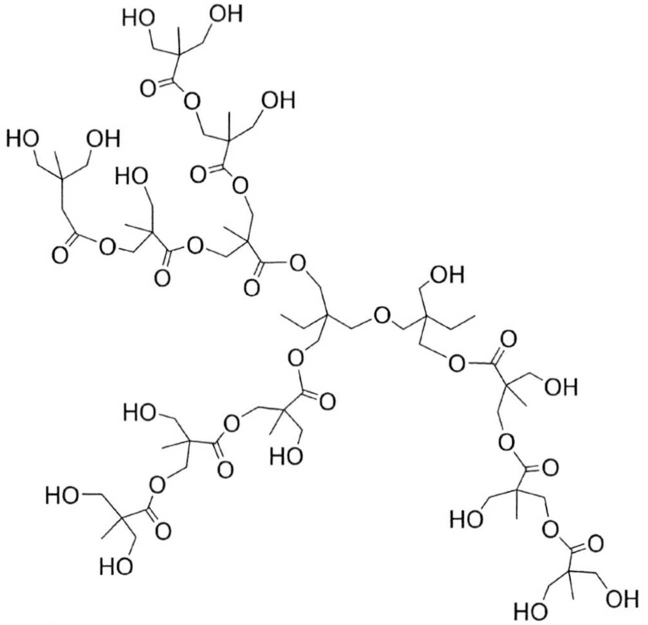

Figure 6. The most probable structure of the sample AHBP-2I

The calculated values of the N_{M+B2}/N_{B4} are in good agreement with the values of the core/monomer ratio used for the synthesis of the different pseudo generation AHBP (Table 1). However, only one part of the used monomers was incorporated into the HB structure in expected manner. This is especially noticeable for the samples of higher pseudo generation. The formation of the HB macromolecules with B_2 core increases with the increasing number of pseudo generation (Table 7) and by changing the procedure for the synthesis from pseudo one-step to one-step. At the same time, the fraction of the HB species with DiTMP core molecule, x_{B4}, decreases, due to the self-condensation of the bis-MPA. When larger amount of monomer is present in the reaction mixture (during the synthesis of the higher pseudo generation samples or during the one-step procedure), the probability for this side reaction is increased. Because of that, the total number of macromolecules in the reaction mixture increases, as well as the number of lower molar mass branches. This finally leads to the considerable decrease of the M_n in comparison to M_{theor} (especially for the samples of higher pseudo generation; Table 1). The increase of the molar mass only up to the sixth pseudo generation for the samples of series I was also confirmed by limiting viscosity number measurements, which were performed using N-methyl-2-pyrrolidinon or 0.7 mass % solution of LiCl in N,N-dimethylacetamide as solvents [22,23].

As it was mentioned before, intramolecular etherification also occurs during the synthesis of AHBP. The amount of the formed ether groups was calculated from ^{1}H NMR spectra by comparing the integrals of the –OH and –CH$_2$OH groups. Since methylene protons of the –CH$_2$–O–CH$_2$– group resonate at the same location as the protons of the –CH$_2$OH group, ½ I(CH$_2$OH) is for some samples not proportional to the real amount of the –OH groups in the structure, i.e. it is bigger than I(OH) [15]. This implies that etherification as a side reaction occurred during the synthesis of these polymers. In the calculation of the amount of ether groups it is necessary to include the presence of the ether groups which belongs to the DiTMP core. Protons of the methylene groups near the ether group in DiTMP resonate at around 3.2 ppm. For the samples of the lower pseudo generation this signal is relatively good separated from the signal of the –CH$_2$OH groups (3.4 ppm) and therefore, it was not included for the calculation of I(CH$_2$OH). Finally, the amount of the formed ether groups, x_{ether}, in the samples from second till forth pseudo generation was calculated using the equation (14):

$$x_{ether} = \left[1 - \frac{I(OH)}{\frac{1}{2}I(CH_2OH)} \right] \times 100 \tag{14}$$

On the other side, in the ^{1}H NMR spectra of the higher pseudo generation AHBP the signal of the methylene protons near the ether group from the DiTMP core is not possible to distinguish from the signal of the –CH$_2$OH groups. Because of that, the percentage of the ether groups for the higher pseudo generation AHBP was calculated in the following manner:

$$x_{ether} = \left[1 - \frac{I(OH)}{\frac{1}{2}\left(I(CH_2OH - \frac{1}{4}I(CH_2)_{B4} \right)} \right] \times 100 \tag{15}$$

The results calculated using equations (14) and (15) are presented in Table 7. Percentage of ether groups decreases from 3.4 % for the second to 0.1 % for the fifth pseudo generation. The probable reason for such behaviour is the fact that lower pseudo generations AHBP have the highest percentage of the –OH groups per molecule (Table 3), increasing in this way the possibility for the ether formation. For the samples AHBP-4II and AHBP-6II, synthesized by one-step procedure, the amount of the formed ether groups is higher compared to adequate samples obtained with pseudo one-step procedure. The calculated amount of the ether groups is for all samples relatively small (1-5 %) and much lower than values determined for poly(bis-MPA) without the core molecule (1-12 %) [14]. Similar results for sample BH-4 were also obtained by Žagar et al. [15]. Due to the low amount of the formed ether groups in BH-4 in comparison to the results presented in ref. [14], these authors have concluded that ether groups were probably formed during intramolecular etherification (cyclization). The same conclusion can be made for the AHBP investigated in this work.

Determined $(M_n)_{VPO}$ values (Table 1) are lower than calculated $(M_n)_{NMR}$ values (Table 7) as a consequence of the formation of cyclic structures through the ester bonds, which were not included in the evaluation of the $(M_n)_{NMR}$ from ¹H NMR data [13]. The comparison of the $(M_n)_{VPO}$ and $(M_n)_{NMR}$ values was used to calculate the extent of cyclization, ξ, in examined AHBP (Table 7) [13]. Value of ξ increases up to the sixth pseudo generation and ξ of the sample AHBP-8II is higher than for the sample AHBP-8I. Combination of ¹H NMR and VPO results shows that for samples AHBP-3I, AHBP-4I and AHBP-5I cyclization occurred through ester and ether bonds, which is quantified in the value of ξ. On the other side, in samples AHBP-6I, AHBP-8I and AHBP-10I the presence of the cycles formed by intramolecular etherification was not detected in ¹H NMR spectra. Therefore, it is possible to suggest that for these samples cyclization occurred only as a consequence of intramolecular esterification.

The presence of the side reaction products in the AHBP was also confirmed from the results obtained by MALDI-TOF MS. Mass spectra of these samples (Figure 7) are consisted of regularly spaced series of peaks (series 1), separated by 116 mass units, corresponding to the molar mass of the monomer unit. Each peak in these spectra represents a molecule of different degree of polymerization. Second series of signals (series 2) with lower intensity was also observed in the mass spectra of the samples AHBP-3I and AHBP-4I (Figure 7b and 7c). The position of this series of signals is 18 mass units lower than the main series of peaks. Finally, in the mass spectrum of the sample AHBP-4I additional series of lower intensity signals was also found, 18 mass units after the main series (series 3). The molar mass of a particular molecule was calculated by consideration of the polymerization process and using the equations developed by Frey and co-workers, since HB polymers which they have examined are of similar structure as AHBP investigated in this work [13]. According to the Frey's equations, the major series of signals (series 1) observed for the AHBP corresponds to molecules formed in the desired successive condensation reaction, i.e. they represent acyclic HB molecules containing a core unit [13]. The second series of signals (series 2) is attributed to the loss of the single water moiety in order to form cyclic structures. Molecules presented with this series of peaks were formed through intramolecular esterification or by hydroxy-ester interchange and represent macro-cyclic bis-MPA macromolecules without a DiTMP core. On the other hand, signals of the series 2 can also be assigned to the cycles formed by intramolecular etherification. Molar mass of these cycles is simply the molar mass of the signal from the series 1 reduced by 18 mass units, since these cyclic species also contain a core unit. However, in MALDI-TOF mass spectra of AHBP it is not possible to distinguish between cycles formed by intramolecular etherification and esterification. It was excluded that the loss of the water took place during the mass spectral ionization, since the presence of the cyclic structures in the AHBP was proved by ¹H NMR and VPO measurements. In the MALDI-TOF mass spectrum of sample AHBP-2I (Figure 7a) a second series of signals was not detected. This shows that the extent of cyclization during the synthesis of this sample was much lower than for the AHBP samples of higher pseudo generation. Since the intensity of the second series of signals increases with increasing number of pseudo generations, it was concluded that the extent of cyclization has the same trend.

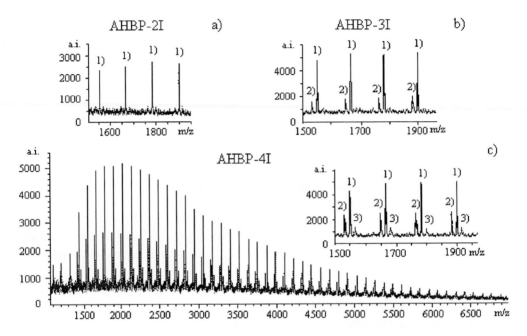

Figure 7. MALDI-TOF mass spectra of a) AHBP-2I, b) AHBP-3I and c) AHBP-4I

In the MALDI-TOF mass spectrum of sample AHBP-4I (Figure 7c) another sub-distribution of signals can be observed (series 3). Taking into the account results obtained by NMR spectroscopy, this series of peaks is attributed to the acyclic molecules with unreacted –COOH groups, i.e. HB molecules without DiTMP core unit and with B_2 core. Furthermore, signals of the cyclic structures formed from the acyclic HB species without DiTMP core are overlapping with signals from the HB species with a core unit. The explanation of the MALDI-TOF mass spectrum of the commercial sample BH-4 (Figure 8) is the same as for the AHBP-4I. Furthermore, the mass spectrum of this sample implies that the amount of the cyclic structures is much higher than in the sample AHBP-4I.

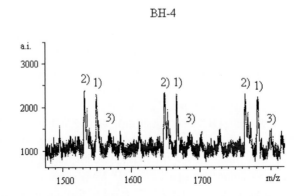

Figure 8. MALDI-TOF mass spectra of BH-4

The identification of different molecules in the MALDI-TOF mass spectra of higher pseudo generation AHBP is somewhat different than for the lower pseudo generation samples. From the NMR results it was determined that AHBP from fifth till tenth pseudo generation have higher x_{B2} than x_{B4} (Table 7). Therefore, it is assumed that in the case of higher pseudo generation AHBP the third series of peaks corresponds to acyclic HB molecules containing a DiTMP core unit, while the first series of signals is attributed to the acyclic molecules with unreacted –COOH group (Figure 9). A second series of signals is assigned to the macrocyclic bis-MPA macromolecules with lack of the DiTMP core, while cyclic structures formed by the loss of the single water moiety from the third distribution of molecules are overlapping with the first series of signals. It was also observed that the relative proportion of the signals, belonging to cyclic structures, increases with the increasing size of the molecules, i.e. with increasing degree of polymerization. This is especially pronounced in the mass spectra of the higher pseudo generation samples. The same was obtained for the samples synthesized by one-step procedure (samples AHBP-6II and AHBP-8II). For these two samples intensities of the signals corresponding to the cyclic structures are slightly higher than for the analogue AHBP samples synthesized by pseudo one-step procedure.

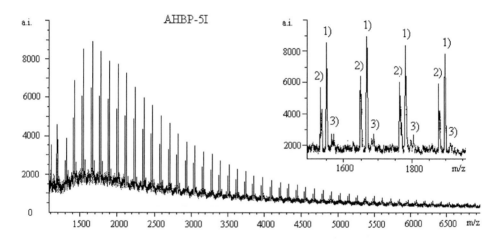

Figure 9. MALDI-TOF mass spectra of AHBP-5I

CONCLUSION

According to the experimental investigation of the influence of the type of procedure (pseudo one-step or one-step) of the acid-catalyzed esterification of bis-MPA (AB_2 monomer) and the DiTMP (B_4 functional core) in melt on molecular structure of AHBP, the following results were obtained:

- Around 50 % of all –OH groups present in the molecule of AHBP are situated on the linear units, which indicates that during the synthesis reactivity of different –OH groups was not the same. As a consequence of that, the degree of branching of all investigated samples is lower than 0.50.

- Samples synthesized by one-step procedure have somewhat lower values of the degree of branching and higher portion of the linear units than samples synthesized by pseudo one-step procedure. Percentage of end –OH groups per molecule decreases with increasing number of pseudo generation.
- The most important side reactions, which occurred during the synthesis of AHBP are the formation of the poly(bis-MPA). In this manner, HB structures without the B_4 (DiTMP) core molecule were formed. The amount of the HB macromolecules without B_4 core increases with the increasing number of pseudo generation and it is more pronounced in one-step procedure. The main consequences of this side reactions are a significant lowering of the M_n of the AHBP compared with theoretical values and increase of the molar mass only up to the sixth pseudo generation.
- Formation of the cycles by intramolecular etherification and esterification are additional side reactions. The extent of cyclization increases up to the sixth pseudo generation and it is slightly higher for the samples synthesized by one-step procedure. The amount of the formed ether groups decreases with increasing number of the pseudo generation. Up to the fifth pseudo generation (samples of series I), cyclization occurred through ester and ether bonds (intramolecular), while for the higher pseudo generation samples cycles were formed mainly as a consequence of the intramolecular esterification. During the synthesis of the commercial samples, the extent of the mentioned side reactions was higher than for the adequate samples synthesized in this work.
- No matter which synthetic procedure was used for the investigated AHBP (pseudo one-step or one-step), it was not possible to prevent the occurrence of the side reactions and to obtain AHBP of the higher molar mass. The deviation from the ideal dendrimer structure for the samples synthesized by pseudo one-step procedure is around 10 % lower than for the samples synthesized by one-step procedure.

REFERENCES

[1] Vogtle, F.; Wehner, W.; Buhleier, E. *Synthesis* 1978, *55*, 155-158.

[2] Tomalia, D. A.; Baker, H.; Dewald, J.; Hall, M.; Kallos, G.; Martin, S.; Roeck, J.; Ryder, J.; Smith, J. *Polymer J.* 1985, *17*, 117-132.

[3] Newkome, G. R.; Yao, Z. Q.; Baker, G. R.; Gupta, V. K. *J. Org. Chem.* 1985, *50*, 2003-2004.

[4] Fréchet, J. M. J. *J. Macromol. Sci. Pure Appl. Chem.* 1996, *A33(10)*, 1399-1425.

[5] a) Kim, Y. H.; Webster, O. W. *J. Am. Chem. Soc.* 1990, *112*, 4592-4593; b) Johansson, M.; Hult, A. *J. Coat. Technol.* 1995, *67*, 35-39; c) Massa, D. J.; Shriner, K. A.; Turner, S. R.; Voit, B. I. *Macromolecules* 1995, *28*, 3214-3220; d) Zhang, Y.; Wang, L.; Wada, T.; Sasabe, H. *Macromo.l Chem. Phys.* 1996, *197*, 667-676; e) Griebel, T.; Maier, G. *Polym. Prepr. (ACS, Polymer Division)* 2000, *41*, 89-90.

[6] Malmström, E.; Johansson, M.; Hult, A. *Macromolecules* 1995, *28*, 1698-1703.

[7] a) Radke, W.; Litvinenko, G.; Müller, A. H. E. *Macromolecules* 1998, *31*, 239-248; b) Yan, D.; Zhou, Z.; Müller, A. H. E. *Macromolecules* 1999, *32*, 245-250; c) Hölter, D.;

Frey, H. *Acta. Polym.* 1997, *48*, 298-309; d) Hanselmann, R.; Hölter, D.; Frey, H. *Macromolecules* 1998, *31*, 3790-3801.

[8] Sunder, A.; Hanselmann, R.; Frey, H.; Mülhaupt, R. *Macromolecules* 1999, *32*, 4240-4246.

[9] a) Nunez, C. M.; Chiou, B.-S.; Andrady, A. L.; Khan, S. A. *Macromolecules* 2000, *33*, 1720-1726; b) Rogunova, M.; Lynch, T.-Y. S.; Pretzer, W.; Kulzick, M.; Hiltner, A.; Baer, E. *J. Appl. Polym. Sci.* 2000, *77*, 1207-1217; c) Hsieh, T.-T.; Tiu, C.; Simon, G. P. *Polymer* 2001, *42*, 1931-1939; d) Mackay, M. E.; Carmezini, G. *Chem. Mater.* 2002, *14*, 819-825.

[10] Luciani, A.; Plummer, C. J. G.; Nguyen, T.; Garamszegi, L.; Månson, J.-A. E. *J. Polym. Sci. Part B: Polym. Phys.* 2004, *42*, 1218-1225.

[11] Arndt, K. F.; Müller, G. *Polymer-charakterisierung*; Carl Hanser Verlag: München - Wien, 1996; 100-103.

[12] Dušek, K.; Šomvársky, J.; Smrčková, M.; Simonsick Jr., W. J.; Wilczek, L. *Polymer Bulletin* 1999, *42*, 489-496.

[13] Burgath, A.; Sunder, A.; Frey, H. *Macromol. Chem. Phys.* 2000, *201*, 782-791.

[14] Komber, H.; Ziemer, A.; Voit, B. *Macromolecules* 2002, *35*, 3514-3519.

[15] Žagar, E.; Žigon, M. *Macromolecules* 2002, *35*, 9913-9925.

[16] Žagar, E.; Žigon, M.; Podzimek, S. *Polymer* 2006, *47*, 166-175.

[17] Chikh, L.; Tessier, M.; Fradet, A. *Polymer* 2007, *48*, 1884-1892.

[18] Malmström, E.; Hult, A. *Macromolecules* 1996, *29*, 1222-1228.

[19] Magnusson, H.; Malmström, E.; Hult, A. *Macromolecules* 2000, *33*, 3099-3104.

[20] Hawker, C. J.; Lee, R.; Fréchet, J. M. J. *J. Am. Chem. Soc.* 1991, *113*, 4583-4588.

[21] Hölter, D.; Burgath, A.; Frey, H. *Acta Polym.* 1997, *48*, 30-35.

[22] Vuković, J.; Lechner, M. D.; Jovanović, S. *Macromol. Chem. Phys.* 2007, *208*, 2321-2330.

[23] Vuković, J.; Lechner, M. D.; Jovanović, S. *J. Serb. Chem. Soc.* 2007, *72*, 1493-1506.

In: Modern Trends in Macromolecular Chemistry
Editor: Jon N. Lee

ISBN: 978-1-60741-252-6
© 2009 Nova Science Publishers, Inc.

Chapter 3

FUNCTIONAL PLATINUM-CONTAINING MACROMOLECULES WITH TUNABLE TRIPLET ENERGY STATES

Cheuk-Lam Ho and Wai-Yeung Wong [*]

Department of Chemistry and Centre for Advanced Luminescence Materials,
Hong Kong Baptist University, Waterloo Road, Kowloon Tong, Hong Kong, P.R. China.

ABSTRACT

This chapter discusses recent progress in the development of metallopolyyne polymers of platinum(II) with a vast range of triplet and bandgap energies. Considerable focus is placed on the design concepts toward the structural tuning of the optical bandgap and triplet state of this family of functional metallopolymers. Various applications of such class of metal-containing macromolecules in optoelectronic and photonic devices are critically discussed. The future challenges and perspectives of this research area are also highlighted.

Keywords: Luminescence; Macromolecules; Metallopolymers; Platinum; Triplet states

ABBREVIATIONS AND SYMBOLS

E_g	Bandgap
eV	Electron-volt
ITO	Indium tin oxide
$(k_{nr})_P$	Non-radiative phosphorescence decay rate
$(k_r)_P$	Radiative phosphorescence decay rate
ΔE_{S-T}	Energy gap for the T_1-S_0 transition

[*] Correspondence to: Wai-Yeung Wong, Department of Chemistry, Hong Kong Baptist University, Waterloo Road, Kowloon Tong, Hong Kong, P.R. China. E-mail: rwywong@hkbu.edu.hk

P3HT	Poly(3-hexylthiophene)
PEDOT:PSS	Poly(3,4-ethylenedioxythiophene):poly(styrenesulfonate)
PCBM	[6,6]-Phenyl C_{61}-butyric acid methyl ester
PL	Photoluminescence
PPV	Poly(p-phenylenevinylene)
S_0	Singlet ground state
S_1	Singlet excited state
T_1	Triplet excited state
TPBI	1,3,5-Tris[N-(phenyl)benzimidazole]benzene

1. INTRODUCTION

Conjugated polymer-based materials have emerged as one of the most exciting areas of scientific endeavor in this decade.[1] They are finding broad applicability in electronic and electro-optical devices, including chemo- and biosensing,[2] electroluminescence,[3] and optical limiting.[4] Although much is known about the properties of the singlet state in conjugated polymers, comparatively less is known about the lowest triplet state.[5] This is clearly an important issue, since the triplet exciton plays an important role in electroluminescence applications,[6] however, phosphorescence is rarely observed due to very low singlet to triplet intersystem crossing yields in organic conjugated materials.[7] Therefore, this has been motivating scientists to investigate the properties of organometallic conjugated materials since the 1950s.[8] This interest derives from the fact that incorporation of heavy metals into an organic conjugated framework can elicit large effects on the electronic and optical properties of the materials.[8] One of them is the quantum spin statistics prediction of a population of singlets and triplets from charge recombination in a ratio of 1:3. The incorporation of heavy metal into the backbone of conjugated organics allows for mixing of singlet and triplet states and thus increases the population of the triplet excited state.

With reference to the Jablonski energy level diagram in a simple photoluminescence (PL) system (Figure 1), there are two radiative decay processes after the absorption of photons by a molecule from the ground state (S_0). They are termed fluorescence ($S_1 \rightarrow S_0$) and phosphorescence ($T_1 \rightarrow S_0$). The relative positions of the lowest singlet (S_1) and triplet (T_1) excited states strongly affect the intersystem crossing rate into the triplet manifold. This provides a major nonradiative decay mechanism for organic systems, which tends to reduce the PL efficiency in purely fluorescent molecules.[1,9]

Polymers derived from alkynyl building blocks have further widened the territory of functional polymers, which are structurally analogous to poly(phenyleneethynylene)s and have high intersystem crossing efficiencies so that their photophysics are dominated by long-lived and phosphorescent $^3\pi-\pi^*$ excited states,[10] both in solution and in the solid state.[10a,b,d] Figure 2 shows the general skeleton of transition metal polyyne polymer (i.e. polymetallayne). Fine-tuning of the desired physical and functional properties can be achieved by appropriate structural variations of the polymetallaynes involving the exploitation of different metal groups (M), auxiliary co-ligands (L) around the metal coordination spheres or bridging spacers (R).

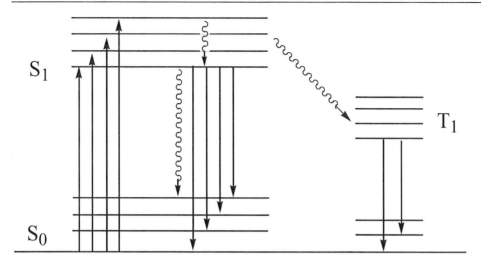

Figure 1. The Jablonski diagram for a classical photoluminescence (PL) system.

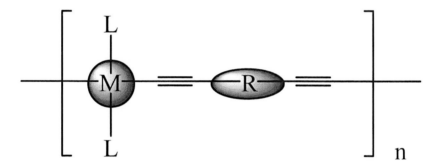

Figure 2. The general skeleton of metal polyyne polymer.

In this chapter, we summarize results on this research area with an emphasis on the influence of organic linkers to the triplet and bandgap energies of the resulting Pt(II) polymers. It is believed that the overlap extent between the metal and ligand conjugated system, the size of the organic spacer and the degree of conjugation in the ligand, and donor-acceptor interaction[11] between Pt(II) atom and the ligands would affect the resultant optical gap in these polymers. It would therefore be very useful to understand what controls the relative positions of the singlet and triplet energy levels in order to chemically tailor the singlet-triplet gap. A concise consideration and a good compromise of the triplet photophysics and the associated bandgap in the polymer are essential for designing highly efficient and wavelength-tunable luminescent and photovoltaically-active materials. There are numerous reports on the experimental and theoretical investigations on the energy levels of S_1 and T_1 states in metallopolyynes to date.[12] In general, the triplet excited states in Pt-containing conjugated polymers comply with the energy gap law, in which the nonradiative decay rate for phosphorescence $(k_{nr})_P$ from the triplet state increases exponentially with decreasing triplet-singlet energy gap according to the expression $(k_{nr})_P \propto \exp(-C\Delta E_{S-T})$, where ΔE_{S-T} is the energy gap for the T_1–S_0 transition and C is a term controlled by the molecular parameters and vibrational modes.[10e] There is always a trade-off between the bandgap and the rate of phosphorescence in such metallopolymeric system. In other words, a high-bandgap polymer with a high T_1 state will favor the observation of phosphorescence whereas the low-gap counterpart will not be triplet-emitting even at low temperatures. There is a strong dependence of the phosphorescence radiative decay rates $(k_r)_P$ on the structural and temperature variations of such macromolecules. A comprehensive account of the effect of the central spacer group on the degree of π-conjugation as well as the spatial extent of the S_1 and T_1 states of these metal polyynes is clearly given.

2. Platinum (II) Polyyne Polymers with Tunable Triplet and Bandgap Energies

The synthesis of Pt(II) acetylenic polymers involves a polycondensation reaction between platinum(II) dichloride and diterminal alkynes via a cuprous halide catalyzed dehydrohalogenation process. The reactions are typically performed in an amine solvent to afford high-molecular-weight polymers. Scheme 1 shows a series of Pt(II) polymers that are going to be discussed in this chapter. The molecular geometries of the dinuclear platinum(II) model complexes of selected molecules were elucidated and the X-ray crystal structures are shown in Figure 3 for **8b**, **14** and **16**, in which the Pt groups are end-capped by the phenyl ring.

Depending on the organic spacers, the optical gap is shifted over about 1.93 eV. Platinum polyyne **1** has been proven an effective model system for the study of triplet excitons in conjugated polymers which was first described by Hagihara et al.[13] In this polyyne, the spin-orbit coupling associated with the heavy element enables efficient intersystem crossing, thus making emission from the triplet manifold directly observable at room temperature. According to the study by Wittmann et al., the absolute intersystem crossing efficiency of **1** is close to 1 at about 16 K.[14] As compared to the polymers with organic spacer naphthalene **2** and anthracene **3**,[15] very weak triplet emission was observed in the former but was nearly unobservable in the latter even at low temperature. While the radiative decay rate for phosphorescence $(k_r)_P$ should be similar in these three organometallic polyynes **1–3** since the Pt(II) atom mainly determines the spin-orbit coupling, k_{nr}, however, increases exponentially with decreasing T_1 energy and is expected to be three orders of magnitude larger in **3** than in **1**, so that the detection of the phosphorescence becomes very difficult. On the other hand, the bandgap decreases as the size of the aromatic linker group increases (**1**: 2.90 eV; **2**: 2.80 eV; **3**: 2.35 eV), as well as the triplet energy $(T_1 \rightarrow S_0)$, consistent with an increase in donor-acceptor interaction between the Pt(II) center and the conjugated ligand in going from benzene, through naphthalene to anthracene as the linker unit.

The Pt(II) polymer **4** shows different phosphorescence energy. Replacing the phenylene spacer by an acetylenic linkage in **4** increases the extent of electron conjugation so that the E_g is reduced to 2.6 eV (relative to 2.9 eV in **1**).[16]

As found for related polymers where the phenyl units are replaced by thiophene ring(s), the optical gap is reduced from 2.80 (**5**) to 2.55 (**6**) and 2.40 (**7**) eV with increasing thiophene content (Figure 4).[10c] This is mainly attributed to an increased delocalization of π electrons along the polymer backbone. However, the overall effect on the bandgap reduction is less pronounced relative to the effect of fused ring in polyynes **1–3** (Figure 4). There would be little benefit in increasing the number of thiophene units above three. As they are cooled down to 18 K, triplet emissions were easily detectable in **5** at 2.06 eV, but there is a sudden disappearance of the triplet emission intensity on going from **5** to **7**. The weak triplet peaks appear at 1.67 and 1.53 eV for **6** and **7**, respectively. This can be rationalized from the notion that increasing the number of thiophenes of the ligand reduces the influence of the heavy metal center, which is mainly responsible for the intersystem crossing.

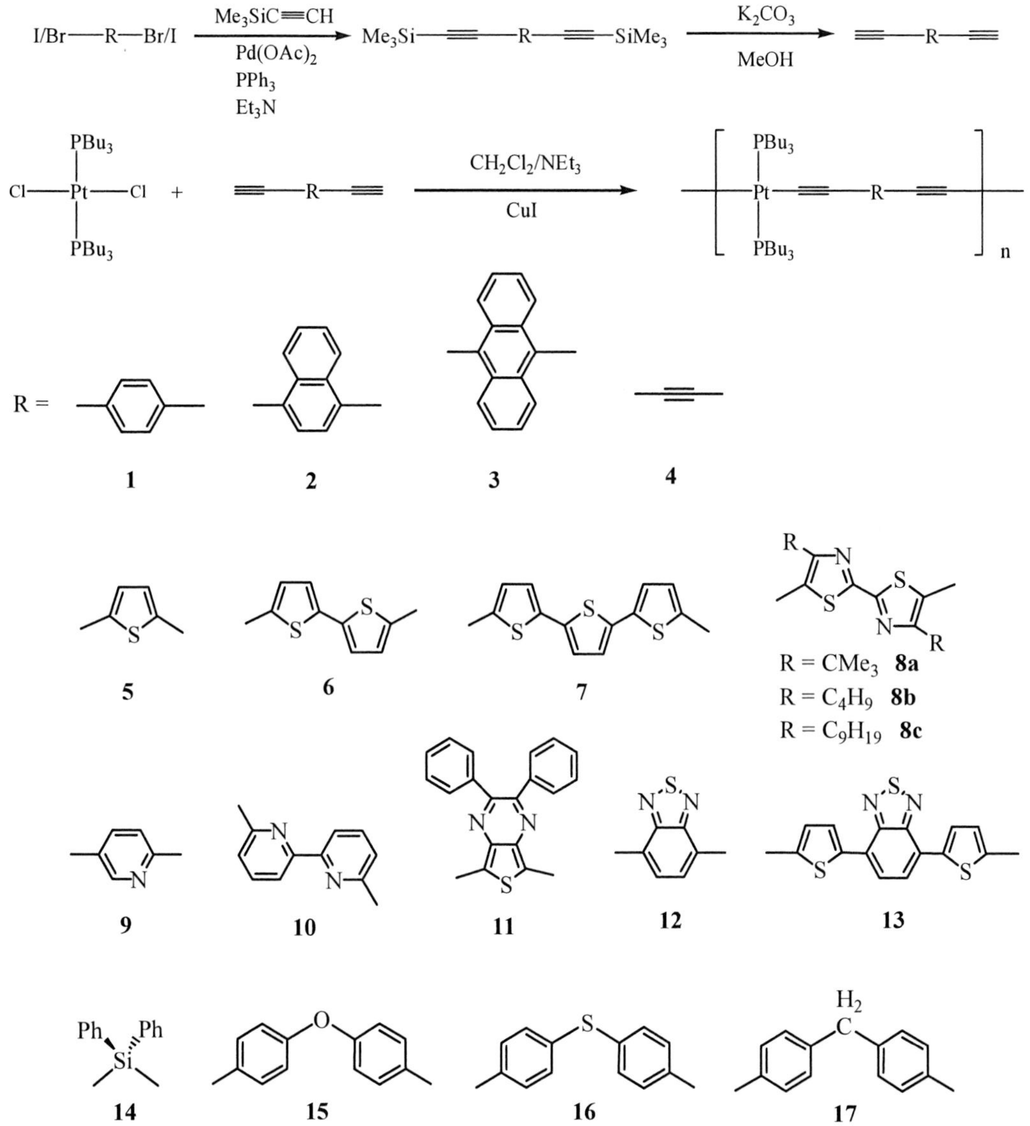

Scheme 1. Synthesis of platinum metallopolyynes with various spacers.

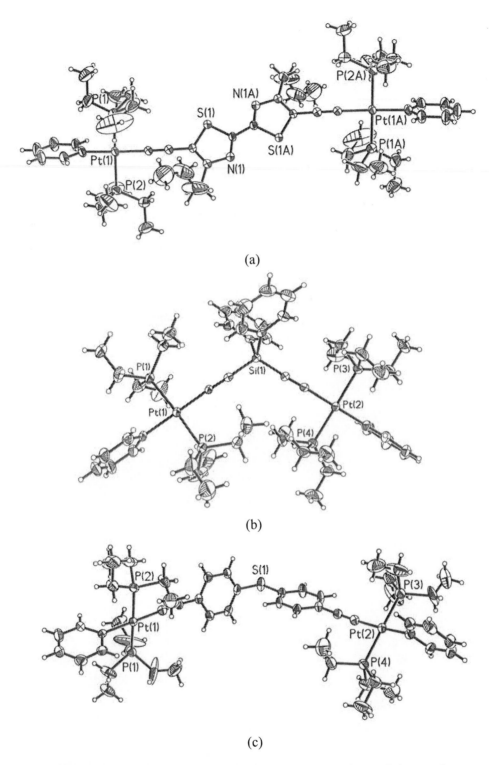

(a)

(b)

(c)

Figure 3. X-ray crystal structures of the model complexes of **8b**, **14** and **16**. Labels on carbon atoms are omitted for clarity.

Polymers **8a–8c**, by virtue of the electron-donating and -withdrawing properties of the thiazole ring in the main chain which are generally regarded as a hydrid of the thiophene and pyridine groups are also known.[17,18] Due to the presence of the electron-withdrawing imine nitrogen atoms, a red shift of about 0.15–0.20 eV of optical gap was observed as compared to the analogous polymer containing 2,2'-bithiophene-5,5'-diyl unit in **6**.[18] In **6**, weak phosphorescence was seen even at 10 K, but only singlet emission with a small Stokes shift from the strongest band in the absorption spectra became apparent in **8a–8c**. It is believed that this emission feature in **6** is attributed to the phosphoresence signal caused by interchain interaction that is similar to other Pt(II) polyynes.

Adding a nitrogen atom into the phenyl ring (i.e. 2,5-diethynylpyridine) leads to an increase of the first transition energy in **9**. The E_g follows the order **9** (3.00 eV) > **1** (2.90 eV) > **5** (2.80 eV) (Figure 5). By comparing with the phenylene unit which shows little donor-acceptor interaction along the rigid backbone of the polymer, the electron rich thiophene unit enhances conjugation.[10c,19] On the other hand, the electron deficient pyridine unit reduces conjugation and increases the electron affinity of the derived polymer. Evidently, reducing conjugation can increase the intersystem crossing, altering their spatial extent of singlet and triplet excited states. From a careful examination of the intensity of the triplet emission to singlet emission at low temperature, the relative efficiency of intersystem crossing was found to increase in the order **5** < **1** < **9**. The energy of triplet emission also experiences a red shift as the conjugation of linker spacer becomes more extensive.

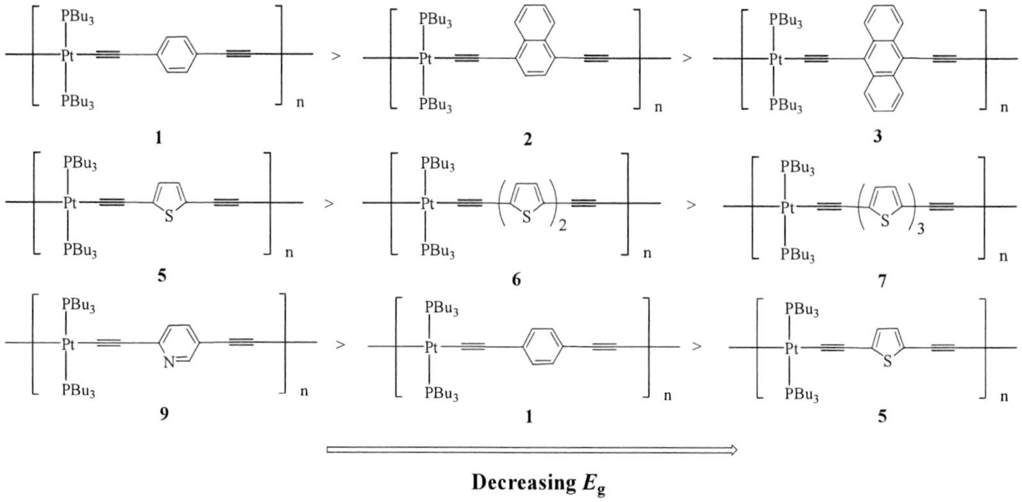

Figure 4. Effect of spacer ring on E_g in various metallopolyyne systems.

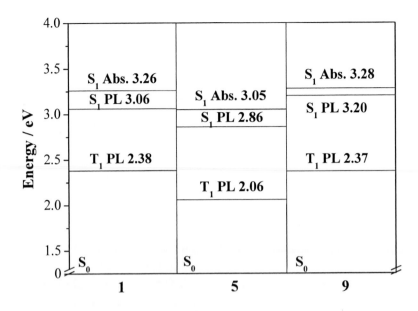

Figure 5. Spatial extent of the lowest excited states for **1, 5** and **9**.

Different from the fully conjugated 2,5-diethynylpyridine-containing polymer **9**, a blue-shifted onset of absorption (ca. 0.3 eV) was observed in **10**. This may be attributed to the fact that the alkynyl groups occur at the 6,6'- position in **10** hindering conjugation between the pyridine rings and may also be due to an increase of the number of pyridine units which enhance the electron-accepting ability of the polymers.[20] Recent observations demonstrate that the intersystem-crossing rate from the singlet excited state to the triplet excited state is notably enhanced when conjugation is decreased.[21] There is no fluorescence but only the phosphorescence band at 10 K for the kinked polyyne **10**. The reduced conjugation shifts the phosphorescence band of **10** to the blue by 0.31 eV as compared to the linear pyridine-containing analogue 9, so that the vibrational peaks of the 10 K emissions are located around 2.68 eV. In the pyridyl class of materials, this is one of the highest energy at which phosphoresce has been observed so far and correspondingly, according to the energy gap law, it should be associated with the lowest rate of non-radiative decay from the T_1 state at 10 K.[10e]

With the aim of preparing some Pt(II) polyynes with bandgap smaller than 2.0 eV, polymers **11**[11a] and **13**[22] consisting of thieno[3,4-*b*]pyrazine and benzothiadiazole, respectively, are also known. The E_g values for **11** and **13** are 1.77 and 1.85 eV, respectively, and these polymers are isolated as deep blue to purple solids. Owing to the presence of a more extended π-electron delocalized system throughout the chain and the creation of an alternate donor-acceptor chromophore based on highly electron-accepting benzothiadiazole rings in combination with the electron-donating thiophene rings, the bandgap of **13** is significantly lowered by 0.7 eV relative to the purely electron-rich bithienyl **6** or electron-deficient benzothiadiazole linked counterpart **12**, both appearing yellow in the solid and solution phases. Both of the polymers exhibit an intense fluorescence band only but no emission from the triplet excited state over the measured spectral window even at low temperature. The measured lifetime for **13** is very short and lies in the nanosecond range, which is in contrast to other Pt(II) polyynes with a higher bandgap. In accordance with the energy gap law for Pt-containing conjugated polymers, there is always a trade-off between the bandgap and the rate of phosphorescence in such a system. Therefore, it is not the triplet state but mostly a charge-transfer excited state that contributes to the efficient photo-induced charge separation in the energy conversion for **13**, which is different from **5** based blends where charge separation occurs via the triplet state of the polymer.[23] From these results, we ascribe the localized state centred at 2.22 eV to a strong donor-acceptor charge-transfer type interaction.

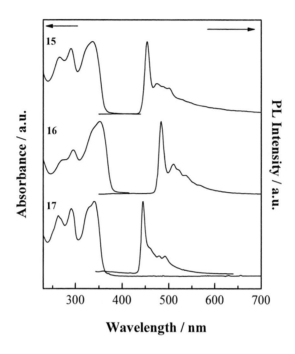

Figure 6. Room-temperature absorption and low-temperature emission spectra of **15–17** in CH$_2$Cl$_2$.

In order to maximize the efficiency of optoelectronic light-emitting devices, triplet excited states must be harvested for light emission. Therefore, considerable research works have been focused on synthesizing some materials with high-energy triplets to avoid competition with non-radiative decay. A series of organometallic polyyne polymers with weakly or non-conjugated organic linkers have been reported. Polymeric Pt(II) species **14** containing silylacetylene units greatly improves the phosphorescence properties which exhibit strong triplet emission with negligible fluorescence intensity at low temperature.[24] The reduced conjugation in the presence of the SiPh$_2$ unit shifts the optical gap and the phosphorescence band to the blue as compared to other Pt(II) polyynes with other aromatic spacers. As compared to **1** and **5**, the phosphorescence band of **14** shifts to the blue by 0.06 and 0.38 eV, respectively. Another approach to achieve high-energy triplet becomes available by using novel conjugated polymers with non-π-conjugated group 16 heteroatoms and CH$_2$ in the polymer chains. The O, S and CH$_2$ functionalities can act as the conjugation-breaking unit to interrupt the electronic π-conjugation. With reference to the similarly prepared bi(*p*-phenylene)-linked polymer,[25] the E_g values of the three polyynes **15–17** are blue shifted,[26,27] and follow the order **17** > **15** > **16**. This suggests that the energy of the S$_1$ state can be modified simply by varying the electronic properties of the spacer group; this is vital for governing the efficiency of the triplet-state emission. We observe that the triplet energies $E(T_1\text{–}S_0)$ follow the order CH_2 (2.79 eV) > O (2.73 eV) > S (2.56 eV), indicating that CH_2 is less effective than the other two in electronic conjugation in the present system (Figures 6 and 7). This observation attests to the importance of designing high-energy gap materials through the use of heteroatoms and CH_2 conjugation-interrupter along the main chain. Moreover, these complexes possess a rigid structure and a large bandgap; all of these properties should favor good blue-light-emitting properties.[28]

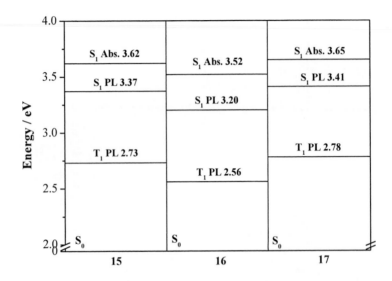

Figure 7. Electronic energy level diagram of Pt polyynes with main group heteroatom elements and CH_2 group based on the absorption and PL data. All energy values are quoted in eV.

Table 1. Photophysical data for platinum-containing polyyne polymers 1–17.

Complex	Organic spacer R	E_g (eV)[a]	Triplet energy T_1 (eV)
1		2.90	2.38
2		2.8	1.96
3		2.35	b
4		2.6	1.88, 2.09, 2.37
5		2.80	2.06
6		2.55	1.67

7		2.40	1.53
8a		2.35	b
8b		2.40	b
8c		2.40	b
9		3.0	2.37
10		3.3	2.68
11		2.2	1.49
12		1.77	b
13		1.85	b
14		3.70	2.44
15		3.40	2.73
16		3.26	2.56
17		3.42	2.79

[a] Estimated from the onset wavelength of the optical absorption.

[b] No triplet emission was observed.

3. OPTOELECTRONIC APPLICATIONS

New perspectives on luminescent organometallic polyyne polymers have resulted in exciting prospects for their applications in PLEDs, photovoltaic or solar cells. A solar cell is a wide area electronic device that converts solar energy into electricity by the photovoltaic effect. Therefore, polymer photovoltaics bring promise as a future renewable source of energy,[29] due to the attractive cost and environmental benefits gained using this technology.

We have studied the photocurrent action of some Pt(II) polyynes. It was shown that 5–7 are good photoconductors in suitably fabricated configurations.[10c] The photocurrent spectra of the Au/5/Al, ITO/6/Al and ITO/7/Al photocells show two peaks, one at the absorption onset [2.92 (5), 2.64 (6) and 2.43 eV (7)] and the other at higher energies [3.81 (5), 3.56 (6) and 3.38 eV (7)]. The second photocurrent peaks are probably caused by absorption into the higher-lying absorption bands. Polymers 5–7 show a short-circuit quantum efficiency of ~0.04% at the first photocurrent peak, which is a common value for single-layer devices. We found no great difference in the maximum photocurrent quantum efficiencies with variation of the thiophene content. The photocurrent quantum efficiency of the second peak is different among 5–7 and is air-sensitive.[10c] The photocells show open-circuit voltages at the first peak of 0.50, 0.75 and 0.47 V and fill-factors of 0.32, 0.35 and 0.30 for 5–7, respectively. The longer lived triplet excitons can migrate to dissociation centers and can thus contribute to the photogeneration of charge carriers. The presence of air can increase the photocurrent in 5-based cell by enhancing exciton dissociation in the polymer film but is reduced after annealing under vacuum. Such event can be recycled.[30] While the low-bandgap polymer 12 is not phosphorescent in the visible wavelength region, it can still show an unusually high photocurrent quantum efficiency of up to 1% at 400 nm for single-layer photovoltaic cells in air. For 1, the excited states are mostly strongly bound triplet excitons upon photoexcitation and these are confined to one monomer unit, while quantum efficiencies of ~1–2% could be achieved for these cells with the addition of 7 wt.-% of C_{60}.[31]

Polymer solar cells were fabricated by using 13 as an electron donor and [6,6]-phenyl C_{61}-butyric acid methyl ester (PCBM) as an electron acceptor (Figure 8). The hole collection electrode consisted of indium tin oxide (ITO) with a spin-coated layer of poly(3,4-ethylenedioxythiophene):poly(styrenesulfonate) (PEDOT:PSS), while Al served as the electron collecting electrode. For 13:PCBM blends, cells with a 1:4 ratio resulted in the best performance with a high power-conversion efficiency of 4.10% under illumination of an AM1.5 solar simulator. The open-circuit voltage obtained for the best cell was 0.82 V, the short-circuit current density was 15.43 mA/cm^2 and the fill factor was 0.39 (Figure 9).[22] The open-circuit voltage is also higher compared with P3HT:PCBM cells, which is probably due to the lower HOMO level of 13 (−5.37 eV, compared with −5.20 eV for P3HT). We observed that the extinction coefficient of 13:PCBM blend is higher above 550 nm, leading to improved absorption of the incident solar radiation. The 13:PCBM blends give comparable hole and electron mobilities as revealed by the time-of-flight technique and space-charge limited current modelling. As a result, 13:PCBM blends exhibit balanced charge transport, as required for efficient solar cell performance, and the mobility data are comparable to those measured in P3HT:PCBM cells with optimal (1:1) blend ratio.[32] The generally good solubility of Pt polyynes over their organic counterparts, viz. poly(heteroaryleneethynylene)s, favors the utilization of the former polymers in the advance of polymer photovoltaic devices. The novel approach reported here can provide a useful avenue to PCE enhancement by extending the absorption to longer wavelengths (Figure 10) and metal incorporation without the need or involvement of the triplet excitons in the high-efficiency energy conversion. A recent report shows an external quantum efficiency of ~9% in **5**:PCBM bulk heterojunction cells, resulting in a power conversion efficiency of only 0.27%.[23] It absorbed only in the blue-violet spectral region with a maximum efficiency at 400 nm, and consequently the efficiency was low due to low coverage of the solar spectrum, although the charge separation was shown to occur via triplet state of the polymer.

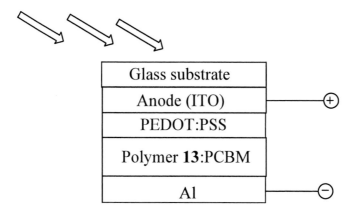

Figure 8. A typical bulk-heterojunction solar cell based on **13**.

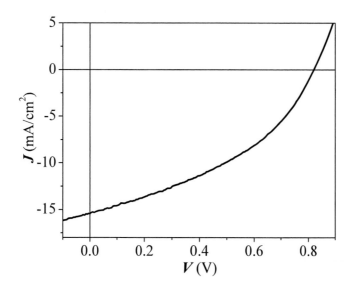

Figure 9. Current density (*J*)-voltage (*V*) characteristic of bulk-heterojunction solar cell based on **13**.

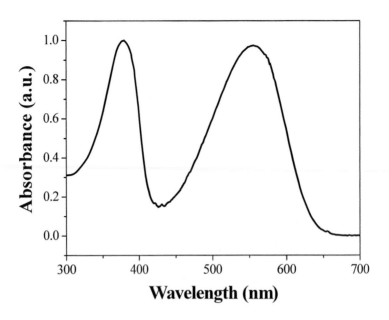

Figure 10. Absorption spectum of polymer **13** in CH₂Cl₂.

The current worldwide interest in electroluminescence using conjugated polymers (e.g. poly(*p*-phenylenevinylene), PPV) is largely driven by the perceived potential for the low-cost fabrication of large-area LEDs through simple processing methods. In such devices, light emission arises from radiative decay of singlet excitons, created by electron-hole recombination within the polymer layer. Triplet excitons are also produced (either by direct recombination of the charge carriers, or indirectly, through intersystem crossing from the singlet to the triplet manifold) but are not utilized for harvesting the triplet energy. The characterization of both the singlet and triplet excitations is thus of prime importance for understanding the mechanisms of LEDs based on conjugated polymers and thereby improving their macroscopic emission characteristics. Since the triplet states constitute the majority of electrogenerated excited states, the successful use of the triplet manifold to produce light should undoubtedly increase the overall device efficiency. Transition metal-containing polyynes were widely employed as a research vehicle for studying the triplet excited states in conjugated polymers directly in which the heavy metals increase spin-orbit coupling, rendering the phosphorescence allowed.[12] Thus, in contrast to hydrocarbon conjugated polymers, the triplet excited state is readily accessible to experiment. The emitting properties of Pt polyynes have recently been successfully applied in the development of electroluminescent devices. Electrophosphorescence from the T_1 state can be observed for a PLED made from a Pt polyyne **18** with a 2,3-diphenylquinoxaline-5,8-diyl moiety in a structure of ITO/PEDOT:PSS/**18**/Ca:Al (Figure 11).[33] The average singlet generation fraction of 57 ± 4% for **18** was determined, suggesting that a spin-dependent process, favoring singlet formation, predominates in the polymer film. The value is more than double the value expected from simple spin statistics and is above the lower limit for the singlet generation fraction of 35–50% measured for PPV derivatives. The results demonstrate that singlets are favored over triplets and recombination of holes and electrons is spin-independent.

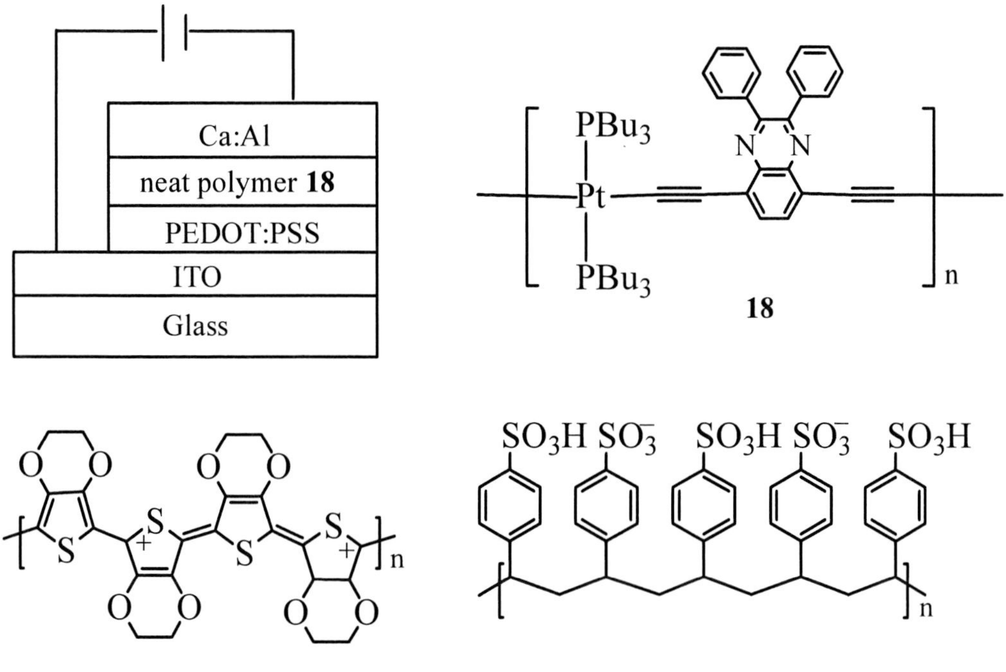

Figure 11. Polymer LEDs derived from **18**.

Conventional multi-layer polymer LED devices with the ITO/PEDOT:PSS (20 nm)/**8b** (70 nm)/TPBI (30 nm)/LiF (0.5 nm)/Al (50 nm)/Ag (50 nm) configuration (TPBI = 1,3,5-tris[N-(phenyl)benzimidazole]benzene) were also fabricated using **8b**.[34] Figure 12 depicts the configuration of the devices and the molecular structures of the compounds used in these devices. PEDOT:PSS only acts as the hole injection layer, TPBI as the hole-blocking/electron-transporting layer, LiF (0.5 nm) as the electron injection layer and Al:Ag as the bilayer cathode. The TPBI layer was adopted for the device to confine excitons within the emissive zone. Figure 13 shows the EL spectrum of **8b**, which is similar to its solution PL spectrum and that of the polymer thin film. The close resemblance between the PL and EL spectra indicates the absence of aggregation or π-stacking and hence both processes arise from the same excited state or the same type of exciton. Polymer **8b** emitted green light with a prominent EL maximum at 500 nm with a shoulder at 522 nm and the Commission Internationale de L'Eclairage (CIE) chromaticity coordinates lie at x = 0.26, y = 0.58 at 8 V. There are no undesirable emission peaks from other layers upon electrical excitation. The polymer LED had a turn-on voltage for light emission at a brightness of 1 cd/m^2 of 8.8 V and the luminance reached 37 cd/m^2 at 16 V. The relatively high driving voltages relative to other typical vacuum-evaporated devices may be due to the thicker emissive layer used and the large hole barrier at the PEDOT:PSS/**8b** interface (HOMOs of **8b** ~5.7 eV versus PEDOT ~5.0 eV). With increasing applied voltage, the current began to increase rapidly, accompanied by a rapid increase in luminance. The maximum luminance efficiency of the device is moderate at 0.11 cd/A at 9 V.

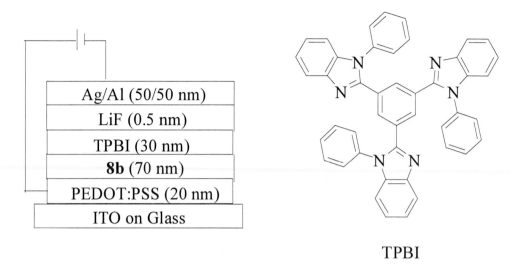

TPBI

Figure 12. Polymer LED structure based on **8b**.

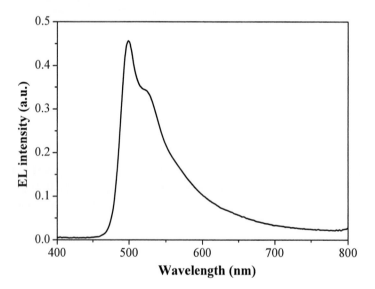

Figure 13. EL spectra of **8b** at 12 V.

4. CONCLUSIONS

In summary, a rational design of an interesting series of organometallic poly(aryleneethynylene)s containing heavy platinum element allows a facile tuning of the optical bandgaps and their phosphorescent energy states. These new polymer systems clearly enable a systematic study of the structure-property-function relationship of these conjugated metallated polymers and such an investigation is desirable for optoelectronic applications that harvest the solar energy for energy conversion and the triplet excited state for efficient light emission. In general, a high-bandgap polymer with a high triplet-state energy will normally favor the observation of triplet emission, whereas the low-bandgap congener will probably not be phosphorescent even at low temperatures. Remarkably, metalated conjugated polymer **13** with a low bandgap value (1.85 eV) has demonstrated a good promise in the creation of high-efficiency polymer solar cells. It is believed that lower bandgaps could increase the efficiency by harvesting a larger proportion of the sunlight. It is likely that this research direction will continue to shed light on the up-and-coming optoelectronic applications based on these functional metallopolymeric materials.

5. REFERENCES

[1] Skotheim, T. A.; Elsenbaumer, R. L.; Reynolds, J. R. *Handbook of Conducting Polymers*, Marcel Dekker, New York, 1998.

[2] McQuade, D. T.; Pullen, A. E.; Swager, T. M. *Chem. Rev.* 2000, *100*, 2537.

[3] (a) Burroughes, J. H.; Bradley, D. D. C.; Brown, A. R.; Marks, R. N.; Mackay, K.; Friend, R. H.; Burn, P. L.; Holmes, A. B. *Nature* 1990, *347*, 539; (b) Braun, D.; Heeger, A. J. *App. Phys. Lett.* 1991, *58*, 1982; (c) Friend, R. H.; Gymer, R. W.; Holmes, A. B.; Burroughes, J. H.; Marks, R. N.; Taliani, C.; Bradley, D. D. C.; Dos Santos, D. A.; Brédas, J. L.; Lögdlund, M.; Salaneck, W. R. *Nature* 1999, *397*, 121.

[4] Spangler, C. W. *J. Mater. Chem.* 1999, *9*, 2013; (b) McKay, T. J.; Bolger, J. A.; Staromlynska, J.; Davy, J. R. *J. Chem. Phys.* 1998, *108*, 5537.

[5] (a) Walters, K. A.; Ley, K. D.; Schanze, K. S. *Chem. Commun.* 1998, *10*, 1115; (b) Burrows, H. D.; Seixas de Melo, J.; Serpa, C.; Arnaut, L. G.; Monkman, A. P.; Hamblett, I.; Navarathnam, S. *J. Chem. Phys.* 2001, *115*, 9601; (c) Monkman, A. P.; Burrows, H. D.; Hamblett, I.; Navarathnam, S.; Svensson, M.; Andersson, M. R. *J. Chem. Phys.* 2001, *115*, 9046; (d) Monkman, A. P.; Burrows, H. D.; Hartwell, L. J.; Horsburgh, L. E.; Hamblett, I.; Navaratnam, S. *Phys. Rev. Lett.* 2001, *86*, 1358.

[6] (a) Baldo, M. A.; O'Brien, D. F.; You, Y.; Shoustikov, A.; Sibley, S.; Thompson, M. E.; Forrest, S. R. *Nature* 1998, *395*, 151; (b) Sandee, A. J.; Williams, C. K.; Evans, N. R.; Davies, J. E.; Boothby, C. E.; Köhler, A.; Friend, R. H.; Holmes, A. B. *J. Am. Chem. Soc.* 2004, *126*, 7041; (c) Baldo, M. A.; Thompson, M. E.; Forrest, S. R. *Nature* 2000, *403*, 7503; (d) Lamansky, S.; Djurovich, P.; Murphy, D.; Abdel-Razzaq, F.; Lee, H. E.; Adachi, C.; Burrows, P. E.; Forrest, S. R.; Thompson, M. E. *J. Am. Chem. Soc.* 2001, *123*, 4304.

[7] Köhler, A.; Wilson, J. S.; Friend, R. H.; Al-Suti, M. K.; Khan, M. S.; Gerhard, A.; Bässler, H. *J. Chem. Phys.* 2002, *116*, 9457.

[8] Kingsborough, R. P.; Swager, T. M. *Prog. Inorg. Chem.* 1999, *43*, 123.

[9] Brédas, J. L.; Cornil J.; Heeger, A. J. *Adv. Mater.* 1996, *8*, 447; (*b*) Sariciftci, N. S. (Ed.), *Primary Photoexcitations in Conjugated Polymers: Molecular Exciton vs Semiconductor Band Model*, World Scientific, Singapore, 1997.

[10] (a) Wittman, H. F.; Friend, R. H.; Khan, M. S. *J. Chem. Phys.* 1994, *101*, 2693; (b) Beljonne, D.; Wittmann, H. F.; Köhler, A.; Graham, S.; Younus, M.; Lewis, J.; Raithby, P. R.; Khan, M. S.; Friend, R. H.; Brédas, J. L. *J. Chem. Phys.* 1996, *105*, 3868; (c) Chawdhury, N.; Köhler, A.; Friend, R. H.; Wong, W.-Y.; Lewis, J.; Younus, M.; Raithby, P. R.; Corcoran, T. C.; Al-Mandhary, M. R. A.; Khan, M. S. *J. Chem. Phys.* 1999, *110*, 4963; (d) Wilson, J. S.; Köhler, A.; Friend, R. H.; Al-Suti, M. K.; Al-

Mandhary, M. R. A.; Khan, M. S.; Raithby, P. R. *J. Chem. Phys.* 2000, *113*, 7627; (e) Wilson, J. S.; Chawdhury, N.; Al-Mandhary, M. R. A.; Younus, M.; Khan, M. S.; Raithby, P. R.; Köhler, A.; Friend, R. H. *J. Am. Chem. Soc.* 2001, *123*, 9412; (f) Yam, V. W.-W. *Acc. Chem. Res.* 2002, *35*, 555; (g) Yam, V. W.-W.; Tao, C. H.; Zhang, L. J.; Wong, K. M. C.; Cheung, K. K. *Organometallics* 2001, *20*, 453; (h) Ohshiro, N.; Takei, F.; Onitsuka, K.; Takahashi, S. *J. Organomet. Chem.* 1998, *569*, 195.

[11] (a) Younus, M.; Köhler, A.; Cron, S.; Chawdhury, N.; Al-Mandhary, M. R. A.; Khan, M. S.; Lewis, J.; Long, N. J.; Friend, R. H.; Raithby, P. R. *Angew. Chem. Int. Ed. Engl.* 1998, *37*, 3036; (b) Havinga, E. E.; Tenhoeve, W.; Wynberg, H. *Synth. Met.* 1993, *55*, 299.

[12] Wong W.-Y.; Ho, C.-L. *Coord. Chem. Rev.* 2006, *250*, 2627.

[13] (a) Fujikura, Y.; Sonogashira K.; Hagihara, N. *Chem. Lett.* 1975, 1067; (b) Sonogashira, K.; Takahashi S.; Hagihara, N. *Macromolecules* 1977, *10*, 879; (c) Takahashi, S.; Kariya, M.; Yatake, T.; Sonogashira K.; Hagihara, N. *Macromolecules* 1978, *11*, 1063.

[14] Wittmann, H. F. Ph.D. Thesis, University of Cambridge, 1993.

[15] Khan, M. S.; Al-Mandhary, M. R. A.; Al-Suti, M. K.; Al-Battashi, F. R.; Al-Saadi, S.; Ahrens, B.; Bjernemose, J. K.; Mahon, M. F.; Raithby, P. R.; Younus, M.; Chawdhury, N.; Köhler, A.; Marseglia, E. A.; Tedesco, E.; Feeder N.; Teat, S. J. *Dalton Trans.* 2004, 2377.

[16] Johnson, B. F. G.; Kakkar, A. K.; Khan, M. S.; Lewis, J.; Dray, A. E.; Friend R. H.; Wittmann, F. *J. Mater. Chem.* 1991, *1*, 485.

[17] (a) Yamamoto, T.; Suganuma, H.; Maruyama, T.; Inoue, T.; Muramatsu, Y.; Arai, M.; Komarudin, D.; Ooba, N.; Tomaru, S.; Sasaki, S.; Kubota, K. *Chem. Mater.* 1997, *9*, 1217; (b) Cao, J.; Kampf, J. W.; Curtis, M. D. *Chem. Mater.* 2003, *15*, 404; (c) MacLean, B. J.; Pickup, P. G. *J. Mater. Chem.* 2001, *11*, 1357; (d) Politis, J. K.; Curtis, M. D.; Gonzalez, L.; Martin, D. C.; He, Y.; Kanicki, J. *Chem. Mater.* 1998, *10*, 1713; (e) Masui, K.; Mori, A.; Okano, K.; Takamura, K.; Kinoshita, M.; Ikeda, T. *Org. Lett.* 2004, *6*, 2011.

[18] Wong, W.-Y.; Chan, S.-M.; Choi, K.-H.; Cheah, K.-W.; Chan, W.-K. *Macromol. Rapid Commun.* 2000, *21*, 453.

[19] Lewis, J.; Long, N. J.; Raithby, P. R.; Shields, G. P.; Wong, W.-Y.; Younus, M. *J. Chem. Soc. Dalton Trans.* 1997, 4283.

[20] Khan, M. S.; Al-Mandhary, M. R. A.; Al-Suti, M. K.; Hisahm, A. K.; Raithby, P. R.; Ahrens, B.; Mahon, M. F.; Male, L.; Marseglia, E. A.; Tedesco, E.; Friend, R. H.; Köhler, A.; Feeder N.; Teat, S. J. *J. Chem. Soc. Dalton Trans.* 2002, 1358.

[21] Chawhury, N.; Köhler, A.; Friend, R. H.; Younus, N.; Long, N. J.; Raithby, P. R.; Lewis, J. *Macromolecules* 1998, *31*, 722.

[22] Wong, W.-Y.; Wang, X.-Z.; He, Z.; Djurišić, A. B.; Yip, C. T.; Cheung, K.-Y.; Wang, H.; Mak C. S. K.; Chan, W.-K. *Nat. Mater.* 2007, *6*, 521.

[23] Guo, F.; Kim, Y. G.; Reynolds, J. R.; Schanze, K. S. *Chem. Commun.* 2006, 1887.

[24] Wong, W.-Y.; Wong, C.-K.; Lu, G.-L.; Cheah, K.-W.; Shi J.-X.; Lin, Z. *J. Chem. Soc. Dalton Trans.* 2002, 4587.

[25] Liu, L.; Poon, S.-Y.; Wong, W.-Y. *J. Organomet. Chem.* 2005, *690*, 5036.

[26] Poon, S.-Y.; Wong, W.-Y.; Cheah, K.-W.; Shi, J.-X. *Chem. Eur. J.* 2006, *12*, 2550.

[27] Wong, W.-Y.; Poon, S.-Y. *J. Inorg. Organomet. Polym. Mater.* 2008, *18*, 155.

[28] Ng, S.-C.; Lu, H.-F.; Chan, H. S. O.; Fujii, A.; Laga, T.; Yoshino, K. *Macromolecules* 2001, *34*, 6895; (b) Kauffman, J. M.; Litak, P. T.; Novinski, J. A.; Kelley, C. J.; Ghiorghis, A.; Qin, Y. *J. Fluoresc.* 1995, *5*, 295.

[29] Li, G.; Shrotriya, V.; Huang, J.; Yao, Y.; Moriarty, T.; Emery, K.; Yang, Y. *Nat. Mater.* 2005, *4*, 864.

[30] Chawdhury, N.; Younus, M.; Raithby, P. R.; Lewis, J.; Friend, R. H. *Opt. Mater.* 1998, *9*, 498.

[31] Köhler, A.; Wittman, H. F.; Friend, R. H.; Khan M. S.; Lewis, J. *Synth. Met.* 1996, *77*, 147.

[32] Huang, J., Li, G..; Yang, Y. *Appl. Phys. Lett.* 2005, *87*, 112105-1; (b) von Hauff, E.; Parisi, J.; Dyakonov, V. *J. Appl. Phys.* 2006, *100*, 043702-1.

[33] Wilson, J. S.; Dhoot, A. S.; Seeley, A. J. A. B.; Khan, M. S.; Köhler, A.; Friend, R. H. *Nature* 2001, *413*, 828.

[34] Wong, W.-Y.; Zhou, G.-J.; He, Z.; Cheung, K.-Y.; Ng, A. M.-C.; Djurišić, A. B.; Chan, W.-K. *Macromol. Chem. Phys.* 2008, *209*, 1319.

In: Modern Trends in Macromolecular Chemistry   ISBN: 978-1-60741-252-6
Editor: Jon N. Lee   © 2009 Nova Science Publishers, Inc.

Chapter 4

TOWARDS NOVEL POLYMERIC POROUS MATERIALS

Chrystelle Egger
Keele University, Keele, Staffordshire, UK

The field of porous materials is not new, chemists have been exploring it for decades. Its origin can be associated with the discovery by the Swedish mineralogist Cronstedt of the "boiling stones" also known as zeolites in the late XVIIIth century[1] or perhaps it may be more realistically linked to the pioneering work started by Barrer in the 1930's on zeolites synthesis[1],[2]. Since then, thanks to a sudden re-emergence of interest in the early 1990's with the synthesis of mesoporous materials [3],[4] (pore size ranging from 2 to 50nm), porous materials have been prepared with various pore volume and pore size distribution, from numerous precursors, as bulk materials, powders or thin films; some of them are now commercially available from major chemical suppliers, e.g. the mesoporous silica MCM-41 [5]. However, the study of porous materials, instead of dying off as it could have been expected, is still very alive amongst the international community and generates quite a large number of collaborations with major industrial partners.

The rationale behind incorporating pores within a network can be related to the creation of low-density or low-dielectric materials (typically for thermal or electric insulation purposes, respectively), but it is primarily in most cases a way to create excellent mass transfer associated with high surface area within the bulk of a material, for recognition or catalysis puprposes. Indeed, interfaces promote or enhance chemical phenomena in heterogeneous media; they are created by successfully dispersing small particles in a medium (colloidal solution), by coating large areas (thin films), or by incorporating pores into a matrix (organic or inorganic).The driving force behind the design of novel porous materials is two-fold: (i) there is a direct pulling effect from the need for innovations by the industrial sector, always looking for new applications requiring materials with novel or combined properties; (ii) there is also a need for a more detailed and deep understanding of fundamental aspects in such materials. Indeed, despite the regular commercialisation on a large scale of novel porous materials, the understanding that the community has gathered on the formation mechanisms of these solids, or on their structural features, remain in some instance surprisingly very poor (to the extent that some materials found in Nature still have no synthetic counterpart [6]). The challenge in studying such solids is that both their preparation and their characterisation are very much dependent on the pore dimensions which are sought. Porosity can be implemented in the materials following one or more of the four strategies available: sacrificial templating [3],[4], sol-gel methods [7],[8], nanolithography techniques [9], or through a gassing process [10],[11]. Pore dimensions are usually tuneable according to which approach or combination of approaches is followed (with a clear increase from a few nanometers to a few hundred microns from the templating to the gassing option), and the associated characterisation techniques (e.g. sorption, porosimetry, scattering and imaging techniques) are often used in combination in order to access a more comprehensive profile of the materials.

The use of polymers as precursors for preparing porous solids offers interesting advantages over that of inorganic species: indeed, soft amorphous matter could be preferred over crystalline oxides in order to easily accommodate a high degree of curvature (mesoporous materials prepared from a templating strategy have first been successfully reported with silica, an amorphous inorganic polymer [3],[4]); plus, the ease of processing soft matter *vs.* crystals is non negligible as usually milder conditions are used, with the possibility to shape the materials through moulding for instance. Moreover, such materials also have much milder working conditions (temperature and pressures) allowing for novel applications at the interface with biochemistry/biology (molecular recognition) or for novel technologies (low-temperature fuel cell). The post-war advent of polymer chemistry has very likely contributed opening new paths towards materials. For instance, polymers formed by radical polymerisation such as polystyrene (PS), polyvinylbenzene (PVB), polyacetonitrile (PAN) or polymethyl metacrylate acid (PMMA) have also been prepared as porous materials. Polymerisation through polycondensation (nylon 6.6) is also very important and if carried out in a solvent, it is the so-called sol-gel process: a soluble monomer is dissolved in a solvent (sol); upon addition of a catalyst, the monomer molecules start polycondensing and form a 3D network; as the polymeric structure forms, the solubility of the oligomers or polymeric aggregates decreases which leads to the precipitation of a cross-linked phase in which the solvent is entrapped (gel-state). Upon successful drying, cavities initially occupied with the solvent are emptied and generate the pore structure (solid-state). Control over pore size and pore volume is both thermodynamic (change of enthalpic and entropic free energy) and kinetic (competitive kinetics of precipitation and polymerisation). Changing synthesis parameters such as temperature, nature of solvent or monomers, will affect the thermodynamics/kinetics of the gels formation. One advantage of using sol-gel methods over radical polymerisation is that the solubility of the precursors used in sol-gel syntheses is usually much better in polar or aqueous solvents, therefore not precluding any applications in bio-related medium. Plus, sol-gel processes are still unique in their ability to accommodate new reactants and the control they allow over experimental parameters. Both organic and inorganic porous polymers have been prepared by sol-gel methods, although silicates and silanes were originally used to prepare aerogels (materials with pore volume>90% and pore sizes< 50nm)[8].

Recently, however, organic aerogels have also been prepared from phenol-type or amino-type resins. Some of them can be successfully pyrolised leading to carbon aerogels with large surface areas[12],[13]. These resins offer the great advantage of versatility: similar but not identical monomers can be used within the same synthesis pot, expanding the variety of materials that can result. As an example, a library of monomers available, based on the phenol-type or amino type motif, is given in Figure 1.

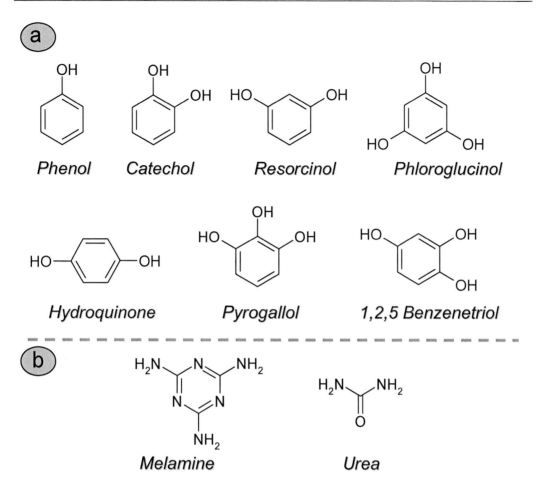

Figure 1. a) Various phenol-type precursors which can be cross-linked with formaldehyde. The number and location of the hydroxy groups affect the overall material. b) Common amino precursors which are cross-linked with formaldehyde. The amount of formaldehyde added affects the number of amino groups which reacts, and consequently the properties of the final materials. Melamine and urea are often used together with formaldehyde to add flexibility into the final network.

In the case of the phenol-type, the number of hydroxo functions present on the benzene ring affects the number of cross-linkers (formaldehyde) which can be added. As for the amino precursors, formaldehyde substitutes one or more hydrogen atom of the amino groups. In both cases, the degree of cross-linking and the number of functions (hydroxyl or amino) left in the network after polymerisation strongly affect the resulting material. For instance, phenol-type or amino-type resins are susceptible to polycondensation after the addition of the cross-linker (formaldehyde) as schematised in Figure 2. The number of cross-linkers on the aromatic ring can be varied experimentally leading to monomers which exhibit one to six cross-linking groups per molecule, an incredibly large and diversified amount of bridging units. Porous melamine-formaldehyde networks have been prepared by various synthesis strategies with different porosities and pore size distributions[14],[15],[16],[17]. Solid-state NMR experiments have recently been carried out on melamine-formaldehyde gels leading to the conclusion that the type of bridges (methoxy- or ethylene-, see Figure 2, right) is directly influenced by the initial amount of formaldehyde in the resin [18].

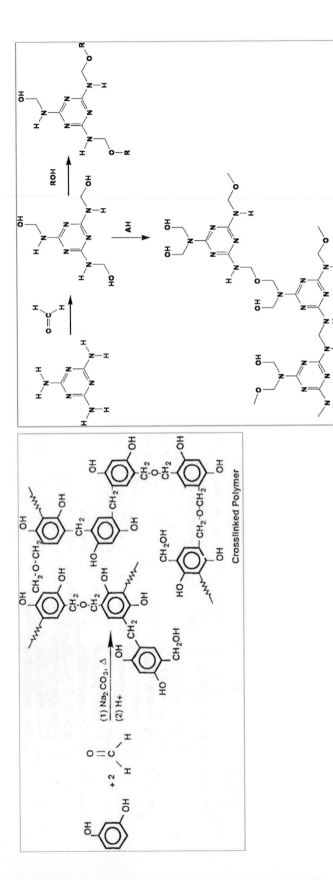

Figure 2. Schematic representation of the reaction of resorcinol with formaldehyde (left [14]) and melamine with formaldehyde (right [18]), followed by an acid-catalysed polycondensation, leading to a cross-linked network of resorcinol-formaldehyde or melamine-formaldehyde.

For given chemical reactions to occur within pores, the cavities should contain the appropriate functions (nature) and coverage (density). In the case of sol-gel processes, the function is brought into the network through the use of a functional monomer. Unless the functional monomer is used solely, it is impossible to achieve 100% coverage on the wall surface and instead, a statistical distribution of functions occurs within the network with a small percentage only located at the surface of the pores. In order to enhance the wall coverage, the order/time of addition of constituents could be altered and a 2-step strategy could be employed (post-treatment methods). For instance, chemical engineering work can be carried out inside the pores: e.g. melamine-based dendrimers have been grown on the walls of mesoporous silica [19]. The importance of such dendrimers in the pores is that they do not bring only one function but several, the number varying with dendrimer generation. Dendrimers are rather large molecules and the inconvenience can be that part of the porosity is lost. In the recent work of Acosta et al.[19] in which melamine-based dendrimers have been grafted on the mesoporous silica SBA-15, pore diameters are decreased from 8.3 to 4.5nm with increasing dendrimer generation from 1 to 5, and an optimum dendrimer size to pore size is found to be 0.5. Above this ratio, larger dendrimers cannot be grown. Indeed, the size of the dendrimers which can be grafted in pores is very much dependent on the curvature of the pore, and there is a dendrimer dimension for which the curvature of the wall is no longer negligible. However, it seems difficult to state whether the growth of dendrimers in a pore modifies the surface area or not, as it is reported that during nitrogen adsorption, nitrogen could be adsorbing into the dendrimers [19], expanding the BET model to cases for which it may no longer valid (the interactions dendrimers-nitrogen are not yet understood). A reference such as a non-porous surface grafted with the same type of dendrimers is required for direct comparison. The remarkable potential for such hybrid materials lies in the incredibly large number of amino groups available within the pores. However, the engineering of dendrimers in a given porous structure is costing as the synthesis is not straight forward and requires multi-step organic synthesis, usually difficult to reproduce at a large scale. Plus, dendrimers are fragile molecules and it could be that the overall thermal stability of the hybrid network is not optimum for utilisation in heterogeneous catalysis for instance.

As pressure from governmental health organisations is increasing towards (porous) materials, coatings or membranes for large scale industrial applications with zero formaldehyde emission (formaldehyde is suspected to be a human carcinogen), it seems that the use of resins (phenol- or melamine-based) could be compromised. Even if different aldehydes are used as cross-linkers [20],[21], they would not be as effective as their reactivity is lower; plus, melamine has recently had very bad publicity since the scandal in September 2008 about contaminated milk products coming from China [22]. There is a clear challenge in synthesising new resins which are as good as the ones currently in use, but with a safer impact on human health or on the environment for the standards set in the new 21st century. Some have already taken up the challenge and have reported soy-flour based resin with no formaldehyde [23]. From such novel resins, new materials will result that will impact on the design of well-defined porous networks with desired properties. However, despite the changes in regulations, materials with formaldehyde as a cross-linker are still under study. The reason for this is that the versatility of the chemistry of such resins is unique, leading to a wide range of new materials. For instance, triazine herbicide, barbituric acid and creatinine sensors based on molecularly imprinted melamine-based polymers have been reported in 1997 [24], 2005 [25] and 2006 [26] respectively.

From the preparation of inorganic aerogels or mesoporous materials to that of mesoporous silica with enantiomeric surfaces [27] (Figure 3), more than a decade has been necessary, illustrating the magnitude of the challenges that materials chemists are facing and the time it requires to overcome the difficulties.

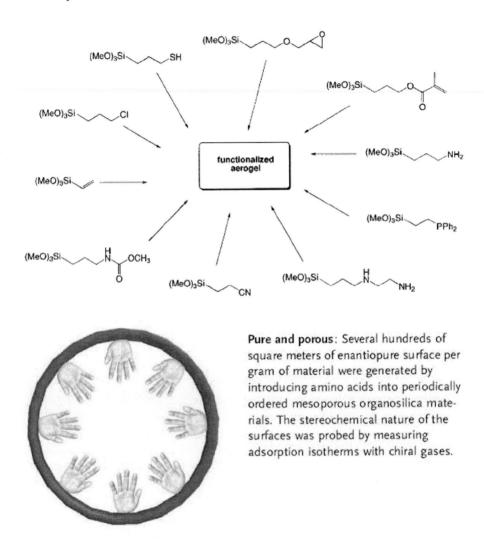

Figure 3. Top- Variety of organosilanes precursors used in preparing functional silica aerogels [8]. Bottom- Enantiomeric surfaces within mesoporous organosilica [27].

With the advent of macromolecular chemistry, novel materials will continusouly emerge and more complex structures will be generated. Hybrid organic-inorganic materials are currently being explored and may be the solution to stable materials, with well-defined pore structures and with high surface areas coupled to high chemical reactivities. However, the challenges that still lie ahead are not only in the synthesis of such materials but also in the thorough characterisation of their chemical and structural features with various physical techniques. Advances in materials science will always be linked to the detailed understanding experimentalists have of the final networks they create, and also to the knowledge they gather about formation mechanisms.

REFERENCES

[1] Payra P, Dutta PK, In: Auerbach SM, Carrado KA, Dutta PK (eds), 2003, Handbook of zeolite science and technology. Marcel-Dekker, NY.

[2] Vaughan D. E. W., Studies Surf. *Sci. Cat.*, 2007 , 170 (B), pp. 87-95.

[3] Yanagisawa T, Shimizu T, Kuroda K, Kato C, *Bull Chem Soc Japan* 1990, 63(4), 988–992 .

[4] Kresge CT, Leonowicz ME, Roth WJ, Vartuli JC, *Beck JS, Nature*, 1992, 359,710–712.

[5] MCM-41 (hexagonally mesostructured porous silica first prepared in 1992) can be now bought from Sigma-Aldrich (*www.sigmaaldrich.com*).

[6] Perry C.C., Biomineralization, 2003, 54, 291-327.

[7] Barton TJ, Bull LM, Klemperer WG, Loy DA, McEnaney B, Misono M, Monson PA, Pez G, Scherer GW, Vartuli JC, Yaghi OM, *Chem. Mater.*, 1999, 11(10), 2633-2656.

[8] Hüsing N, Schubert U, Angew Chem Int Ed, 1998, 37-22.

[9] Xia Y., Rogers J. A., Paul K. E. and Whitesides G. M., *Chem Rev.*, 1999, 99, 1823-1848.

[10] Park CB, Baldwin DF, Suh NP, *Polym Eng Sci* , 1995, 35(5), 432-440.

[11] Handa YP, Zhang Z, Wong B, *Cell Polym*, 2001, 20(1), 1-16.

[12] Kosonen H., Valkama S., Nykänen, Toivanen M., ten Brinke G., Ruokolainen J., Ikkala O., *Adv. Mater.*, 2006, 18, 201-205.

[13] Mulik S., Sotiriou-Leventis C., Leventis N., Chem Mater., 2008, DOI 10.1021/cm801428p.

[14] Pekala R. W., *J. Mater. Sci.*, 1989, 24, 3221-3227.

[15] C.C. Egger, C. du Fresne, D. Schmidt, J. Yang, V. Schädler, J. *Sol-Gel Sci.* Tech., 2008, 48, 86-94.

[16] C.C. Egger, P. Mueller-Buschbaum, V. Raman, S. Roth, C. du Fresne, V. Schädler, T. Frechen, *Langmuir*, 2008, 24, 5877-5887.

[17] du Fresne von Hohenesche C., Schmidt D. F., Schädler V., *Chem. Mater.*, 2008, 20 (19), 6124-6129.

[18] C. C. Egger, V. Schädler, J. Hirschinger, J. Raya, B. Bechinger, *Macromol. Chem. Phys.*, 2007, 20(21), 2204-2214.

[19] Acosta E. J., Carr C. S., Simanek E. E., Shantz D. F., *Adv. Mater.*, 2004, 16(12), 985-989.

[20] Šebenik A., Osredkar U., Zigon M., *Polymer Bulletin*, 1989, 22, 155-161.

[21] Šebenik A., Osredkar U., Lesar M., *Polymer*, 1990, 31, 130-134.

[22] http://www.food.gov.uk/news/newsarchive/2008/sep/melamine

[23] Amaral-Labat G.A., Pizzi A., Gonçalves A. R., Celzard A., Rigolet S., Rocha G. J. M., J. *Applied Polymer Sci.*, 2008, 108, 624-632.

[24] Matsui J., Okada M., Tsuruoka M., Takeuchi T., *Analytical Comm.*, 1997, 34, 85-87.

[25] Prasad B. B., Lakshmi D., *Electroanalysis*, 2005, 17 (14), 1260-1268.

[26] Lakshmi D., Prasad B. B., Sharma P. S., *Talanta,* 2006, 70, 272-280

[27] Kuschel A., Sievers H., Polarz S., Angew. *Chem. Int. Ed.*, 2008, DOI: 10.10002/anie.200803405

In: Modern Trends in Macromolecular Chemistry
Editor: Jon N. Lee,

ISBN: 978-1-60741-252-6
© 2009 Nova Science Publishers, Inc.

Chapter 5

MOLECULAR DESIGN OF ORGANIC-INORGANIC HYBRID GELS SYNTHESIZED BY MEANS OF A HYDROSILYLATION REACTION BETWEEN SI CONTAINING CROSSLINKING MONOMERS AND BI-FUNCTIONAL VINYL MONOMERS

Naofumi Naga*[1], Sanae Fujii[2], Emiko Oda[2], Akinori Toyota[2] and Hidemitsu Furukawa*[3]

[1]Department of Applied Chemistry, Materials Science Course, College of Engineering, Shibaura Institute of Technology, Toyosu, Kohtoh-ku, Tokyo, Japan
[2]Department of Organic and Polymer Materials Chemistry, Faculty of Technology, Tokyo University of Agriculture and Technology, Naka-Cho, Koganei, Tokyo, Japan
[3]Division of Biological Science, Graduate School of Science, Hokkaido University, Kita-ku, Sapporo, Japan

ABSTRACT

Organic-inorganic hybrid gels have been synthesized by means of a hydrosilylation reaction of multiple Si-H functional crosslinking monomers with bi-functional vinyl spacer monomers using a Pt catalyst. We select 1,3,5,7-tetramethylcyclotetrasiloxane (TMCTS), tetrakis(dimethylsilyloxy)silane (TDSS), or silsesquioxane, 1,3,5,7,9,11,13,15-octakis(dimethylsilyloxy)pentacyclo-[9,5,1,1,1,1]octasilsesquioxane (POSS) as the crosslinking monomers. Two series of the bi-functional spacer monomers are applied in the organic-inorganic hybrid gels. One is divinyl or diallyl monomers containing Si, divinyldimethylsilane (VMS), diallyldimethylsilane (AMS), diallydiphenylsilane (APS), 1,3-divinyltetramethyldisiloxane (VMSO), or 1,4-bis(dimethylvinylsilyl)benznene (VMSB), and the other is bi-functional polar monomers, 1,6-hexanediol divinylether (HDVE), 1,6-hexanediol diacrylate (HDA), and 1,6-hexanediol dimethacrylate (HDMA). Network structures of the resulting gels have been quantitatively characterized by means of a novel scanning microscopic light scattering system. Both the structures of the crosslinking monomers and spacer monomers affect the critical gelation concentration (minimum monomers concentration which attains formation of the gel), reaction time for formation of the gels, and network structure of the gels.

INTRODUCTION

Organic-inorganic hybrid polymers having network structures have been developed due to their characteristic properties; high transparency, high thermal stability, good mechanical strength, excellent solvent resistance, low dielectric constant, and so on. The organic-inorganic hybrid polymers have been prepared with some effective methods.[1-13[Hydrosilylation reaction of multi-functionalized corsslinking reagents containing Si-H or vinyl group is one of an effective methods to yield organic-inorganic network polymers.[14-24] The authors have recently developed organic-inorganic hybrid gels using hydrosilylation reaction of cyclic siloxane or cubic silsesquioxane with α,ω-nonconjugated dienes, and quantitatively characterized the network structures of the gels by means of a characteristic analytical method, namely scanning microscopic light scattering (SMILS) system.[25] The chemically synthesized gels generally have defect and/or mesh-size distribution in their network structures.[26-31] The SMILS analysis of the resulting gels from cyclic siloxane or cubic silsesquioxane with α,ω-nonconjugated dienes cleared the extremely narrow distribution of mesh size in the gels. Further more, the mesh size of the gels were correspond with the length of the α,ω-nonconjugated dienes used. However, the gels with the narrow mesh-size distribution were obtained only in the specific monomer concentration, namely optimal gelation concentration. Flexible methylene chain in α,ω-nonconjugated diene should cause the inhomogeneity of network structure due to unexpected intermolecular crosslinking and/or intramolecular cyclization in the gels which were prepared in other monomer concentrations.[25]

We have been challenging tailored synthesis and accurate characterization of the organic-inorganic hybrid gels using this approach to produce a well-defined hybrid network structure. We also have been developing highly functionalized gels. Various types of divinyl or diallyl compounds are applicable instead of α,ω-nonconjugated dienes as spacer monomers. In this study, we have selected divinyl or diallyl monomers containing Si, and bi-functional polar monomers. This chapter reports synthesis of a series of organic-inorganic hybrid gels from multiple Si-H functional crosslinking monomers, 1,3,5,7-tetramethylcyclotetrasiloxane (TMCTS), tetrakis(dimethylsilyloxy)silane (TDSS), or silsesquioxane, 1,3,5,7,9,11,13,15-octakis(dimethylsilyloxy)pentacyclo-[9,5,1,1,1,1]octasilsesquioxane (POSS), with divinyldimethylsilane (VMS), diallyldimethylsilane (AMS), diallydiphenylsilane (APS), 1,3-divinyltetramethyldisiloxane (VMSO), 1,4-bis(dimethylvinylsilyl)benznene (VMSB), 1,6-hexanediol divinylether (HDVE), 1,6-hexanediol diacrylate (HDA), or 1,6-hexanediol dimethacrylate (HDMA), by means of hydrosilylation reaction with a Pt catalyst, as shown in Scheme 1 or 2. The network structures of the resulting gels were investigated with SMILS to study the effect of the monomer structure or concentration on the gelation behavior, mesh size, and mesh-size distribution of the resulting gels.

1,3,5,7-tetramethylcyclotetrasiloxane
TMCTS

tetrakis(dimethylsilyloxy)silane
TDSS

R=OSiH(CH₃)₂

1,3,5,7,11,13,15-octakis(dimethylsilyloxy)
pentacyclo(9.5.1.1.1.1)octasiloxane
POSS

Scheme 1. Crosslinking monomers

divinyldimethylsilane
DVDMS

1,3-divinyltetramethyldisiloxane
VTMDS

diallyldimethylsilane
DADMS

DADPS

diallyldiphenylsilane

1,4-bis(dimethylvinylsilyl)benzene
DMVSB

1,6-hexanediol divinylether
HDVE

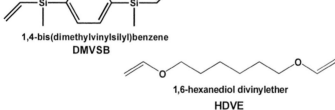

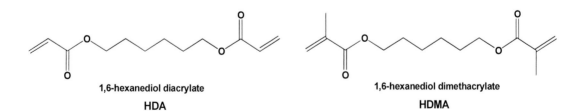

1,6-hexanediol diacrylate
HDA

1,6-hexanediol dimethacrylate
HDMA

Scheme 2. Spacer monomers

SYNTESIS AND STRUCTURE OF ORGANIC-INORGANIC HYBRID GELS

1. Experimental

Materials

VMS (Chisso Co. Ltd.), AMS (Tokyo Chemical Industry Co. Ltd.), APS (Chisso Co. Ltd.), VMSO (Aldrich Chemical Co. Ltd.), and VMSB (Shin-Etsu Chemical Co., Ltd.) were used as received. HDVE, HDA, and HDMA were commercially obtained from Aldrich Chemical Co. Ltd., and purified it through activated alumina under nitrogen atmosphere. TMCTS (Chisso Co. Ltd.), TDSS (Aldrich Chemical Co. Ltd.) and POSS (Aldrich Chemical Co. Ltd.) were used without further purification. Platinum-divinyltetramethyldisiloxane complex (catalyst **1**) were purchased from Chisso Co. Ltd., and used without purification. Toluene was dried over calcium hydride under refluxing for 6 h and distilled before use. The catalyst **1** was diluted to $2.2 \cdot 10^{-3}$ mol/L with distilled toluene and stored under nitrogen. The molar ratio of vinyl or diallyl group in spacer reagents to Si-H group in crosslinking reagents was adjusted to 1.0. The concentration of the catalyst **1** was $1.5 \cdot 10^{-3}$ mol/L. Samples were prepared with special care for getting rid of dust and measured in as-prepared state in toluene at 25 ˚C.

Gelation of TMCTS with VMS

Run 2 (monomer concentration = 10 wt.-%): In a sample tube of 4 mm diameter, 12.0 mg (0.05 mmol) of TMCTS and 11.2 mg (0.10 mmol) of VMS were dissolved in a 110 µL of toluene at room temperature. Then a 132 µL toluene solution of the catalyst **1** was added and shaking the sample tube several times to mix the reaction reagents. It was placed without stirring and a cleared-brown gel was generated. The gelations of different monomer concentration or different spacer monomers containing Si were also conducted with the same procedure.

Gelation of TDSS with AMS

Run 21 (monomer concentration = 16 wt.-%): In a sample tube of 4 mm diameter, 13.2 mg (0.04 mmol) of TDSS and 11.2 mg (0.08 mmol) of TDSS were dissolved in 72.4 µL of toluene. Then 87.3 µl of toluene solution of catalyst **1** was added and shaking the sample tube several times to mix the reaction reagents. It was placed without stirring and a cleared-brown gel was generated. The gelations of different monomer concentration or different spacer monomers containing Si were also conducted with the same procedure.

Gelation of POSS with AMS

Run 39 (monomer concentration = 13 wt.-%): In a sample tube of 4 mm diameter, 10.5 mg (0.0103 mmol) of POSS and 5.8 mg (0.041 mmol) of POSS were dissolved in 56.9 µL of toluene. Then 68.6 µl of toluene solution of catalyst **1** was added and shaking the sample tube several times to mix the reaction reagents. It was placed without stirring and a cleared-brown gel was generated. The gelations of different monomer concentration or different spacer monomers containing Si were also conducted with the same procedure.

Gelation of POSS with HDVE

Run 62 (monomer concentration = 20 wt.-%): In a sample tube of 4 mm diameter, 7.5 mg (0.074 mmol) of POSS and 5.0 mg (0.296 mmol) of HDVE were dissolved in a 44 μL of toluene at room temperature. Then a 14 μL toluene solution of the catalyst **1** was added and shaking the sample tube several times to mix the reaction reagents. It was placed without stirring and a cleared gel was generated. The gelations of different crosslinking monomers or different bi-functional polar spacer monomers were also conducted with the same procedure.

Measurements

Scanning Microscopic Light Scattering (SMILS)

Quantitative determination of minute mesh size of the gels was performed with the SMILS system, [27,31] which was recently developed for the detailed characterization of network structure in polymer gels. Conventional light scattering systems is hardly used to quantitatively characterize the network structure in polymer gels since the static inhomogeneities that are inevitably memorized in the network. The inhomogeneities sometimes cause undesired large interference, which prevents the exact characterization. The developed SMILS system enable us to scan and measure at many different positions in an inhomogeneous gel, in order to rigorously determine a time- and space-averaged, i.e. ensemble-averaged, (auto-)correlation function of fluctuating mesh size in the gel. Analysis of the ensemble-averaged function makes it possible to quantitatively characterize the mesh-size distribution of network structure in inhomogeneous gels. Scanning measurement was performed at more than 25 points for each sample to determine ensemble-averaged dynamic structure factor. The determined correlation function was transformed to the distribution function of relaxation time by using numerical inversed Laplace transform calculation. For these gels, a few peaks of relaxation modes were observed in the distribution function. Based on the observation of scattering-angle, q, dependence of the relaxation modes, the observed modes usually have q^2-dependence, which correspond to translational diffusion. In the following, all the results were determined at the scattering angle fixed at 90°. The observed modes, as assigned to the cooperative diffusion of gel network, were used for the determination of radius of mesh (mesh size) (ξ; m) with Einstein-Stokes formula.

$$\xi = \frac{16\pi n^2 \tau_R K_B \sin^2 \frac{\theta}{2}}{3\eta\lambda^2}$$

where n, τ_R, K_B, θ, η, and λ are refractive index of toluene, Ensemble-averaged relaxation time (s), Boltzmann constant (1.38×10^{-23} JK^{-1}), scattering angle (90°), viscosity coefficient of toluene at 298 K (5.6×10^{-4} N m^2 s^{-1}), wave length of incident ray (4.42×10^{-7} m), respectively.

Solution NMR Analysis

^{1}H NMR spectra of sol samples were recorded on a JEOL AL-300 spectrometer operated at 300 MHz in the pulse Fourier Transform mode. The pulse angle was 45° and 32 scans were accumulated in 7 s of the pulse repetition. The spectra were recorded at room temperature in CDCl$_3$ solution.

2. Synthesis of Organic-Inorganic Hybrid Gels with Divinyl or Diallyl Monomers Containing Si

Critical Gelation Concentration

The hydrosilylation reaction of the crosslinking monomers with the divinyl or diallyl monomers containing Si as spacer monomers was investigated with the catalyst **1** under various monomer concentrations in toluene at room temperature. In the reaction of TMCTS with VMS, the corresponding gels were obtained in the reaction with higher monomer concentrations than 6.0 wt.-% (wt.-% of the summation of TMCTS and VMS in the reaction system). On the other hand, no gel was generated in reactions with less than 6.0 wt.-% of monomer concentration. We will use the term "critical gelation concentration" to refer to this border in the monomer concentration between gelation and non-gelation. The critical gelation concentration of the gels is summarized in Table 1. The gels with TMCTS tend to show low critical gelation concentration, whereas the gels with TDSS show high critical gelation concentration, except the gel with VMSB. Both TMCTS and TDSS crosslinking monomers have four Si-H groups per molecule. The reaction conversions, which were determined by ^{1}H NMR at the sol state in the reaction less than the critical gelation concentration, of TMCTS with VMS, AMS, or VMSO for 1 h were 25.6 %, 39.4 %, or 17.8 %, respectively, and were smaller than those of TDSS (37.0 % with VMS, 70.3 % with AMS, 23.9 % with VMSO). The similar results were previously reported in the reaction with α,ω-nonconjugated dienes.[25] The flexible and/or tetrahedral-like structure of TDSS may prevent from forming an infinite network. The gels with VMSB as a spacer monomer showed particular low critical gel concentrations independent of the crosslinking monomer used. The difference should be derived from the molecular size and/or structure (rigid or flexible) of the spacer monomers. The long and rigid molecular structure of VMSB should promote the formation of network, even the reaction with TDSS.

Table 1 Critical gelation concentration of the gels with divinyl or diallyl monomers containing Si (wt.-%)

	VMS	AMS	APS	VMSO	VMSB
TMCTS	6.0	3.0	6.0	9.0	4.0
TDSS	14.0	16.0	50.0	16.0	3.0
POSS	5.0	7.5	13.0	9.0	4.0

Synthesis and Structure of Gels from TMCTS

The ensemble-averaged relaxation time $[P_{en}(\tau_R)]$ distributions as a function of the relaxation time of the gels from TMCTS as a crosslinking monomer have been corrected with SMILS, and the structures of the gels are summarized in Table 2. The ensemble-averaged relaxation time $[P(\tau_R)]$ is quantitatively relevant to mesh size of the gels, as shown in

Einstein-Stokes formula. The mesh-size distribution of the gels has been evaluated by the standard deviation of the peak of ensemble-averaged relaxation time distribution (σ). Figure 1 shows the ensemble-averaged relaxation time distribution as a function of the relaxation time of TMCTS-VMS gels. All the resulting gels formed uni-modal mesh-size distribution independent of the monomer concentration. The monomer concentration gave little effect on the mesh size of the gels. The gels from TMCTS with AMS or APS showed the same results, and typical ensemble-averaged relaxation times distributions as a function of the relaxation time in those gels are shown in Figure 2, indicating homogeneous mesh-size distribution of the gels. In the case of the gels of TMCTS-α,ω-nonconjugated diene, the monomer concentration strongly affected the mesh-size distribution and mesh size, and formed the gels with homogeneous mesh-size distribution and the small mesh size only in the optimal gelation conditions, which was a little higher than the critical gelation concentration.[25] The difference should be derived from the rigidity of the spacer monomers containing Si atom with methyl or phenyl substituents.

Figure 3 illustrates the ensemble-averaged relaxation time distributions as a function of the relaxation time in the TMCTS-VMSO gel. The gel prepared in the low monomer concentration (9.0 wt.-%) showed broad peak and bimodal distribution. The result indicates structural defects in the network. Whereas the gels obtained in the reaction of high monomer concentration (13.0 – 17.0 wt.-%) showed a uni-modal distribution of $P(\tau_R)$ indicating homogeneous network structure. The gel obtained in the 13.0 wt.-% of monomer concentration showed the smallest mess size and the narrowest mesh-size distribution, and the monomer concentration was the "optimal gelation concentration" of the gel.

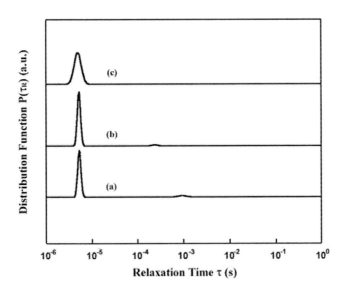

Figure 1. Ensemble-averaged relaxation time distributions as a function of the relaxation time of TMCTS-VMS gels prepared using various monomer concentrations; (a) 6.0 wt.-% (Run 1), (b) 10.0 wt.-% (Run 2), and (c) 12.0 wt.-% (Run 3).

Table 2.   Structure of TMCTS-VMS, AMS, APS, VMSO, or VMSB gels

Run	Spacer monomer	Monomer concentration wt.-%	$\tau_R \cdot 10^{-6\ a)}$ s	$\sigma^{\ b)}$	Mesh size nm
1	VMS	6.0	5.3	0.036	1.9
2	VMS	10.0	5.2	0.036	1.8
3	VMS	12.0	4.9	0.078	1.7
4	AMS	5.0	5.0	0.028	1.8
5	AMS	7.0	4.4	0.034	1.5
6	AMS	11.0	4.6	0.043	1.6
7	APS	6.0	4.4	0.039	1.6
8	APS	10.0	4.1	0.037	1.5
9	APS	12.0	4.3	0.056	1.5
10	VMSO	9.0	4.0	0.068	1.4
11	VMSO	13.0	3.8	0.029	1.4
12	VMSO	15.0	4.6	0.035	1.6
13	VMSO	17.0	4.7	0.037	1.7
14	VMSB	6.0	4.7	0.039	1.7
15	VMSB	8.0	4.5	0.037	1.6
16	VMSB	12.0	4.5	0.038	1.6

a) Relaxation time of the gels determined by SMILS

b) Standard deviation of a peak of the ensemble-averaged relaxation time distribution.

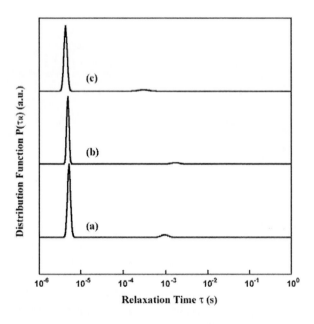

Figure 2. Ensemble-averaged relaxation time distributions as a function of the relaxation time of TMCTS-spacer monomers gels prepared under the optimal gelation concentration; (a) TMCT-VMS 6.0 wt.-% (Run 1), (b) TMCTS-AMS 5.0 wt.-% (Run 4), and (c) TMCTS-APS 6.0 wt.-% (Run 7).

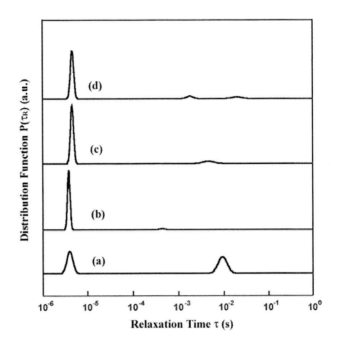

Figure 3. Ensemble-averaged relaxation time distributions as a function of the relaxation time of TMCTS-VMSO gels prepared using various monomer concentrations; (a) 9.0 wt.-% (Run 10), (b) 13.0 wt.-% (Run 11), (c) 15.0 wt.-% (Run 12), and 17.0 wt.-% (Run 13).

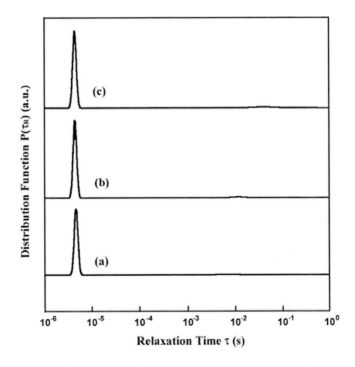

Figure 4. Ensemble-averaged relaxation time distributions as a function of the relaxation time of TMCTS-VMSB gels prepared using various monomer concentrations; (a) 6.0 wt.-% (Run 14), (b) 8.0 wt.-% (Run 15), and (c) 12.0 wt.-% (Run 16).

Network structure of the TMCTS-VMSB gel was also investigated with SMILS. The ensemble-averaged relaxation time distributions as a function of the relaxation time in TMCTS-VMSB gel are shown in Figure 4. The gels formed homogeneous mesh-size distribution independent of the monomer concentration. The result is different from that of TMCT-VMSO gel. The structure of the substituent in the two Si atoms, oxygen in VMSO or 1,4-phenyl in VMSB, should cause the monomer concentration effect on mesh size and mesh-size distribution.

Synthesis and Structure of Gels from TDSS

Synthesis and SMILS analysis of the gels with TDSS as a crosslinking monomer were also investigated, and the structures of the gels are summarized in Table 3. The typical ensemble-averaged relaxation time distributions as a function of the relaxation time of the TDSS-VMS and TDSS-AMS gels are illustrated in Figure 5. The gels obtained from TDSS with VMS or AMS formed homogeneous mesh-size distribution, and the mesh size of the gels was almost independent of the monomer concentration, as observed in the TMCTS-VMS and TMCTS-AMS gels. The SMILS analysis of TDSS-APS gels was impossible due to the muddiness of the resulting gels, which were formed in the high monomer concentration more than 50 wt.-%. The gels with VMSO which were prepared in the monomer concentration with a little higher (16.0 wt.-%) or much higher (22.0 wt.-%) than the critical gelation concentration showed bimodal distribution of the mesh-size, as shown in Figure 6. The flexible molecular structure of VMSO should induce the unexpected intermolecular crosslinking and/or intramolecular cyclization. The gels with rigid spacer monomer VMSB (Runs 29-31) formed the network with narrow mesh-size distribution ($\sigma = 0.037 - 0.039$) of the mesh size 1.6 - 1.7 nm independent of the monomer concentration.

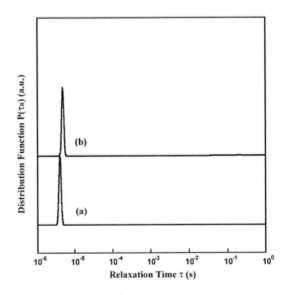

Figure 5. Ensemble-averaged relaxation time distributions as a function of the relaxation time of TDSS-spacer monomers gels prepared under the optimal gelation concentration; (a) TDSS-VMS 20.0 wt.-% (Run 19), and (b) TDSS-AMS 22.0 wt.-% (Run 23).

Table 3.   Structure of TDSS-VMS, AMS, APS, VMSO, or VMSB gels

Run	Spacer monomer	Monomer concentration wt.-%	$\tau_R \cdot 10^{-6\ a)}$ s	$\sigma^{b)}$	Mesh size nm
17	VMS	14.0	4.8	0.041	1.7
18	VMS	16.0	4.3	0.036	1.5
19	VMS	20.0	4.1	0.030	1.4
20	VMS	22.0	4.5	0.042	1.6
21	AMS	16.0	3.9	0.029	1.4
22	AMS	18.0	4.2	0.031	1.5
23	AMS	22.0	4.9	0.027	1.7
24	AMS	24.0	4.9	0.024	1.7
25	VMSO	16.0	6.6	0.064	2.3
26	VMSO	18.0	6.9	0.034	2.4
27	VMSO	20.0	6.9	0.039	2.4
28	VMSO	22.0	7.2	0.069	2.6
29	VMSB	8.0	5.9	0.059	2.1
30	VMSB	10.0	5.8	0.044	2.0
31	VMSB	16.0	5.6	0.047	2.0

a) Relaxation time of the gels determined by SMILS

b) Standard deviation of a peak of the ensemble-averaged relaxation time distribution.

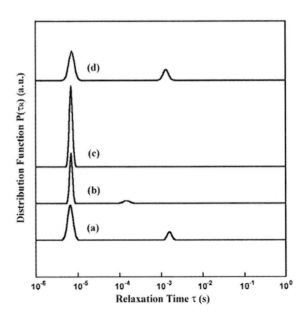

Figure 6. Ensemble-averaged relaxation time distributions as a function of the relaxation time of TDSS-VMSO gels prepared using various monomer concentrations; (a) 16.0 wt.-% (Run 25), (b) 18.0 wt.-% (Run 26), (c) 20.0 wt.-% (Run 27), and 22.0 wt.-% (Run 28).

Synthesis and Structure of Gels from POSS

SMILS analysis of the gels with POSS as a crosslinking monomer was also investigated, and the structures of the gels are summarized in Table 4. The ensemble-averaged relaxation time distributions as a function of the relaxation time of typical gels with VMS (Run 35), AMS (Run 38), and APS (Run 40) are illustrated in Figure 7. The mesh size of the main structure of the gels was almost independent of the monomer concentration. Those gels showed bimodal distribution of the $P(\tau_R)$, derived from small mesh-size $(4 - 6 \cdot 10^{-6}$ s$)$ and large mesh-size $(10^{-3} - 10^{-2}$ s$)$. The similar results were reported in the gels of POSS-α,ω-nonoconjugated dienes. The structure of POSS, the three dimensional rigid cubic structure and/or the number of Si-H per one crosslinking monomer, should inevitably form the network defects independent of the spacer monomers used.

The SMILS analysis of POSS-VMSO gels was also conducted. The ensemble-averaged relaxation time distributions as a function of the relaxation time of the gels are shown in Figure 8. All the gels showed bimodal distribution of the $P(\tau_R)$, derived from small mesh-size and large mesh-size. The gels obtained in the monomer concentration with a little higher (11.0 wt.-%) or much higher (17.0 wt.-%) than the critical gelation concentration showed broad distribution of the main mesh-size. The gel obtained in the monomer concentration of 15.0 wt.-% formed the smallest mesh size and narrowest mesh-size distribution among the prepared gels. The gels with VMSB showed similar tendency of the gels with VMSO.

Table 4.   Structure of POSS-VMS, AMS, APS, VMSO, or VMSB gels

Run	Spacer monomer	Monomer concentration wt.-%	$\tau_R \cdot 10^{-6}$ [a] s	σ [b]	Mesh size nm
32	VMS	5.0	5.1	0.133	1.8
33	VMS	7.0	5.1	0.074	1.8
34	VMS	9.0	5.1	0.049	1.8
35	VMS	10.0	5.7	0.063	2.0
36	AMS	8.0	5.5	0.035	1.9
37	AMS	9.0	5.3	0.038	1.9
38	AMS	11.0	4.7	0.031	1.7
39	AMS	13.0	4.9	0.032	1.7
40	APS	14.0	4.1	0.095	1.4
41	APS	15.0	4.1	0.082	1.5
42	APS	17.0	4.3	0.042	1.5
43	VMSO	11.0	5.6	0.067	2.0
44	VMSO	15.0	4.9	0.027	1.7
45	VMSO	17.0	5.1	0.044	1.8
46	VMSB	4.0	6.9	0.047	2.5
47	VMSB	8.0	6.2	0.040	2.2
48	VMSB	10.0	7.3	0.055	2.6

a) Relaxation time of the gels determined by SMILS

b) Standard deviation of a peak of the ensemble-averaged relaxation time distribution.

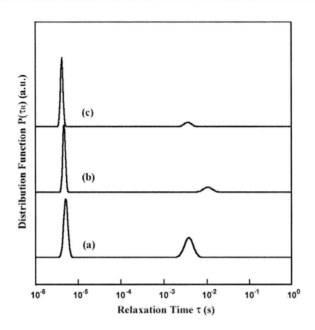

Figure 7. Ensemble-averaged relaxation time distributions as a function of the relaxation time of POSS-spacer monomers gels; (a) POSS-VMS 10.0 wt.-% (Run 35), (b) POSS-AMS 11.0 wt.-% (Run 38), and POSS-APS 14.0 wt.-% (Run 40).

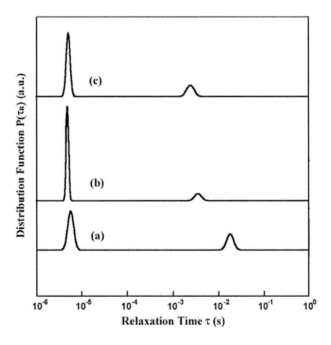

Figure 8. Ensemble-averaged relaxation time distributions as a function of the relaxation time of POSS-VMSO gels prepared using various monomer concentrations; (a) 11.0 wt.-% (Run 43), (b) 15.0 wt.-% (Run 44), and (c) 17.0 wt.-% (Run 45).

Relationship between Mesh Size and Crosslinking or Spacer Monomers

Mesh size of the gels is summarized in Figure 9 to study the effect of the structure of the crosslinking monomer and/or spacer monomer on the mesh size. The stretched molecular length of VMS is shorter than that of AMS, or APS. In the gels with TMCTS or POSS with VMS, AMS, or APS as spacer monomers, however, the mesh size of the gels did not correspond to the length of the spacer monomer. It seems reasonable to suppose that the small reaction groups, vinyl groups, on Si in AMS induce a slight defects in network structure and form lager mesh size than the expected size with TMCTS.[32] In the case of the gels with the spacer monomers containing two Si atoms (VMSO or VMSB), the gels with VMSB showed larger mesh size than those with VMSO, and the results were correlative with the stretched molecular length of the spacer monomers. The mesh size of the gels with TDSS showed quite different tendency from those with TMCTS or POSS, and the gels with VMSO formed the largest mesh size. The difference should be derived from the peculiar structure, flexible and tetrahedral-like structure, of TDSS.

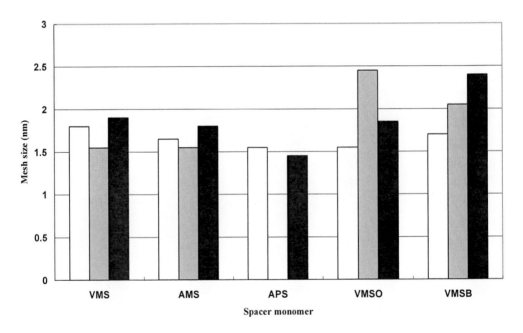

Figure 9. Relationship between spacer monomers and averaged mesh size of the gels with TMCTS (white), TDSS (gray), and POSS (black).

2. Synthesis of Organic-Inorganic Hybrid Gels with Bi-Functional Polar Monomers

Critical Gelation Concentration

The hydrosilylation reaction of the crosslinking monomers (TMCTS, TDSS, or POSS) with the bi-functional polar monomers (HDVE, HDA, or HDMA in Scheme 2) was investigated with the catalyst 1. The critical gelation concentrations of the gels are summarized in Table 5. The gels with HDA and HDMA showed higher critical gelation concentrations than the gel with HDVE. Furthermore, it took very long reaction time (about 3 days) to form the gels with HDA and HDMA. The reaction conversions of the reactions between TMCTS, TDSS, or POSS and HDVE or HDMA were determined by ^{1}H NMR at the sol state in the reactions of low monomer concentrations (less than the critical gelation concentration), and the results are summarized in Table 6. The reaction conversions of all the crosslinking reagents with HDVE for 1 h reaction were higher than those with HDMA. The result indicates that carbonyl group in HDMA should inhibit the hydrosilylation reaction and cause long time induction for the gel formation. The reaction conversions in the reactions of POSS are lower than the other crosslinking monomers. Furthermore, increase of the reaction conversions was not observed in the long time reactions with POSS. All though the reaction conversions in the reaction with of POSS are relative low, the critical gel concentrations of the gels are lower than the gels with other crosslinking monomers. The results indicate that the three dimensional rigid cubic structure and the number of Si-H groups per one crosslinking reagent of POSS should induce effective occupation of the space.

Table 5 Critical gelation concentration of the gels with bi-functional polar monomers (wt.-%)

	HDVE	HDA	HDMA
TMCTS	5.0	12.0	15.0
TDSS	5.0	10.0	15.0
POSS	3.0	10.0	7.5

Table 6 Reaction conversion of crosslinking monomer with bi-functional polar monomer in the sol state

Crosslinking monomer	Bi-functional polar monomer	Reaction conversion (%)	
		1 (h)	96 (h)
TMCTS	HDVE	68	89
TDSS	HDVE	73	94
POSS	HDVE	41	45
TMCTS	HDMA	17	42
TDSS	HDMA	21	89
POSS	HDMA	19	19

Synthesis and Structure of Gels from TMCTS

The ensemble-averaged relaxation time $[P(\tau_R)]$ distributions as a function of the relaxation time of the gels with TMCTS as a crosslinking monomer have been corrected with SMILS, and the structures of the gels are summarized in Table 7. Figure 10 shows the ensemble-averaged relaxation time distributions as a function of the relaxation time of TMCTS-HDVE gels. The monomer concentration strongly affected on the mesh size of the gel. Bimodal distribution of $P(\tau_R)$ is detected in all the gels. The peaks at short relaxation time

$(10^{-4} - 10^{-6}$ s) and long relaxation time $(10^{-2}$ s) are derived from the network of small mesh and large defect, respectively. The mesh size and mesh-size distribution of the small mesh of TMCTS-HDVE gel, which was evaluated by the peaks (σ), decreased with increasing of the monomer concentration. Increase of the monomer concentration should increase the occupied space in the gel with the mesh and induce the homogeneous mesh size distribution. Figures 11 and 12 show the ensemble-averaged relaxation time distributions as a function of the relaxation time of TMCTS-HDA and TMCTS-HDMA gels, respectively. The size of small mesh decreased with increasing of the comonomer concentration, as observed in the formation of TMCTS-HDVE gel.

Table 7. Structure of TMCTS-HDVE, HDA, or HDMA gels

Run	Crosslinking monomer	Spacer monomer	Monomer concentration wt.-%	$\tau_R \cdot 10^{-6\,a)}$ s	$\sigma^{b)}$	Mesh size nm
49	TMCTS	HDVE	5.0	69.8	0.14	12.0
50	TMCTS	HDVE	10.0	13.2	0.085	2.3
51	TMCTS	HDVE	30.0	8.0	0.054	1.4
52	TMCTS	HDA	12.0	6.4	0.068	1.1
53	TMCTS	HDA	15.0	22.7	0.058	3.9
54	TMCTS	HDA	30.0	7.3	0.10	0.9
55	TMCTS	HDMA	15.0	66.2	0.051	11.4
56	TMCTS	HDMA	20.0	1.5	0.054	2.5
57	TMCTS	HDMA	30.0	9.2	0.038	1.6
58	TMCTS	HDMA	40.0	4.1	0.046	0.7

a) Relaxation time of the gels determined by SMILS

b) Standard deviation of a peak of the ensemble-averaged relaxation time distribution.

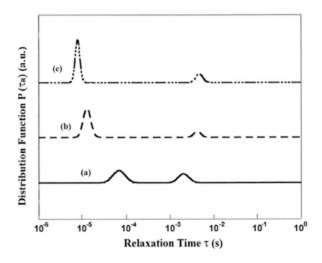

Figure 10 Ensemble-averaged relaxation time distributions as a function of the relaxation time of TMCTS-HDVE gels prepared using various monomer concentrations; (a) 5.0 wt.-% (Run 49), (b) 10.0 wt.-% (Run 50), and (c) 30.0 wt.-% (Run 51).

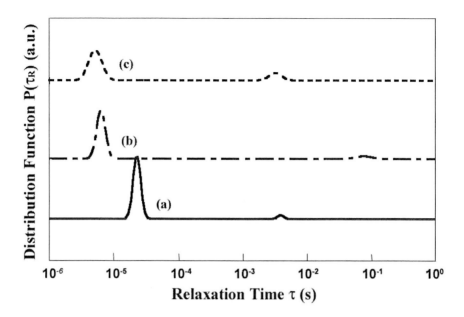

Figure 11. Ensemble-averaged relaxation time distributions as a function of the relaxation time of TMCTS-HDA gels prepared using various monomer concentrations; (a) 12.0 wt.-% (Run 52), (b) 15.0 wt.-% (Run 53), and (c) 30.0 wt.-% (Run 54).

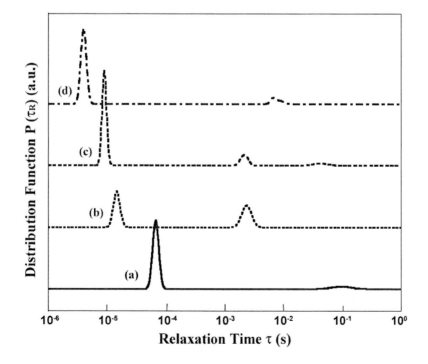

Figure 12. Ensemble-averaged relaxation time distributions as a function of the relaxation time of TMCTS-HDMA gels prepared using various monomer concentrations; (a) 15.0 wt.-% (Run 55), (b) 20.0 wt.-% (Run 56), (c) 30.0 wt.-% (Run 57), and (d) 40.0 wt.-% (Run 58).

Synthesis and Structure of Gels from TDSS and POSS

Figure12 illustrates the ensemble-averaged relaxation time distributions as a function of the relaxation time in the gels with TDSS (20 wt.-%), and the structures of the gels are summarized in Table 8. Although bimodal distribution was observed, the small mesh in all the gels showed narrow size distribution. The mesh size of the TDSS gels derived from the small mesh ranged from 1 to 2 nm. Relatively high reaction conversions in the reactions with TDSS (as summarized in Table 6) would induce the homogeneous mesh size distribution in the gels with TDSS.

Table 8. Structure of TDSS, or POSS-HDVE, HDA, or HDMA gels

Run	Crosslinking monomer	Spacer monomer	Monomer concentration wt.-%	$\tau_R \cdot 10^{-6\ a)}$ s	$\sigma^{b)}$	Mesh size nm
59	TDSS	HDVE	20.0	11.6	0.037	2.0
60	TDSS	HDA	20.0	9.7	0.032	0.9
61	TDSS	HDMA	20.0	11.1	0.046	1.1
62	POSS	HDVE	20.0	15.3	0.072	2.6
63	POSS	HDA	20.0	3.5	0.082	0.6
				122	0.040	20.9
64	POSS	HDMA	20.0	6.9	0.047	2.5
				104	0.061	17.9

a) Relaxation time of the gels determined by SMILS

b) Standard deviation of a peak of the ensemble-averaged relaxation time distribution.

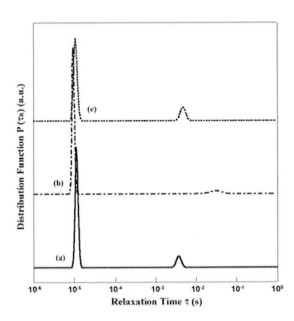

Figure 13. Ensemble-averaged relaxation time distributions as a function of the relaxation time of (a) TDSS-HDVE (Run 59), (b) TDSS-HDA (Run 60), (c) TDSS-HDMA (Run 61) gels (20 wt.-%)

Figure 13 shows the ensemble-averaged relaxation time distributions as a function of the relaxation time in the gels with POSS (20 wt.-%), and the structures of the gels are summarized in Table 8. All the gels with POSS show clear bimodal or multiple mesh size distribution. The low reaction conversion of POSS with the bi-functional polar monomers, as summarized in Table 6, would cause the inhomogeneity of the mesh size distribution. The other reason for the network defects in the gels with POSS would be derived from unexpected intermolecular crosslinking and/or intramolecular cyclization bi-functional monomers.

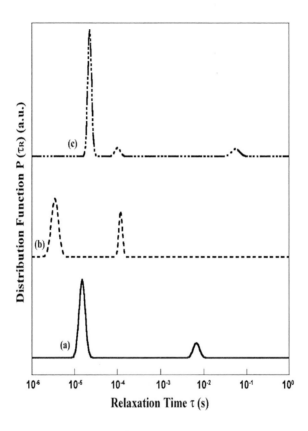

Figure 14. Ensemble-averaged relaxation time distributions as a function of the relaxation time of (a) POSS-HDVE (Run 62), (b) POSS-HDA (Run 63), (c) POSS-HDMA (Run 64) gels (20.0 wt.-%)

CONCLUSION

Organic-inorganic hybrid gels have been successively synthesized by means of a hydrosilylation reaction of multiple Si-H functional crosslinking monomers (TMCTS, TDSS, or POSS) with various bi-functional spacer monomers using a Pt catalyst.

In the first half of this chapter, we have reported the syntheses of the organic-inorganic hybrid gels with divinyl or diallyl monomers containing Si (VMS, AMS, APS, VMSO, or VMSB). The structure of crosslinking monomer and spacer monomer affected formation and network structures of the resulting gels. The gels with TDSS showed higher critical gel concentration than the gels with TMCTS and POSS. The gels from TMCTS or TDSS with Si containing spacer monomers, except VMSO, showed homogeneous mesh-size distribution. On the other hand, bimodal distribution of the mesh size was detected in the gels containing POSS and/or VMSO. A characteristic of these gels prepared in this part is that the mesh size of the gels is almost independent of the monomer concentration and molecular length of the spacer monomers, and ranges from 1.5 to 2.5 nm as summarized in Figure 9. In the case of the gels with α,ω-nonoconjugated dienes, the mesh size of the gels was strongly affected the monomer concentration, and showed the smallest size in the optimal monomer concentration, a little higher than the critical gel concentration as previously reported.[25] Molecular length of the spacer monomer also affected the mesh size, and the mesh size of the gels increased with increasing of the molecular length of α,ω-nonoconjugated dienes. The present synthesis of the organic-inorganic hybrid gels from Si-H functional crosslinking monomers, with Si containing divinyl or diallyl spacer monomer, should be a useful method to synthesize the gels with homogeneous network structure and limited mesh size from 1.5 to 2.5 nm. The kind of hybrid polymer with rigid spacer monomer was effective for the membrane with high selectivity to some pairs of gases. [33] The present hybrid gels with the spacer monomers containing Si should be useful precursor for the functional membrane.

In the latter half of this chapter, we have investigated the syntheses of the organic-inorganic hybrid gels with the bi-functional polar monomers (HDVE, HDA, or HDMA). It took very long time, about 3 days, for formation of the gels with HDA and HDMA. The results indicate that carbonyl groups in the HDA and HDMA monomer should hinder the hydrosilylation reaction. The low reaction conversion of Si-H groups in the croslllinking monomers and HDA or HDMA should support these results. Low reaction conversion and/or unexpected reaction of spacer monomers in the reaction with POSS would cause the inhomogeneity of the mesh size distribution. Although some problems should be solved in this synthetic method, the formation of the gels with bi-functional polar monomers should extend possibility of the wide application of various kinds of spacer monomers.

The next step of our experiment is molecular design and effective synthesis of the organic-inorganic hybrid gels with various spacer monomers to add a variety of functions. Synthesis of some gels with functional spacer monomers, and application of the gels are proceeding and the results will be reported elsewhere.

REFERENCES

[1] J. E. Mark, Y.-C. C. Lee, P. A. Bianconi, Eds.; ACS Symposium Series 585; Hybrid Organic-Inorganic Composites; American Chemical Society: Washington, DC, 1995.

[2] L. Mascia, *Trends Polym. Sci.* 1995, *3*, 61.

[3] B. M. Novak, *Adv. Mater.* 1993, *5*, 442.

[4] A. Provatas, J. G. Matisons, *Trends Polym. Sci.* 1997, *5*, 32.

[5] R. M. Laine, C. Zhang, A. Sellinger, L. Viculis, *Appl. Organomet. Chem.* 1998, *12*, 715.

[6] R. A. Vaia, H . Ishi, E. P. Giannelis, *Chem. Mater.* 1996, *8*, 1728.

[7] M. W. Weimer, H. Chen, E. P. Giannelis, D. Y. Sogah, *J. Am. Chem. Soc.* 1999, *121*, 1615.

[8] T. Saegusa, Y. Chujo, *Makromol. Chem., Macromol. Symp.* 1991, *51*, 1.

[9] C. J. T. Landry, B. K. Coltrain, M. R. Landry, V. K. Long, *Macromolecules* 1993, *26*, 3702.

[10] N. Yamada, I. Yoshinaga, S. Katayama, *J. Mater. Chem.* 1997, *7*, 1491.

[11] K. M. Kim, Y. Chujo, *J. Matr Chem.* 2003, *13*, 1384.

[12] J. Choi, A. F. Yee, R. M. Laine, *Macromolecules* 2003, *36*, 5666.

[13] V. Galiastsatos, P. Subramanian, L. Klein-Castner, *Macromol. Symp.* 2001, *171*, 97.

[14] R. O. Pinho, E. Radovanovic, I. L. Torriani, I. V. P. Yoshida, *Eur. Polym. J.* 2004, *40*, 615.

[15] S. U. A. Redondo, E. Radovanovic, I. L. Torriani, I. V. P. Yoshida, *Polymer* 2001, *42*, 1319.

[16] M. Tsumura, K. Ando, J. Kotani, M. Hiraishi, T. Iwahara, *Macromolecules* 1998, *31*, 2716.

[17] M. Tsumura, T. Iwahara, *Polymer Journal* 1999, *31*, 452.

[18] M. Tsumura, T. Iwahara, *Polymer J.* 2000, *32*, 567.

[19] M.Tsumura, T. Iwaara, *J. Appl. Polym. Sci.* 2000, *78*, 724.

[20] M. J. Michalczyk, W. E. Farneth, A. Vega, *J. Chem. Mater.* 1993, *5*, 1687.

[21] R. Q. Su, T. E. Müller, J. Procházka, J. A. Lercher, *Adv. Mater.* 2004, *14*, 1369.

[22] C. Zhang, F. Babonneau, C. Bonhomme, R. M. Laine, C. L. Soles, H. A. Hristov, A. F. Yee, *J. Am. Chem. Soc.* 1998, *120*, 8380.

[23] R. M. Laine, C. Zhang, A. Sellinger, L. Viculis, *Appl. Organometal. Chem.* 1998, *12*, 715.

[24] G. Pan, J. E Mark, D. W. Schaefer, *J. Polym. Sci.: Part B: Polym. Phys.* 2003, *41*, 3314.

[25] N. Naga, E. Oda, A. Toyota, K. Horie, H. Furukawa, *Macromol. Chem. Phys.* 2006, *207*, 627-635.

[26] H. Furukawa, K. Horie, N. Azuma, Y. Miyashita, *Cellulose Commun.* 2003, *10*, 154.

[27] H. Furukawa, K. Horie, R. Nozaki, M. Okada, *Phys. Rev. E; Statistical, Nonlinear, and Soft Matter Phys.* 2003, *68*, 031406-1.

[28] H. Furukawa, S. Hirotsu, *J. Phys. Soc. Japan* 2002, *71*, 2873.

[29] H. Furukawa, K. Horie, *Kobunshi Ronbunshu* 2002, *59*, 578.

[30] H. Furukawa, M. Okada, *Transactions of the Materials Research Society of Japan* 2000, *25*, 723.

[31] H. Furukawa, M. Kobayashi, Y. Miyashita, K. Horie, *High Performance Polymers* 2006, *18*, 837-847.

[32] Calculated with Chem 3D.

[33] R. O. Pinho, E. Radovanovic, I. L. Torriani, I. V. P. Yoshida, European Polymer Journal 2004, 40, 615-622.

In: Modern Trends in Macromolecular Chemistry
Editor: Jon N. Lee

ISBN: 978-1-60741-252-6
© 2009 Nova Science Publishers, Inc.

Chapter 6

RECENT DEVELOPMENTS IN ASYMMETRIC C-C BOND FORMATION REACTIONS USING HETEROGENEOUS CHIRAL CATALYST ESPECIALLY ON INORGANIC SUPPORTS

Kavita Pathak, Sayed H. R. Abdi[**], Rukhsana I. Kureshy, Noor-ul H. Khan and Raksh V. Jasra

Discipline of Inorganic Materials and Catalysis, Central Salt and Marine Chemicals Research Institute (CSMCRI), Bhavnagar, Gujarat, India.

ABSTRACT

Enantioselective catalytic reactions in which the chirality of an asymmetric catalyst induces the preferred formation of a given product enantiomer have been one of the most important achievements in chemistry in the 20th century. Numerous enantioselective catalytic reactions like C-H, C-C, C-O, C-N, and other bonds have been discovered. In recent years, much research has been devoted on catalytic asymmetric C-C bond formations using various chiral catalysts. Chiral products obtained by asymmetric C-C bond formation reactions are important intermediates for pharmaceuticals, vitamins, agrochemicals, flavors, fragrances and functional materials. Quite often, the cost of sophisticated chiral ligands exceeds the cost of noble metals employed in chiral catalyst. Therefore, catalyst recovery is of paramount importance for the application of enantioselective metal catalysis to large scale processes. This has led researches looking for recoverable and recyclable chiral catalysts. In recent years, inorganic materials have been extensively used as alternative supports for the heterogenization of chiral catalysts due to their excellent thermal and chemical stability, ease of handling and diversity of chemical modifications. Moreover, a large variety of preparative procedures are available leading to materials with diverse morphological properties. They can also be used at both high and low temperatures and at high pressure. Their inflexibility and

---

[*] Correspondence to: Discipline of Inorganic Materials and Catalysis, Central Salt and Marine Chemicals Research Institute (CSMCRI), Bhavnagar- 364 002, Gujarat, India. Tel.: +91 278 2567760; Fax: +91 278 2566970. E-mail: shrabdi@csmcri.org

noncompressibility make silica-tethered catalysts suitable for continuous-flow reactors. In this chapter, we have focused our attention on asymmetric C-C bond formations using chiral metal complexes of 'privileged chiral ligands' as catalysts tethered to inorganic materials by covalent and non-covalent interactions. Our aim is to highlight the great potential and diversity of applications, of immobilized chiral catalysts on inorganic supports in asymmetric C-C bond formations and to discuss the factors of heterogeneous chiral catalysts that influence the asymmetric catalytic performance under heterogeneous reaction conditions.

INTRODUCTION

The demand for optically pure compounds is increasing with every passing year due to the greater awareness with regard to personal health and environment. While various methods are available for selective production of single enantiomers, asymmetric catalysis is unique in the sense that it allows single chiral information embedded in a chiral catalyst to be reproduced several times. This is the most economic way to produce chirally pure compounds, as in principle, one molecule of chiral catalyst can produce millions of chiral molecules (chiral reproduction) [1-4]. Basically, chiral catalysts contain a catalytically active metal center surrounded by a chiral ligand that imparts enantioselectivity to the product. Although, in some cases chiral organic compounds on their own work as chiral catalyst (*e.g.*, organo-catalysts like enzymes). To begin with, practically all chiral catalysts worked under homogeneous condition. Even though homogeneous asymmetric catalysis has the advantages of high enantioselectivity and catalytic activity in a variety of asymmetric transformations under mild reaction conditions [5,6], the high catalyst loadings and the difficulties associated with recovery and the reuse of expensive chiral catalysts severely hampers its practical applications. In addition, sometimes the metal contaminants can leach from the homogeneous catalysts into the products; this contamination is particularly unacceptable for pharmaceutical production [7]. To counter these issues researchers are looking for recoverable and recyclable chiral catalysts [8-14]. Consequently, heterogenization of homogeneous catalysts evolved as a logical alternative to homogeneous catalysis. The potential advantages of heterogeneous catalysis, such as easy separation, efficient recycling, minimization of metal traces in the product, and an improved handling and process control, which finally result in an overall lowering of costs, are well known. However, the journey to heterogenization of homogeneous catalysts is not as smooth as was initially expected. In heterogeneous asymmetric catalysis, there are a large number of variables whose effects on the given system are often not well understood. The adsorption of the modifier is usually complex, with a multiplicity of potential adsorption modes. At the same time, little is known about the role that the support may play.

Various enantioselective catalytic reactions like C-H, C-C, C-O, C-N, and other bonds have been extensively studied. Arguably enantioselective C–C bond formation is perhaps the most important area in terms of constructing complex organic molecules [1,15]. Table 1 illustrates important asymmetric C-C bond formation reactions.

Table 1. Asymmetric C-C bond forming reactions.

Name of reaction	Reaction Scheme	Products
Aldol		β-hydroxy ketone, or 'aldol'
Mukiyama Aldol		β-hydroxycarbonyl compounds
Nitro-Aldol		nitro alcohol (nitroaldol)
Mannich /Nitro-Mannich		2-Nitroamines
Michael Addition		Michael adducts like nitroalcohols
Diels-Alder		Cycloadducts
Hetero Diels-Alder		Cycloadducts like dihydropyrans or tetrahydropyrans
Cyclopropanation		Cyclopropanes
Hydroformylation		Branch and linear aldehydes
Carbonyl-ene reaction		α-Hydroxy carbonyl compounds
Imine-ene reaction		α-Amino carbonyl compounds
Addition of dialkylzinc to aldehydes		Secondary alcohols
Addition of phenylactylene to aldehydes		Propargylic alcohols
Allylation of Aldehydes		Allyl alcohols
Cyanosilylation of Aldehydes		Cyanohydrins
Fridel-Crafts Alkylation		Ketone

STRATEGIES TO SUPPORT CHIRAL HOMOGENEOUS CATALYSTS

The immobilization of chiral catalysts mainly employed two main strategies. First, the homogeneous catalyst is heterogenized by anchoring it onto a solid support (*e.g.*, an inorganic material or polymer) *via* (i) a covalent bonding, (ii) adsorption or ion-pair formation, (iii) encapsulation, or (iv) entrapment. Second, liquid–liquid two phase (biphasic) systems, where the catalyst is retained in one phase (e.g., aqueous phase, fluorous phase, supercritical carbon dioxide (ScCO$_2$) or ionic liquid) and the other phase is used for delivery and/or removal of reactants. Quite recently, a so-called ''self-supported strategy'' has been developed toward efficient immobilization of chiral catalysts [16-18] (Figure 1). However, immobilization usually affects the catalytic performance of the catalyst. Unfortunately, most examples of supported catalysts tend to have inferior catalytic properties compared to their homogeneous counterparts. In addition, most immobilization methodologies need to modify the catalyst structure that raises catalyst costs. However, in some cases, the immobilization results in marked increase in activities, chemoselectivities, and even enantioselectivities. In rare cases, increased activity and selectivity have also been observed upon reuse.

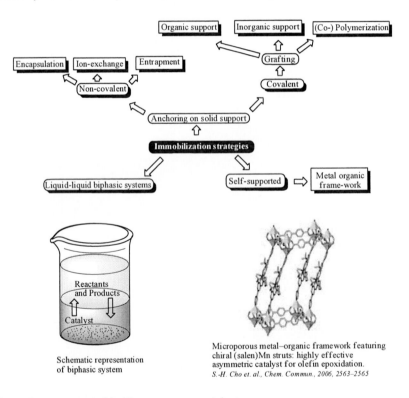

Microporous metal–organic framework featuring chiral (salen)Mn struts: highly effective asymmetric catalyst for olefin epoxidation.
S.-H. Cho et. al., Chem. Commun., 2006, 2563–2565

Schematic representation of biphasic system

Figure 1. Strategies to support chiral homogeneous catalysts.

The scope of the present chapter is to review the progress in recoverable chiral catalysts, especially on inorganic supports, which are effective in enantioselective C-C bond formation reactions. Before we describe catalysts used in specific C-C bond coupling reactions, we have presented a brief account of different types of supports and chiral ligands frequently employed for this purpose.

TYPES OF SOLID SUPPORTS

Two principle supports have been used for the immobilization of chiral homogeneous catalysts: organic (polymeric) and inorganic supports. Both supports have uniques characteristics as per required applications. The organic supports can be soluble or insoluble polymer resins where active catalyst is chemically bound to give a supported catalyst [19-21]. Some times the active catalyst is an intigeral part of the main polymer network. In a few cases, the active catalyst is at the outer most stratum of a dendrimer which serves as a recoverable chiral catalyst. In most cases organic polymers are flexible and swell in solvents used for catalytic runs [20]. In case of immobilization, the support materials need to be thermally, chemically, and mechanically stable during the reaction process. Moreover, the structure of the support needs to be such that the active sites are well dispersed on its surface so that these sites are easily accessible by the reactants. The inorganic supports are generally inert porous structures with highly specific surface areas and there is the opportunity to synthesize a variation in pore size, pore distribution and particle size [22,23]. Since the focus of this chapter is asymmetric C-C bond formations mediated by catalysts supported on inorganic materials, it is pertinent give a brief introduction on inorganic supports widely used in this area (Table 2).

Table 2. Different Inorganic supports.

Types		Attributes
Microporous	Amorphous silica	• Easily available for large-scale applications. • Wieldy use as Filler for paints, plastic, epoxy, glassmaking, papers, seed germination, cosmetics, automotive, storage, chromatographic column treatment, wood treatment, powder metal molding, etc.
	Alumina	• Thermally-stable, high-surface-area forms lead to use as acid or base catalysis or as supports for other catalytic materials (e.g. metals, oxides, sulphides, etc.). • Dehydrated and hydrated forms. • used as a drying agent, catalyst, catalyst support, and column choromatography
	Zeolites	• Crystalline microporous aluminosilicates with molecular-sized interacrystalline channels and cages. • Highly selective adsorbents, can remove minute components of a reaction mixture. • Used as support in "ship in bottle" method for heterogenization of homogeneous catalyst.
	Clay	• Clays are alumino-silicates with a 2-D sheet lattice structure. In contrast, quartz, the common mineral of sand and silt, is a silicate with a three dimensional tetrahedral structure. • The type Montmorillonite - most versatile intermediate charge clay has high cation exchange capacity and good swelling properties allow a wide variety of catalytically active forms to be prepared: acidic cations, metal complexes, photo-catalytically active cations.

Table 2. (Continued)

Types		Attributes
Mesoporous	MCM-41	• Hexagonal arrangement of the mesopores in 1-D, space group *p6mm*. • High surface area (700-1100 m^2/g^{-1}) with narrow pore size distribution. • Most used supports for catalysis and specially in asymmetric catalysis.
	MCM-48	• Cubic arrangement of the mesopores in 3-D, space group *Ia3d*). • High surface area (700-1100 m^2/g^{-1}) with narrow pore size distribution. • Most promising for potential applications in catalysis due to their high thermal stability and high adsorption capacity.
	SBA-15	• Hexagonal arrangement of mesopores in 2-D, space group *p6mm*. • Pore size is much larger than MCM-41 and can be controlled in the range of 6-15 nm.
	MCF	• Synthesized in a similar protocol to SBA-15, but with oil-in-water microemulsions as templates. • MCFs consist of interconnecting cage-like pores with sizes ranging from 20-40 nm, and pore interconnection widths ranging from 8-25 nm.
Macroporous	CPG	• Constitutes an amorphous network of pores of uniform sizes in the range of 4-10^4 nm. • They can be modified to include a variety of functional groups, and the adsorption strength of the glasses can be adjusted over a wide range of values. • Widely used as a stationary phase in chromatography.

Zeolites are one of the earliest inorganic polymers used as supports for the immobilization of chiral catalysts. They are crystalline materials with well-defined pores and channels in the micropore range [24,25]. There are many attributes to this silica support: (i) does not swell in the solvents, (ii) can withstand lower or higher temperatures and higher pressure without undergoing structural changes, (iii) does not allow the aggregation of active catalysts [26], (iv) insoluble in organic solvents [12] and (v) possibility to device stationary chiral phases that is used in a continuous process. In recent years, mesoporous materials have drawn much attention for immobilizing the homogeneous catalysts for chiral synthesis because the size distribution of the mesopore is in the range of 2–20 nm [10,12]. This pore size range is well suited for application in pharmaceuticals and fine chemicals as frequently large sized molecules are involved in the synthesis of these chemicals. Finally, these pores could act as nanoreactors where assembly of reactants and chiral catalyst in these confined media often result in a higher reactivity and selectivity than normally observed in homogeneous catalysis.

TYPES OF CHIRAL LIGANDS

Most used chiral ligands, which were immobilized for asymmetric C-C bond formation reactions, are based on the well-established key ligands as presented in Figure 2. In particular, "Priveliged chiral ligands"[27], BINOL (2,2'-dihydroxy 1,1'-binaphthyl) [28-30], salen (salicyledine ethylenediamine) [31,32], TADDOL [33,34], Box and pybox [35-37], β-aminoalcohols [38,39] rank as the most important chiral catalysts which have successfully been used as chiral backbones for solid-bound catalysts. These chiral ligands having numerous variations of substituents form complexes with a variety of metals. These complexes have been used as homogeneous chiral catalysts in various C-C bond formation reactions, as a consequence, efforts were made to heterogenize these chiral complexes with minimal structural changes made on the original complex so as to retain the efficiency of the homogeneous catalytic system.

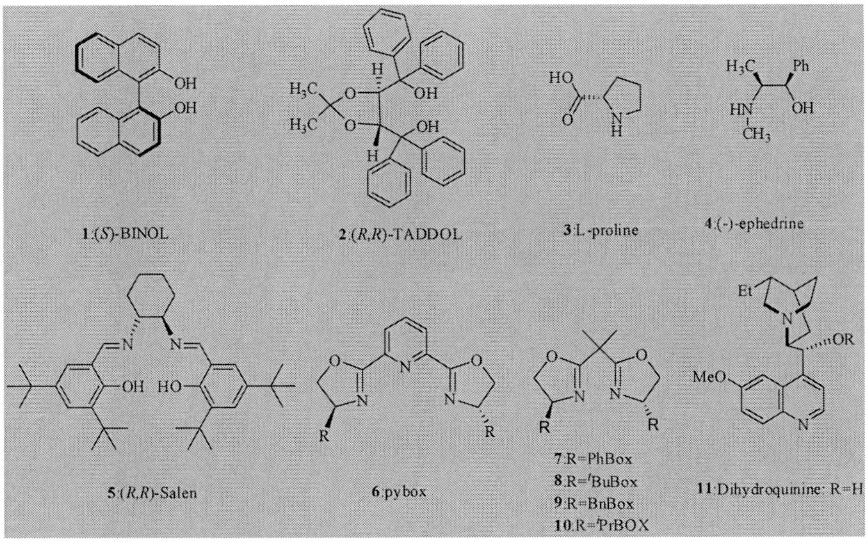

Figure 2. Privileged chiral ligands for asymmetric C-C bond formations.

ALDOL REACTION

The aldol reaction is recognized as being one of the most powerful and popular methods for the construction of C-C bonds and has been widely used in the synthesis of natural products and their synthetic variants [40]. A number of methods based on enzyme catalysts, organo-catalysts and organometallic catalysts have been reported for this transformation [41,42]. In recent years, organocatalysis,[43] in particular, the use of amino acid-proline [44,45] has experienced a renaissance in catalytic asymmetric aldol condensations. The main advantage of this method is that the reaction can be performed in a stereoselective manner, under mild reaction conditions without the need of any metal, this is often preferred in pharma industry to ensure no metal contaminant in the finished drug molecules. Choudhary et al. reported the incorporation of unaltered L-proline and S-BINOL into layered double hydroxides (LDHs) [46]. The heterogeneous catalysts 1-LDH and 3-LDH were used for the aldol reaction (Scheme 1). Under mild reaction conditions, excellent yields were obtained, but with low enantioselectivity (ee < 12%).

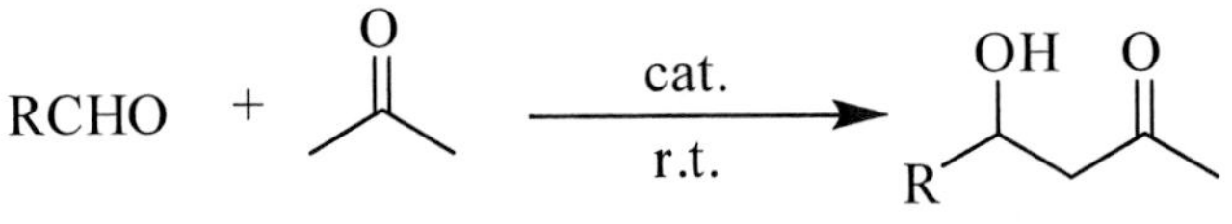

$$RCHO \ + \ \overset{O}{\underset{}{\bigwedge}} \xrightarrow[\text{r.t.}]{\text{cat.}} \underset{R}{\overset{OH \quad O}{\bigwedge\bigwedge}}$$

R= C$_6$H$_5$-CH$_3$, Cl-C$_6$H$_4$-CH$_3$, (CH$_3$)$_2$CH$_2$CH$_2$CH$_3$

with **1**-LDH: 82-90% yield, 3-10% ee
with **3**-LDH: 20-95% yield, <10% ee

Scheme 1. Aldol reaction catalysed by organo-LDHs.

Beadham et al. [47] have reported immobilized proline and benzylpencillin derivatives onto MCM-41 and amorphous silica gel through covalent grafting. The catalysts were tested for direct Aldol reaction between acetone and activated aromatic aldehydes (Scheme 2). MCM-41 supported benzylpenicillin derivative gave higher enantioselectivity than SiO$_2$ supported benzylpenicilin derivative. Moreover, the heterogeneous catalyst could be reused without significant loss of enantioselectivity.

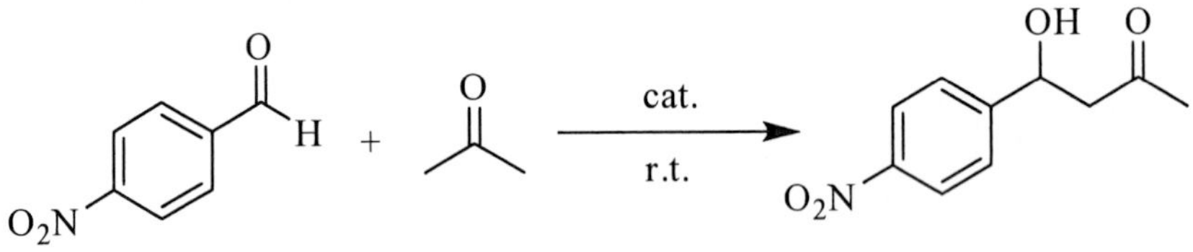

with **3**-MCM-41: 45% yield, 36% ee
with **3**-SiO$_2$: 42% yield, 32%ee
with Benzylpenicilin-MCM-41: 32% yield, <5%ee
with Benzylpenicilin-SiO$_2$: 21% yield, n.d. ee

Scheme 2. Aldol reaction using immobilized proline and benzylpencillin onto silica.

An effective approach for the asymmetric Aldol reaction was developed by Mayoralas et al. [48] using different mesoporous and lamellar siliceous materials with different topologies to heterogenize proline derivatives assisted by heat and microwaves (Scheme 3). The aldol reaction of hydroxyacetone with different aldehydes using thus immobilized proline furnishes 1,2-diols which may be used in the synthesis of carbohydrate analogues as inhibitors of glycosidases and glycosyltransferases. Condensations of aldehydes with hydroxyacetone under thermal heating have resulted in aldol products with moderate to good yields (up to 80%) and stereoselectivities (ee, 70-99%), in some cases leading to the formation of the syn-diol as main product. With the assistance of microwave heating, reaction times were drastically decreased (from 24 h to 10 min) and yields were generally improved (up to 90%). They found that **3**-MCM-41 catalyzed aldol reactions in both hydrophilic and hydrophobic solvents provided products with stereoselectivities comparable with the homogeneous catalyst (ee, >99%). The heterogeneous catalysts could be reused to up to three cycles without significant lost of stereoselectivity.

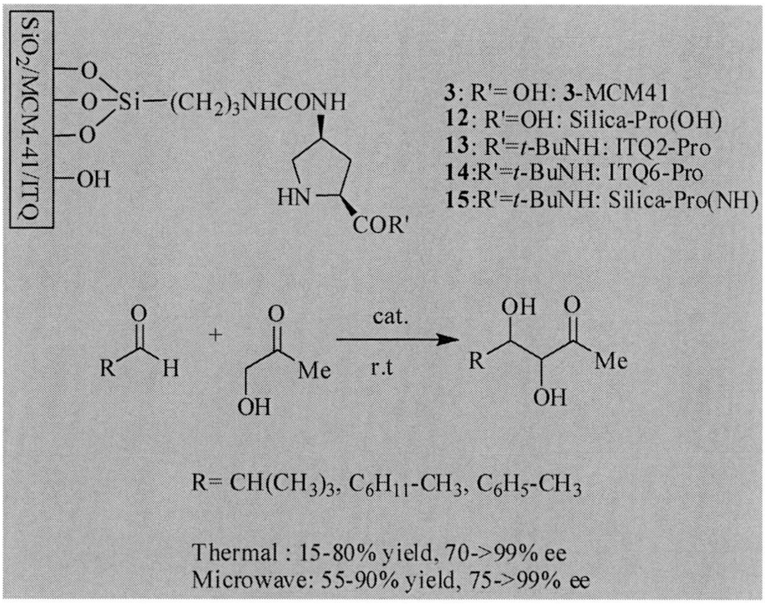

Scheme 3. Immobilized proline derivatives on different silica.

Mayoralas et al. [49] further investigated catalytic efficiency of **3**-MCM-41 in asymmetric aldol reaction of *p*-nitrobenzaldehyde with 2,2-dimethyl-1,3-dioxan-5-one in a wide range of solvents for the synthesis of biologically important chiral molecules (Scheme 4 (a)). They found the progress of the reaction to be dependent on the nature of the solvent. Reactions proceeded more efficiently in hydrophilic polar solvents; however, the addition of a small amount of water (5 equiv.) had a positive effect on the rate and the stereoselectivity of the reaction (conversion, 47% and ee, 78%) performed in hydrophobic toluene. They further extended this reaction under heterogeneous conditions with several chiral aldehydes to give useful intermediates for the synthesis of azasugars (Scheme 4 (b) & (c)). With catalyst **3**-MCM-41, the reactions in formamide gave *anti* aldols as major products in moderate yields but with good diastereoselectivities. These products in turn can be transformed into six- and five membered iminocyclitols by hydrogenation. These are important azasugars where the ring oxygen of a carbohydrate is replaced by nitrogen. Azasugars have shown the ability to act as potent inhibitors of enzymes involved in carbohydrate processing, such as glycosidases and glycosyltransferases.

A heterogeneous catalyst using intercalation of L-proline in Mg–Al LDH was developed by He et al. [50]. The authors found that the thermal stability and optical stability of the immobilized chiral catalytic centers in restricted galleries of LDH materials had improved against thermal treatment and light irradiation (Figure 3). The aldol reaction of benzaldehyde with acetone in the presence of **3**-LDHs as catalyst resulted in corresponding aldol product in very good yield (~90%) and a high enantiomeric excess (ee, 94%). The enantiomeric selectivity of **3**-LDHs was found to be more stable towards thermal pretreatment as compared to pristine L-proline. The possible reaction mechanism was proposed in the light of the mechanism found under homogeneous reaction condition. The effects of chemical composition of **3**-LDHs and other reaction parameters on catalytic efficiency were also investigated. This novel solid catalyst has advantages in terms of high activity under mild liquid-phase conditions, ready separability by simple filtration, and does not produce side products.

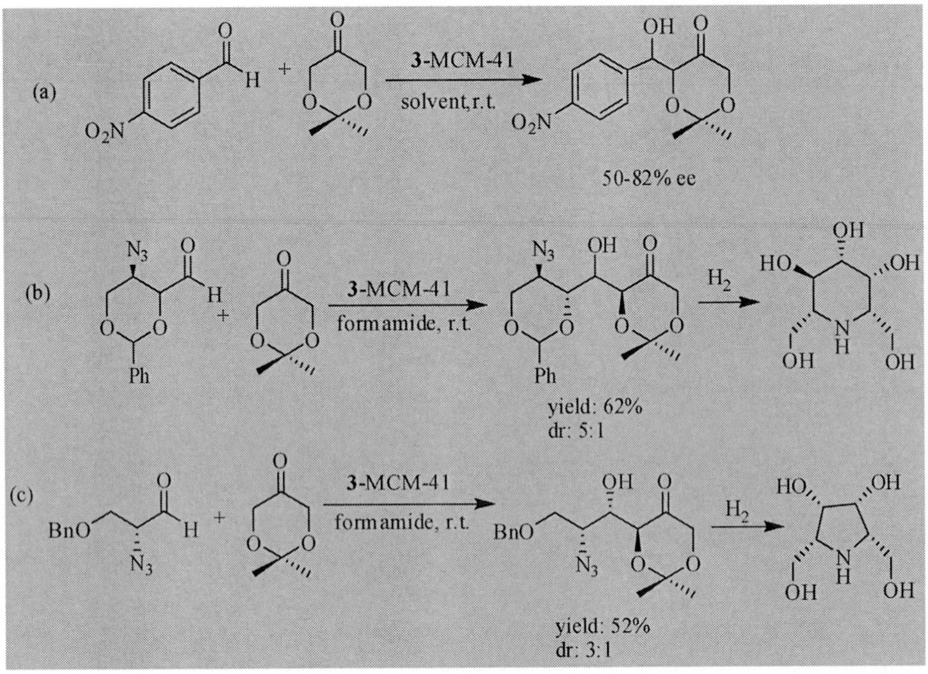

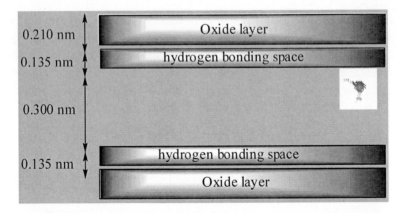

Scheme 4. Asymmetric aldol reaction using MCM-41 supported proline.

Figure 3. The schematic model of L-Proline LDH (3-LDH).

A heterogeneous bifunctional catalyst 16 (Figure 4) for asymmetric Aldol reaction was synthesized by incorporating a chiral amine and various acids onto silica surface [51]. However, the use of different acids like AcOH, CF$_3$COOH, TsOH, tartaric acid showed only modest reactivity and enantioselectivity (yield, 27-69%; ee, 16-40%) for the acyclic and the cyclic ketones with low or no diastereoselectivity in the case of cyclic ketones. For substituted benzaldehyde as the ketone acceptor, p-nitrobenzaldehyde afforded the aldol product with the highest yield (75%) and enantioselectivity (36%) while using AcOH. It is worth noting that synergistic effect between the amine and the acid groups is crucial for the reactivity and the enantioselectivity in these reactions. The highest 67% yield and 60% ee for the product aldol was obtained by the reaction of isobutylaldehyde and acetone under this reaction system.

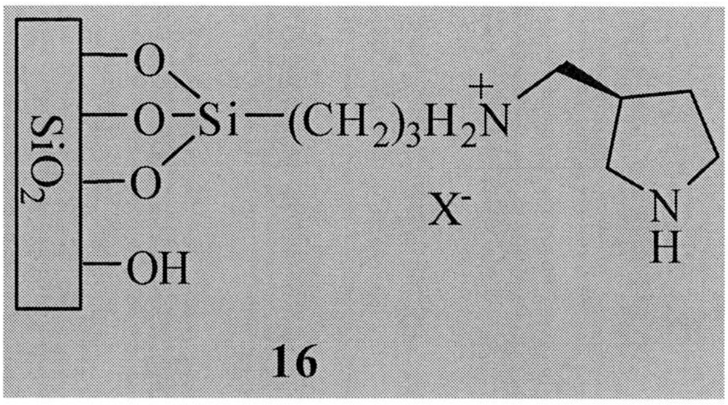

16

Figure 4. Heterognenous bifuntional catalyst.

L-proline being active organocatalyst for enantioselective Aldol reaction, Gruttadauria et al developed new materials for its heterogenization on solid support to facilitate its separation and recycling [52]. The authors supported L-proline on the surface of silica gels modified with a monolayer of covalently attached ionic liquid and used this material alone or with additionally adsorbed ionic liquid as catalysts (Scheme 5). These materials have been applied to the L-proline-catalyzed aldol reaction between acetone and several aldehydes. Good yields (50-99%) and ee values (ee, 63-95%), comparable with those obtained under homogeneous conditions, have been obtained especially with imidazolinium-modified and 4-methylpyridinium-modified silica gels. Moreover, these materials have been easily recovered by simple filtration and used for at least nine cycles with no loss in activity or enantioselectivity. This strategy has showed that ionic liquid modified silica gels could be interesting materials for supporting L-proline and other similar catalysts.

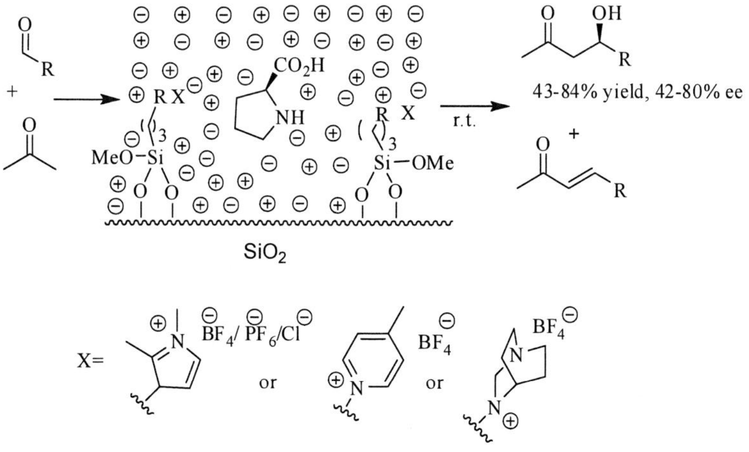

Scheme 5. Supported ionic liquid for asymmetric aldol reaction.

Gruttadauria et al. have further developed ionic liquid-modified silica gels as recyclable materials of L-proline **3** and the tripeptide H–Pro–Pro–Asp–NH₂ 17, by adsorption, onto the surface of modified silica gels functionalized with a monolayer of covalently attached 1,2-dimethyl-imidazolium chloride, tetrafluoroborate or hexafluorophosphate ionic moieties, respectively [53]. Three different linkers were used to attach the ionic liquid moiety to the surface of these supports (Scheme 6). The resulting materials effectively catalyzed the Aldol reaction between acetone and several substituted benzaldehydes. Good yields and enantioselectivities (yield: 99%, ee: 95%), comparable to or better than those obtained under homogeneous reaction conditions were obtained with these catalysts. These catalytic materials were easily recovered by filtration, and reused. Studies performed using L-proline-supported materials have shown that the re-use of these materials is dependent on the nature of the linker. The supported tripeptide H–Pro–Pro–Asp–NH₂ 17 gave higher enantioselectivities than those obtained with supported-proline **3**. Recycling investigations have shown that silica gel supported L-proline has superior recyclability (nine times) than tripeptide-supported materials (four times).

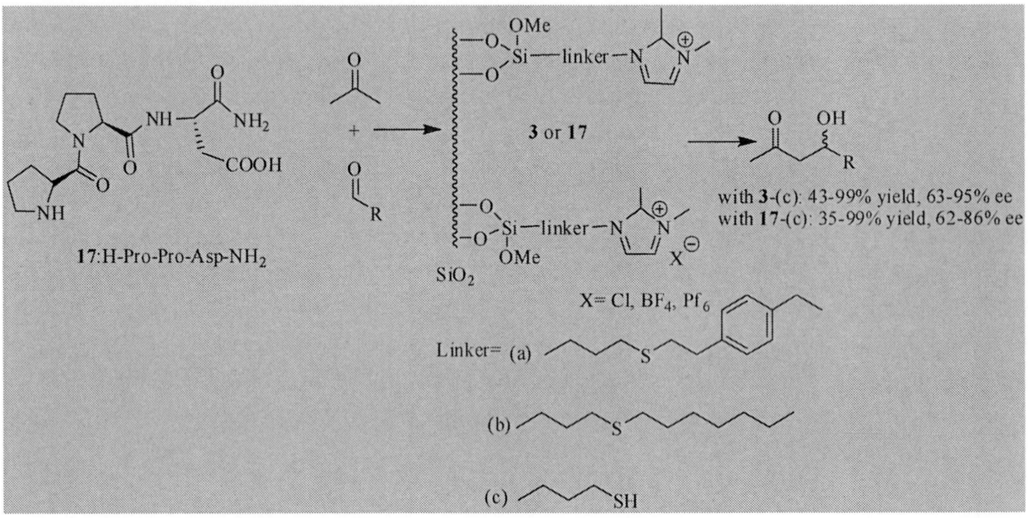

Scheme 6. Supported ionic liquid on silica.

Recently Cheng et al. [54] used polyoxometalate (POM) as support for recoverable asymmetric enamine-based catalysts for aldol reaction (Scheme 7). Polyoxometalates (POMs) belong to a family of metal-oxide clusters that have diverse catalytic applications due to their intrinsic properties such as high acidity and favorable redox potentials [55]. Chiral amine and acids were combined to give highly efficient, stereoselective, and recoverable enamine-based catalysts. Excellent yields and enantioselectivity (51-99% yield, 95 to >99% ee) for Aldol products were achieved with this hybrid catalyst. Less than 1 mol % of chiral amine loading was sufficient for good catalytic activity, and the catalyst could be recovered and reused 6 times with unchanged enantioselectivity and slightly decreased activity. This study provided a new strategy for the development of practical asymmetric organocatalytic synthesis.

A heterogeneous catalyst using intercalation of L-proline in Mg–Al LDH was developed by He et al. [50]. The authors found that the thermal stability and optical stability of the immobilized chiral catalytic centers in restricted galleries of LDH materials had improved against thermal treatment and light irradiation (Figure 3). The aldol reaction of benzaldehyde with acetone in the presence of **3**-LDHs as catalyst resulted in corresponding aldol product in very good yield (~90%) and a high enantiomeric excess (ee, 94%). The enantiomeric selectivity of **3**-LDHs was found to be more stable towards thermal pretreatment as compared to pristine L-proline. The possible reaction mechanism was proposed in the light of the mechanism found under homogeneous reaction condition. The effects of chemical composition of **3**-LDHs and other reaction parameters on catalytic efficiency were also investigated. This novel solid catalyst has advantages in terms of high activity under mild liquid-phase conditions, ready separability by simple filtration, and does not produce side products.

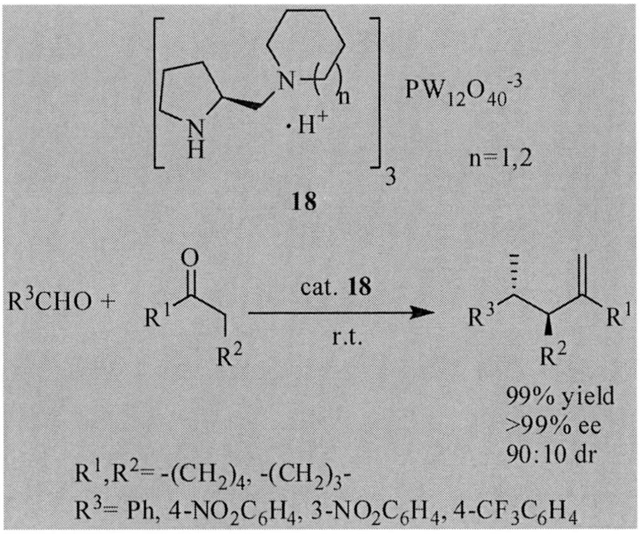

Scheme 7. Chiral amine-POM hybrid catalyst.

Dinuclear pincer-palladium(II) complexes supported on silica were also reported to be effective heterogeneous catalysts for the aldol reaction of aldehydes with methyl isocyanoacetate [56] (Figure 5). To activate the surface of the silica gel it was first treated with excess of thionyl chloride. This procedure produces electrophilic Si–Cl groups at the surface of silica that have been shown to react with alcohol groups to form functionalized surfaces. This method was particularly conducive to catalysts containing a protected vicinal diol that could be easily unveiled by the hydrolysis of the ketal functionality. Two silica-supported catalysts prepared with a bimetallic compound show high catalytic activity but with very minor enantioselectivity (ee: 2-3%). The systems were designed to contain remote chirality in an effort to produce a stereo induction to be realized when both metal centers interact with the substrate.

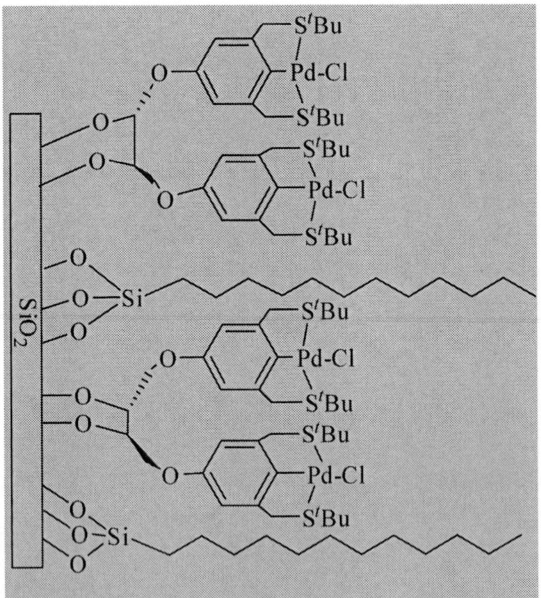

19

Figure 5. Dinuclear pincer-palladium(II) complexes supported on silica.

MUKAIYAMA-ALDOL REACTION

Catalytic asymmetric Mukaiyama-Aldol reaction has emerged as one of the most important carbon–carbon bond forming reactions in the synthesis of optically active β-hydroxycarbonyl compounds [40,57]. Various methodologies of the asymmetric Mukaiyama-Aldol reaction have been studied, which include chiral Lewis acid promoted reaction, [58] Lewis base mediated reaction, [59] and the reaction *via* transition metal enolate intermediates [60]. These chiral Lewis acid catalysts include a number of chirally modified boron complexes that have been prepared and studied as Lewis acid catalysts for the asymmetric reactions including Mukaiyama-Aldol reaction. Few heterogenous chiral catalysts on silica support have also been reported. Fraile et al. reported enantioselective Mukaiyama-Aldol reaction of different α-ketoesters using copper complexes of chiral Box and azaBox ligands in both homogeneous and heterogeneous systems [61]. Results in the homogeneous reactions were greatly influenced by the nature of the ketoester, the chiral ligand, and the reaction solvent. In the case of supported catalysts, the use of strongly coordinating azaBox ligands **20**-laponite prevented the leaching of metal but reduced the Lewis acidity, and thus the catalytic activity, therefore it did not solve the problem of poisoning by strong coordination of products, byproducts, or solvents. The counter-ion effect was also very significant, and electrostatic immobilization was efficient only with Box ligands (up to 67% ee at room temperature), whereas covalent immobilization allowed the use of azaBox ligands (up to 76% ee at room temperature) in the heterogeneous phase (Scheme 8). Recovered deactivated solids could be reused in a reaction.

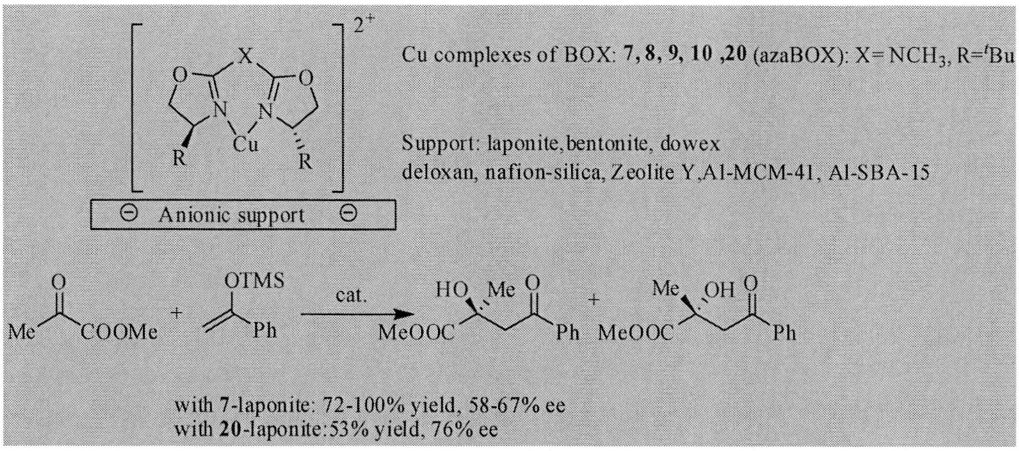

Scheme 8. Laponite clay supported chiral Box and azaBox-Cu catalyzed Mukiyama-aldol reaction.

Recently an interesting work was reported by Doherty et al. using silica and ionic liquid supported Cu-bis(oxazoline) complexes for asymmetric Mukaiyama-Aldol reaction [62]. Lewis acid complexes based on copper(II) and an imidazolium-tagged bis(oxazoline) have been used to catalyse the asymmetric Mukaiyama-Aldol reaction between methyl pyruvate and 1-methoxy-1-trimethylsilyloxypropene under homogeneous and heterogeneous conditions (Scheme 9). Although the ees obtained in ionic liquid were similar to those found in CH_2Cl_2, there was a significant rate enhancement in the ionic liquid with reactions typically reaching completion within 2 min as against only 55% conversion after 60 min. in CH_2Cl_2. However, this rate enhancement was offset by lower chemoselectivity in ionic liquids due to the formation of 3-hydroxy-1,3-diphenylbutan-1-one as a by-product. Supporting the catalyst on silica or imidazolium-modified silica using the ionic liquid or in an ionic liquid-diethyl ether system completely suppressed the formation of this by-product without reducing the enantioselectivity. Although the heterogeneous systems were characterized by a drop in catalytic activity the system could be recycled up to five times without any loss in conversion or ee.

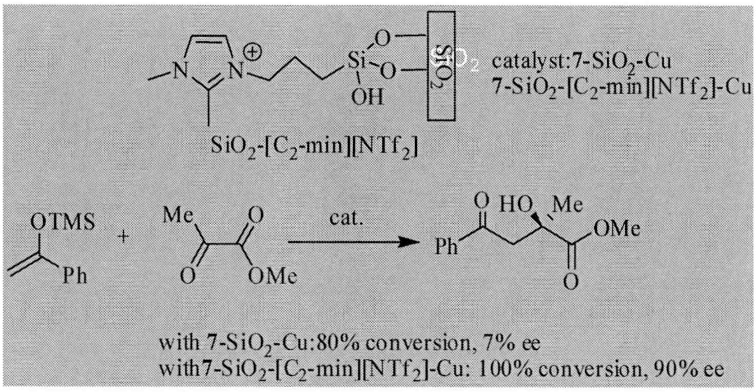

Scheme 9. Bis-oxazoline supported on unmodified and modified silica with ionic liquid.

NITRO-ALDOL REACTION

The nitroaldol or Henry reaction consisting of the addition of a nitroalkane compound to the carbonyl group of an aldehyde or ketone is a powerful C-C bond formation reaction [44,63,64]. The resulting nitro alcohol (nitroaldol) products can be readily transformed into other compound families e.g., to amines by reduction [65], to carbonyl compounds by the Nef reaction [66,67]. Nitro alcohol adducts are also subjected, either to subsequent dehydration reactions to afford conjugate nitroalkenes, which are important building blocks in synthesis [68], or into other derivatives through substitution of the nitro group with various carbon and heteroatom-centred nucleophiles [69]. The catalytic enantioselective version of Henry reaction was first reported by Shibasaki et al., using heterobimetallic lanthanide BINOL catalysts [70,71]. In recent years, catalytic asymmetric nitro-aldol reaction has attracted considerable interest [72]. Both, metal catalysis and organocatalysis, have been successfully applied in nitroaldol reactions. However, there are only few reports on heterogenization of chiral catalyst on inorganic support for nitro-aldol reaction. Choudary et al. [46] used LDH supported proline and BINOL as heterogeneous chiral catalyst for nitro-aldol reaction. Product 2-nitroalkanols were obtained with 100% selectivity and excellent yields albeit with low enantioselectivity (ee: 5-9%) (Scheme 10).

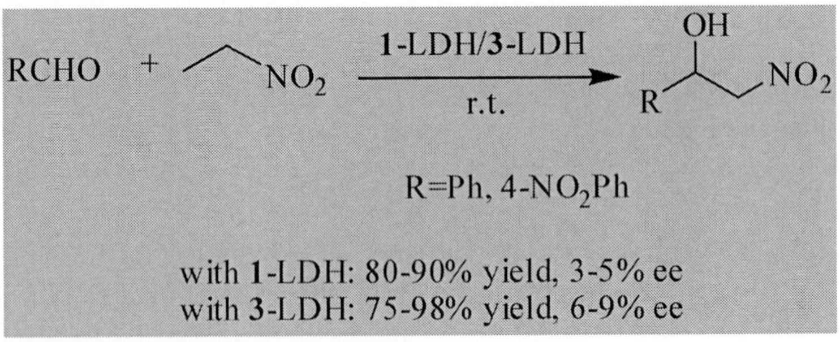

Scheme 10. Henry reaction catalyzed by organo-LDHs.

Choudary et al. [73] used commercially available nanocrystalline magnesium oxide crystals with chiral BINOL as recyclable heterogeneous catalyst for the asymmetric Henry reaction to afford chiral nitro alcohols with excellent yields (up to 95%) and enantioselectivities (90%). When benzaldehydes are substituted at the 4-position with electron-withdrawing groups, higher ee's were observed than the one bearing electron-donating groups at the 4-position. Authors supposed that the surface –OH and O^{2-} of these oxide crystals are expected to trigger this reaction. O^{2-}/O^- of NAP-MgO abstract an acidic proton of the nitromethane, giving a carbanion, which forms a complex with the unsaturated Mg+ site (Lewis acid-type) of NAP-MgO. The asymmetric Henry reaction proceeds *via* dual activation of both substrates (nucleophiles and electrophiles) by NAP-MgO. Thus, the Lewis base (O^{2-}/O^-) of the catalyst activates nitroalkanes, and the Lewis acid moiety (Mg^{2+}/Mg^+) activates the carbonyls of aldehydes (Scheme 11). The chiral auxiliaries, BINOL, bound to NAP-MgO with proper alignment via hydrogen bonds direct the delivery of -CH_2NO_2 stereoselectively to the Mg+ (Lewis acid)-activated carbonyl of aldehydes or chalcones via oxygen coordination to afford the chiral nitro alcohols. The catalyst was reused five times and showed consistent yields and ee's in reaction.

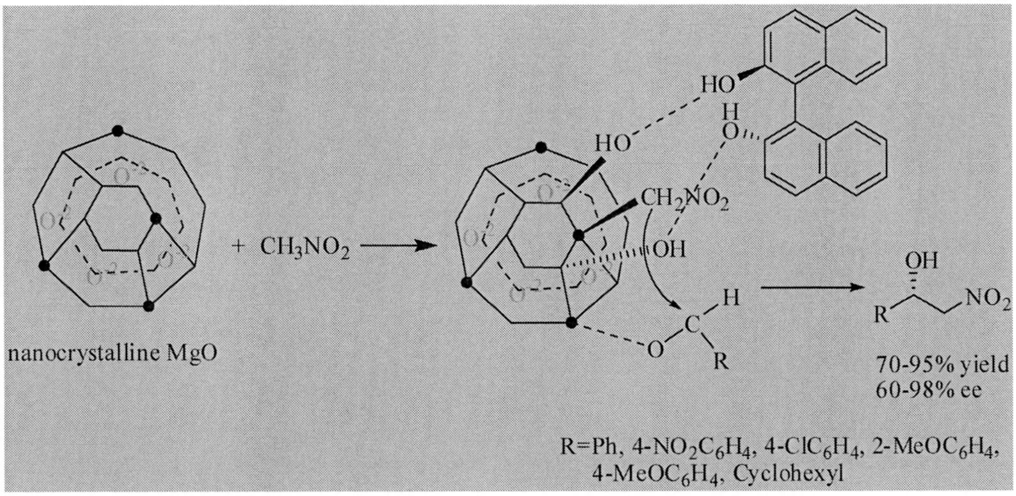

Scheme 11. Asymmetric Henry reaction catalyzed by nanocrystalline MgO.

Abdi and coworkers have reported the immobilization of chiral BINOL ligand on silica and mesoporous MCM-41 through covalent grafting method [74]. Chiral BINOL was supported on two different silica sources, namly amorphous silica and mesoporous MCM-41 through chloropropyl linker (Scheme 12). Catalysts were characterized by various physico-chemical techniques. These immobilized complexes with a lanthanum content of 0.12–0.18 mmol/g have shown promising activity and selectivity in enantioselective Henry reaction. High yield with ee (55–84%) was obtained in the case of aliphatic aldehydes. The catalyst could be separated by simple filtration and reused up to three times without any significant change in enantioselectivity but with some loss in activity.

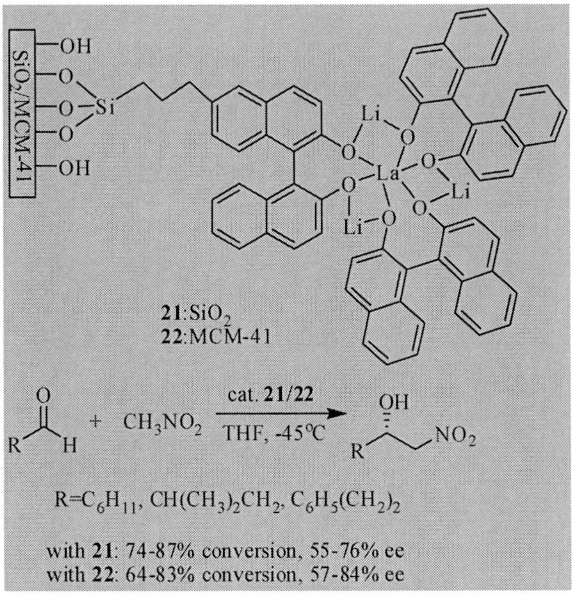

Scheme 12.Chiral La-Li-BINOL supported on silica and MCM-41.

NITRO-MANNICH REACTION

The asymmetric addition of a nitroalkane to imines known as the nitro-Mannich (or aza-Henry) reaction, is a carbon–carbon bond forming process that can result in the generation of two contiguous nitrogen-bearing stereogenic centers. The asymmetric nitro-Mannich reaction is one such reaction producing chiral 2-nitroamines from the reaction of nitroalkanes with aldehydes in presence of a chiral catalyst [75-77]. A variety of nitrogen containing chiral building blocks can be synthesized from 2-nitroamines such as vicinal diamines by reduction of the nitro group [78] and α-amino carbonyl compounds by means of the Nef reaction [79]. Various chiral metal complexes and organocatalysts have been extensively used in this reaction under homogeneous reaction conditions [80-82]. However, there is only one report so far on its heterogeneous version [83]. In this report, chiral bis(oxazoline) ligand with copper (catalyst **23**) was immobilized on mesoporous silica (SBA-15) for asymmetric heterogeneous nitro-Mannich reaction (Scheme 13). Depending upon the size of the alkyl chain in the nitroalkane substrates, enantioselectivities comparable to and higher diastereoselectivities (syn/anti ratio) than those obtained from homogeneous reactions were observed. In the case of the long chain substituted nitroalkane substrate (nitrohexane), the best selectivities (diastereoselectivity: syn/anti = 98/2, and enantioselectivity: 93% and 82% ee's for syn- and anti-isomers, respectively), were observed. The catalyst was easily recovered and reused five times without significant loss of reactivity. However, a gradual reduction in levels of diastereo and enantioselectivities upon each recycling experiment was observed.

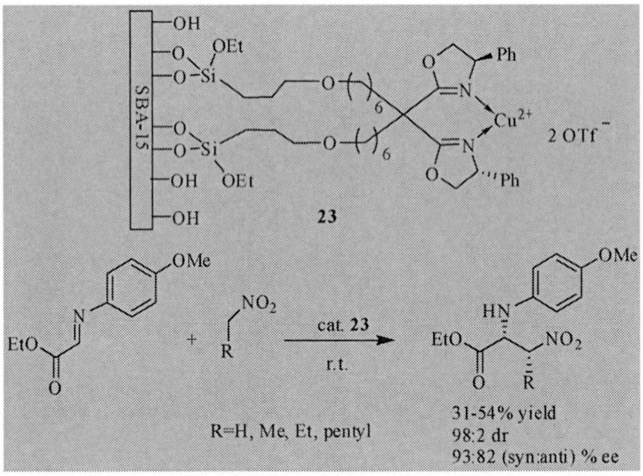

Scheme 13. Asymmetric nitro-Mannich reaction catalyzed by silica-BOX.

MICHAEL ADDITION

Asymmetric Michael addition reaction is among the most powerful and convergent strategies for the enantioselective formation of C–C bonds in organic synthesis [84-86]. However, there are only a few reports on immobilized chiral catalysts for asymmetric Michael reaction. Corma et al. reported the heterogenization of cinchonidine and cinchonine onto MCM-41 and their use as active catalysts for the Micheal addition of ethyl 2-oxocyclopentancarboxylate and methyl vinyl ketones [87]. Immobilized cinchonidine on MCM-41 being more active and enantioselective (ee, 47%) than immobilized cinchonine on MCM-41 (ee, 8%). However in homogeneous reaction condition, in presence of cinchonidine, 98% conversion with 65% ee was achieved while with cinchonine, conversion was 78% and enantioselctivity (ee, 28%). Authors studied that free OH groups of the cinchonidine has a negative impact on the enantioselctivity. When they protected the hydroxyl groups of the heterogenized cinchonidine with carbobenzoxyl group (Scheme 14, catalyst **25**), the catalytic activity increased (99% yield) however the enantioselctivity decreased (ee, 25%) with respect to unprotected catalyst 24.

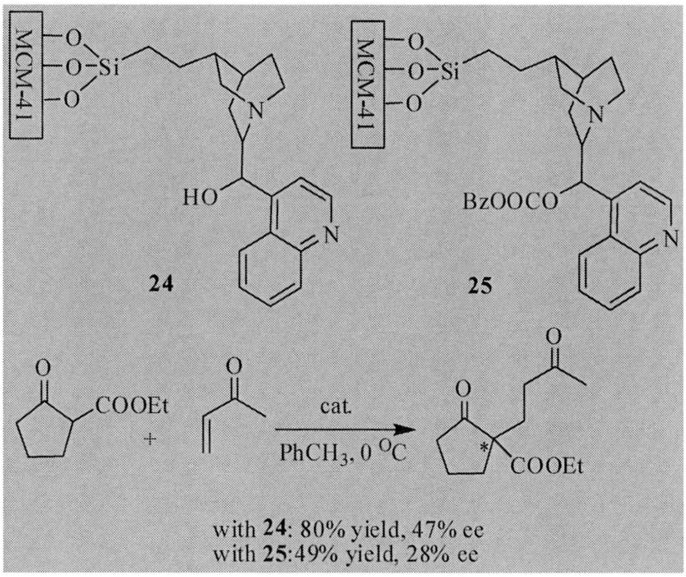

Scheme 14. Asymmetric Micheal reaction catalyzed by MCM-41 supported cinchonidine.

Another interesting result was reported by Choudary et al. [73], where they used commercially available magnesium oxide crystals as recyclable heterogeneous catalyst in the presence of (1R,2R)-(-)-diaminocyclohexane (DAC) as a chiral auxiliary for asymmetric Michael additions of nitromethanes to chalcones. The Brønsted hydroxyls formed hydrogen bridges with DAC and with nitro group directing an asymmetric Michael addition in excellent yields (95 %) and with excellent enantioselectivities (ee, 96 %). Authors reported that this inorganic material contains Lewis acid sites (Mg^{2+}), Lewis basic sites (O^{2-} and O^-) and isolated Brønsted hydroxyls (NAP-MgO). This inorganic material, addressed as Lewis acid sites, promoted electrophylic activation of chalcones carbonyls (Scheme 15). The reactivated catalyst was reused five times and showed consistent yields and ee's for AM additions.

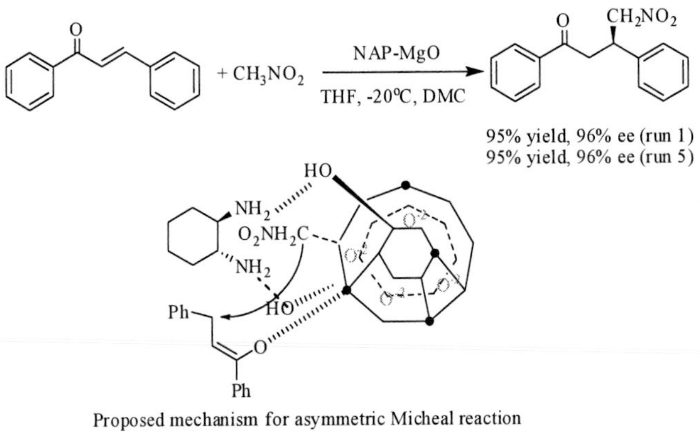

Proposed mechanism for asymmetric Micheal reaction

Scheme 15.Asymmetric Micheal reaction catalyzed by nanocrystalline MgO.

Recently Pitcumani et al. reported the immobilization of L-proline into the inter layers of hydrotalcite (HT) clay [88]. This L-proline-anchored hydrotalcite (HTLP) has good catalytic activity in the asymmetric Michael addition reaction (Scheme 16). The Michael adduct has been prepared by two methods: first, by addition of acetone to β-nitrostyrene (an enamine-type addition) and secondly, by addition of nitromethane to benzylideneacetone (an iminium-type addition). High conversion (100%) has been achieved with the reaction between β-nitrostyrene and acetone using HTLP. L-proline intercalated hydrotalcite catalyst provides improved enantioselectivity (ee, 83.1 %) even better than pure L-proline-catalyzed Michael addition reaction (ee, 46.4%) for the addition of acetone to β-nitrostyrene. Intrestingly inversion in enantioselectivity as compared to L-proline catalysis was obsereved in this case. Authors proposed mechanisms for both routes (Scheme 17 & 18). In an enamine-type addition, non-polar solvent, namely toluene, gave good results as comparied to polar solvent like DMSO and ethanol. Hence the authors concluded that HT not only served as a support for the chiral organocatalyst L-proline, but also functioned as a cocatalyst in the Michael addition between β-nitrostyrene and acetone. This may be attributed to the simultaneous catalysis by organo-inorganic bifunctional (L-proline's amino group and by the surface hydroxyl groups of hydrotalcite) catalysis of HTLP.

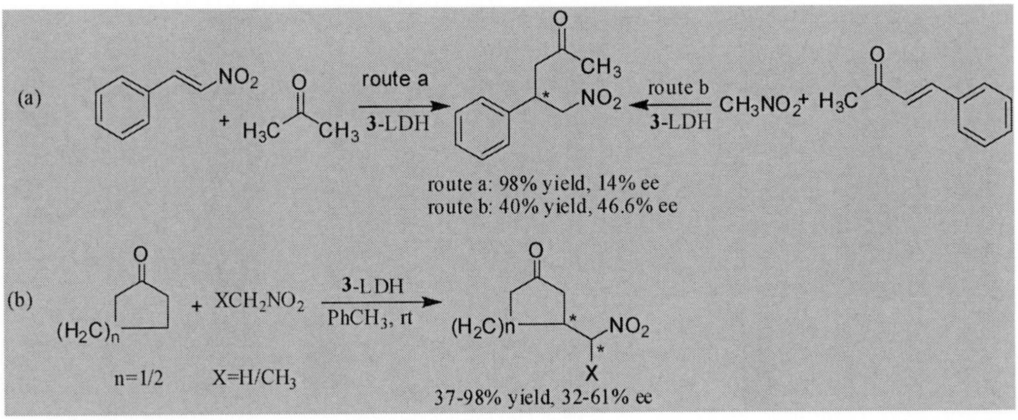

Scheme 16. Asymmetric Micheal addition using proline-LDH.

In a proposed mechanism, the authors postulated that benzylideneacetone is activated by the formation of an iminium intermediate with an amino group of L-proline, which was anchored on HT. It then interacted enantioselectively with the acianion nitromethane, which was stabilized by hydroxyl groups of hydrotalcite to form the Michael adduct (Scheme 20). HTLP catalyst was also used in the addition of nitroalkanes to cyclic enones (Scheme 18(b)). Good enantioselectivity (up to 61%) with high yield (98%) was achieved with addition of cyclopentanone to nitromethane. These heterogeneous catalysts are easy to separate and give consistent yield upon reuse. In addition to this, the solid support HT also acts as a cocatalyst via its basic sites for the above reaction.

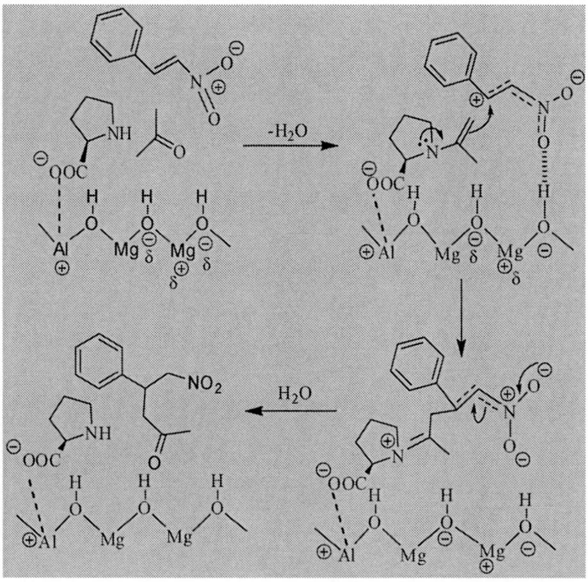

Scheme 17. Proposed enamine-type mechanism for the proline-LDH -catalyzed Michael addition of acetone to β-nitro styrene.

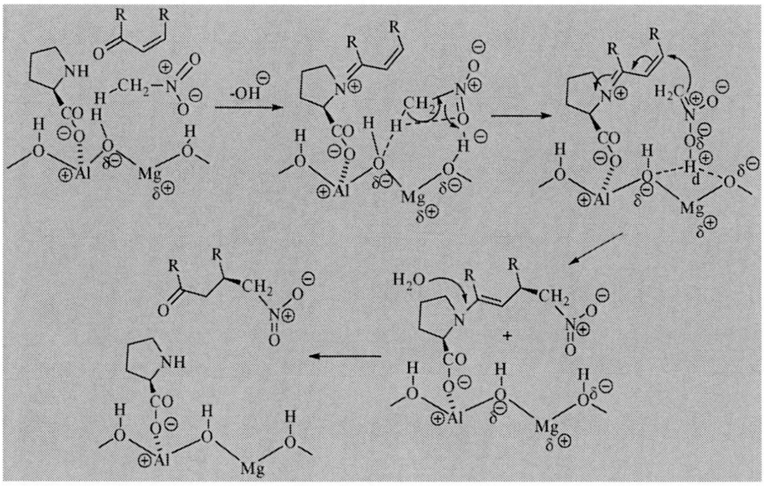

Scheme 18. Proposed iminium ion mechanism for the proline-LDH -catalyzed Michael addition of nitromethane to benzylideneacetone.

DIELS-ALDER REACTION

Diels-Alder reactions are important C-C bond forming reactions, in which up to four new chiral centers can be formed. Many natural products widely used in pharmaceutics and biosynthesis [89,90] can be prepared at an early stage of the synthetic scheme by taking advantage of an asymmetric Diels-Alder reaction. For example loganine has been obtained from a Diels-Alder reaction on a crotonate of a terpene derivative [91]. Complexes of bis(oxazoline) ligands have been shown to be efficient catalysts in asymmetric Diels-Alder reactions [92,93]. Heterogenization of such complexes on to organic and inorganic supports to make this reaction recyclable is of current interest. Some excellent reviews on this topic are already available in literature [94,95]. However, here we will present an up to date account of immobilization of chiral catalysts on inorganic supports for Diels-Alder reaction. Mayoral et al. for the first time [96] reported the synthesis of (S)-tyrosine derivatives supported on silica gel through its phenolic oxygen atom. The material thus obtained was treated with BH₃ to get chiral Lewis acid which was eventually used as catalyst to promote the reaction between methacrolein / bromoacrolein with cyclopentadiene but they did not provide asymmetric induction (Scheme 19, catalyst 26a & 26b) although the conversion was failry good. In the same report, authors supported (S)-Prolinol through the nitrogen atom on silica gel, which when treated with AIEtCl₂ provided catalysts 27 and 28. These solids catalysed the reaction between methacrolein and cyclopentadiene, leading to excellent conversions but again with poor enantiomeric excess (8% ee) for the cycloaddition product.

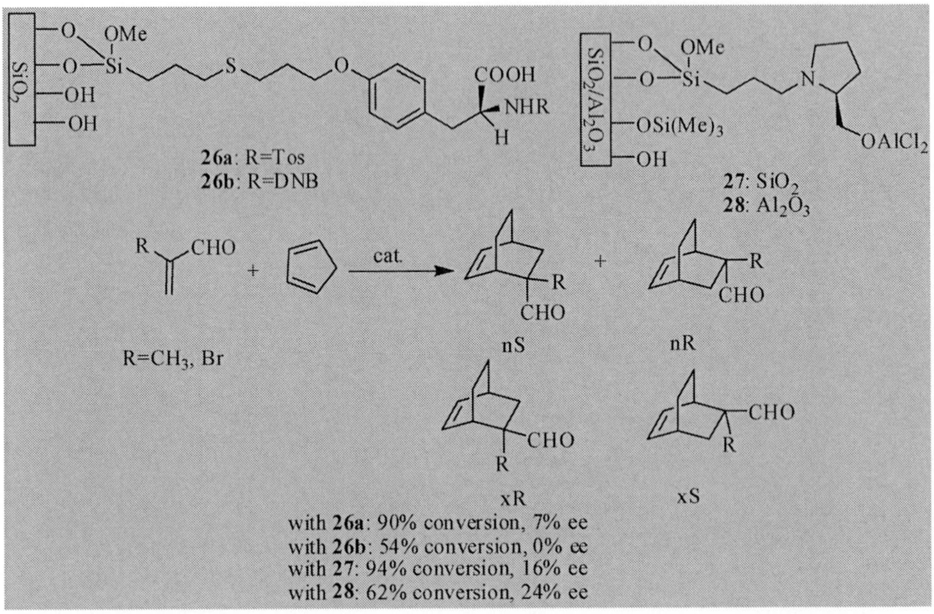

Scheme 19. Chiral Lewis Acids supported on SiO₂ and Al₂O₃, and their use as catalysts in Diels-Alder reaction.

Lemaire et al. [97,98] have published the first example of a Diels–Alder reaction catalyzed by a covalently-supported catalyst based on bis(oxazoline) ligands. An analogous of the IndaBox ligand was functionalized in the methylene bridge with two triethoxysilyl groups via a carbamate linker. This modified ligand was grafted onto activated (acid treated) silica, and complexed with Cu(OTf)$_2$ and Cu(ClO$_4$)$_2$ (Scheme 20). These complexes were tested in the Diels–Alder reaction between cyclopentadiene and 3-acryloyloxazolidin-2-one. The perchlorate complex 29b showed good recoverability during 4 cycles at different temperatures with high conversion (>99%) and enantioselectivities between 65% ee (rt) to 85% ee (−15 °C). This catalyst, on silanization of the silica surface with N-trimethylsilylimidazole showed higher enantioselectivity, (92% ee at −78 °C). Enantioselectivities with the other dienophiles were slightly lower (61-70%). In a later study [99], authors reported a linear relationship between enantioselectivity and silanization degree, measured both by microanalysis and ^{29}Si-CP-MAS-NMR, however the catalyst loading on the support had no significant effect on enantioselectivity. According to the authors, the possible complexation of copper with the free silanol groups which served as non-enantioselective catalytic sites is the main reason for this behavior.

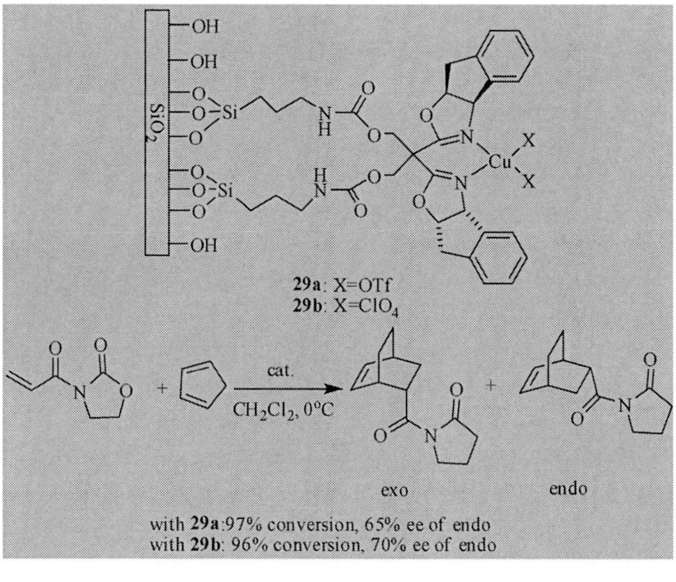

29a: X=OTf
29b: X=ClO$_4$

exo   endo

with 29a:97% conversion, 65% ee of endo
with 29b: 96% conversion, 70% ee of endo

Scheme 20. IndaBox-Cu complex supported on silica.

Kim and coworkers used a similar strategy [100] to immobilize IndaBox on mesocellular foam (MCF). In this case, the starting IndaBox was first dialkylated at the methylene central bridge with protected p-hydroxybenzyl bromide. The deprotection of this material gave free phenol which on alkylation with chloropropyl triethoxysilane provided silyl arm for its grafting on MCF (Scheme 21). The material thus obtained was complexed with Cu(OTf)$_2$, to get active catalyst for Diels–Alder reactions. As compared to the results obtained with the analogous homogeneous catalyst, the initial results with this catalyst (OH of silica protected with TMS) for the DA reaction between cyclopentadiene/cyclohexadiene with different dienophiles like 3-(2-propenoyl)-2-oxazolidinone were not very encouraging. But the optimization of reaction condition and elimination of excess of copper improved the enantioselectivity (up to 75% ee at −78 °C) and endo/exo selectivity (17:1). However in the present case, silanization of free OH groups on silica did not have positive effect on enantioselectivity, as the non-silanized catalyst could afford 78% ee. The best catalyst was used up to 5 cycles with similar enantioselectivities (ee, 70–78%). Even in that case, the analysis showed a considerable amount of copper leaching after each cycle, probably due to the strong complexation with cycloadduct products.

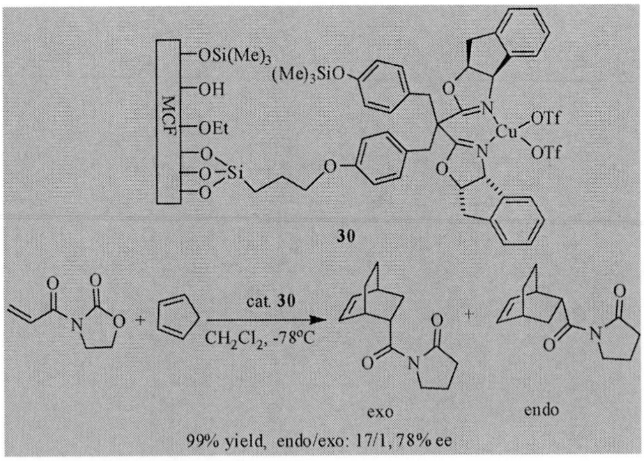

Scheme 21. IndaBox-Cu complex supported on MCF.

Ying et al. [101] have shown the effect of linker and silane groups in the performance of the above type of MCF supported catalysts. Three different linkers and four silanes were tested in the Diels–Alder reaction between cyclopentadiene and 3-acryloyloxazolidin-2-one (Scheme 24). In this case, silanization enhanced enantioselectivity from 47% ee to 70% ee at rt, and from 77% ee to 88% ee at −78 °C. The silanized catalysts were more enantioselective (69-86% ee) than the non-silanized ones (47% ee), as well as the analogous homogeneous catalyst (84% ee). High enantioselctivity (up to 90%) was obtained with MCF supported catalyst when R=(CH$_2$)$_2$PhCH$_2$ and R'=SiMe$_3$ (Figure 6). Catalyst could be recycled up to 5 runs with 100% conversion overall, and displayed high enantioselectivities of 79–90%. Authors also observed that enantioselectivity was also dependent on the nature of the solvent used in the reaction and the catalyst washes. Lower enantioselectivities were associated with the use of anhydrous CH$_2$Cl$_2$ (74% ee), and higher ee value (80%) were obtained when water-saturated CH$_2$Cl$_2$ (1850 ppm of H$_2$O) was used at rt.

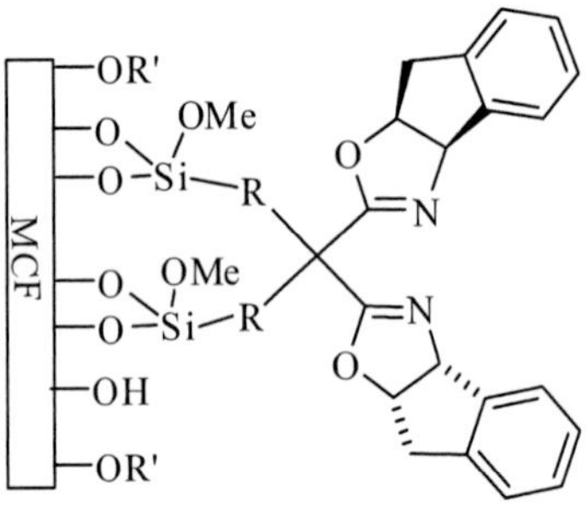

R=-(CH$_2$)$_2$-, -PhCH$_2$-, -(CH$_2$)$_2$-PhCH$_2$-
R'= -SiMe$_3$, -SiPh$_2$Me, -SiMe$_2$(CH$_2$)$_7$CH$_3$, -SiMe$_2$(CH$_2$)$_2$(CF$_2$)$_3$CF$_3$

Figure 6. IndaBox-Cu complex supported on MCF with different linkers.

Iwasawa and coworkers reported [102-104] the immobilization of ^{t}BuBox on silica (Figure 7). The Cu(ClO$_4$)$_2$ complex was much more active than the homogeneous counterpart in the reaction between 3-acryloyloxazolidin-2-one and cyclopentadiene, and at the same time the enantioselectivity was increased from 5 to 15% ee at −10 °C. The silanization of the silanols with different silanes did not produce any significant positive effect, with only one exception in the case where silica was silanized with 3-(trimethoxysilyl) propyl methacrylate, enantioselectivity went up to 65% ee.

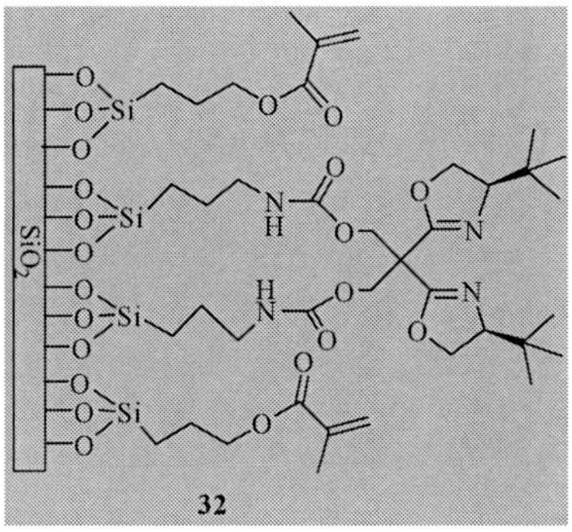

32

Figure 7. Box-Cu complex supported on silica.

Seebach et al. used the TADDOL-Ti catalyst immobilized onto controlled- pore glasses (CPG) (Scheme 22) for [3+2]-cycloaddition of diphenylnitrone to N-crotonyl-1, 3-oxazolidin-2-one [105] with 66% de and 70% ee. Under homogeneous conditions with a comparable TADDOL derivative, similar selectivities (70% de and 85% ee) were found. Good reproducibility was also possible when the catalyst was reused. Only a small loss of activity was observed after four runs.

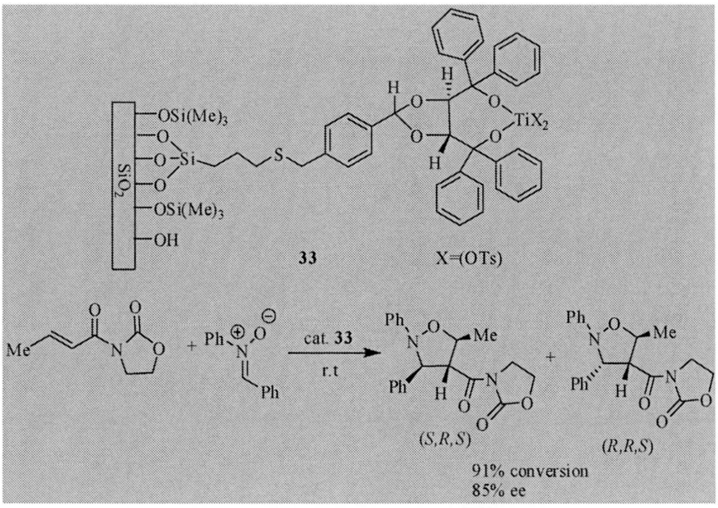

Scheme 22. Immobilized TADDOL-Ti on CPG for [3+2]-cycloaddition.

Asymmetric organocatalysis on solid support combines the environmental advantages of metal-free catalysts and the ease of operation of solid-supported reagents. In this regard, Pihko et al. reported silica supported dervitaive of N-Fmoc-protected (*S*)-phenylalanine **34** (Figure 8) as organocatalyst for enantioselective Diels-Alder reactions [106]. A maximum of 83% yield and 90% ee (for endo isomer) was reported for the cycloaddition product between a diene and a dienophile. The supported catalyst was recoverable and reusable with retion of performance. However, only one reuse data was provided for this study.

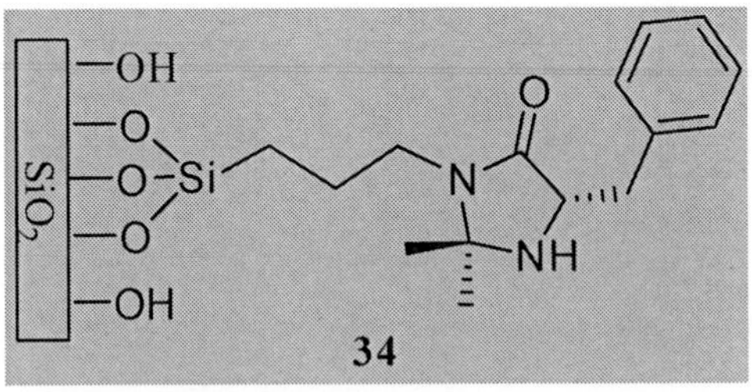

Figure 8. (*S*)-phenylalanine derivative supported on silica.

Ying et al. synthesized imidazolidin-4-one immobilized on siliceous MCF and polymer-coated MCF and successfully demonstrated their highy efficiency and enantioselectivity as heterogenized organocatalyst for Friedel–Crafts alkylation and Diels–Alder cycloaddition [107]. The performance of the supported catalysts in relation to the surface environment of siliceous MCF revealed that partial pre-capping of the MCF with TMS groups is crucial to attain well-dispersed catalysts for optimal performance. However the heterogenized catalysts were slightly inferior to the corresponding homogenous catalyst. Nevertheless, the heterogenized catalyst showed better recyclability (Figure 9).

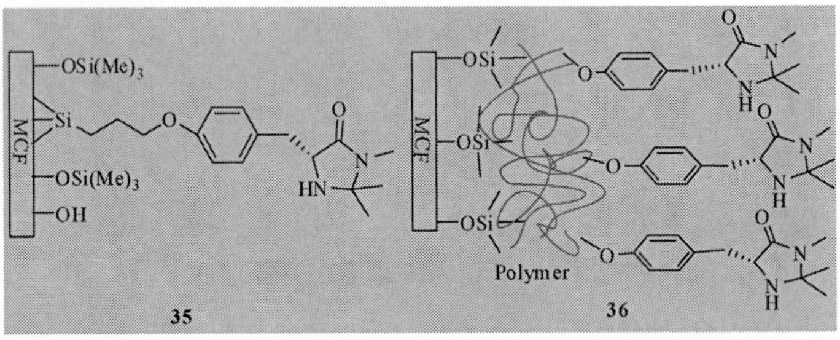

Figure 9. Chiral Imidazolidin-4-one immobilized on siliceous and polymer-coated MCF.

In addition to the covalent immobilization methods, several non-covalent strategies have also been used to immobilize catalysts for Diels–Alder reactions. Cationic exchange method was first used by Fraile et al. [108] to support enantioselective catalyst for Diels–Alder reactions. Cationic complexes of PhBox and Cu, Zn, and Mg salts were exchanged on laponite clay and Nafion-silica nanocomposite (Scheme 23). Overall the results obtained with heterogenized copper and zinc complexes as catalysts in Diels–Alder reactions (R = Me, Scheme 29) were poorer (ee, 10%) than the results obtained in homogeneous phase.

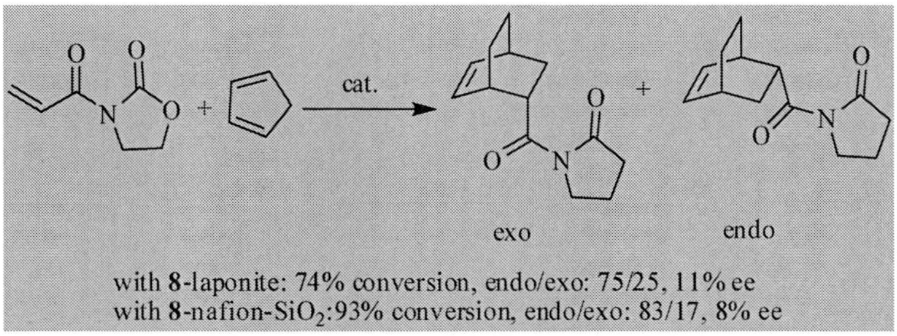

with 8-laponite: 74% conversion, endo/exo: 75/25, 11% ee
with 8-nafion-SiO$_2$:93% conversion, endo/exo: 83/17, 8% ee

Scheme 23. Asymmetric Diels–Alder reaction using bis(oxazoline) supported on laponite and nafion-SiO$_2$.

An alternative method of non-covalent immobilization is the adsorption of Cu(OTf)$_2$ complexes onto silica, where the formation of hydrogen bonds between silanols and triflate anions is proposed (Figure 10) [109].

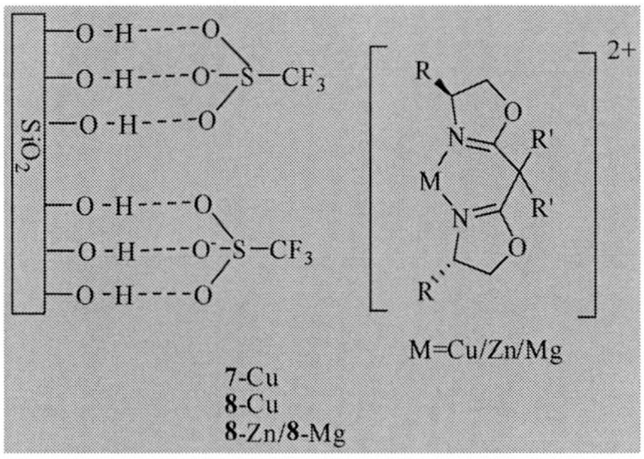

Figure 10. Bis(oxazoline) complexes immobilized on silica *via* electrostatic interactions.

The immobilized complexes were tested as catalysts in the Diels–Alder reaction between 3-acryloyloxazolidin-2-one and cyclopentadiene. Catalyst ^{t}BuBox-Cu 8-Cu (Scheme 30) showed activity and enantioselectivity (57%ee) similar to those observed in homogeneous phase. However, a reversal of the major *endo* enantiomer obtained with catalyst PhBox-Cu (R = H), with regard to the homogeneous phase reaction was noticed. Later on, these results were revisited by Li and coworkers [110] using Cu catalysts, together with analogous of Zn and Mg. Results obtained with supported ^{t}BuBox-Cu were much better than those reported earlier, with enantioselectivities in the range of 85–93% ee at rt. Activity and recoverability were improved by using toluene as the reaction solvent and 3Å MS as water scavenger, allowing three uses with the same activity (conversion, 83-98%) and enantioselectivity (ee, 91% in the three runs). In another study where Li and coworkers [111] used periodic mesoporous ethane-silicas functionalized with sulfonic acid, cesium salts of 12-tungstophosphoric acid, activated silica, and SBA-15 as solid supports for the immobilization of the Cu-bis(oxazoline) catalyst *via* non-covalent interactions (Figure 11). These heterogeneous catalysts have been evaluated for the Diels-Alder reaction between 3-((E)-2-butenoyl)-1,3-oxazolin-2-one and cyclopentadiene. The performance of the catalyst relies on the nature of both the support and the anion on the surface of the support. The small number of the silanol groups on the support led to poor enantioselectivity when SBA-15 was used. Experimental data indicated that the performance of the catalysts depended on the nature of the support as well as on the anion present on the surface of the support. The best activity and enantioselectivity were obtained with activated silica calcined at 300°C as support. In all the cases conversion was fairly good (77-100%) but enantioselectivity was not up to the mark (ee, 15-42%)

Figure 11. PhBOX-Cu complex immobilized via non-covalent interactions.

HETERO-DIELS–ALDER REACTIONS

The hetero Diels-Alder reaction of electron-rich alkenes with α, β-unsaturated carbonyl compounds is an efficient method for the preparation of dihydropyrans [112,113]. These cycloadducts and their derived tetrahydropyrans, are prevalent structural subunits in numerous important natural products [37, 114]. There exist only few examples in literature on supported chiral catalysts for Hetero-Diels–Alder reactions. Anwander et al. reported hetero-Diels-Alder reaction using organic-inorganic hybride materials [115]. The authors reacted MCM-41 silica with Y[N(SiHMe$_2$)$_2$]$_3$(THF)$_2$, followed by its end-capping with HN(SiHMe$_2$)$_2$ with various chiral ligands (Scheme 24). Among them the L-(-)-3-(perfluorobutyryl)-camphor **3** derivative [MCM-41]Y(38)$_x$(THF)$_y$ produced the highest *de* and *ee* (de, 67%; ee 37%).

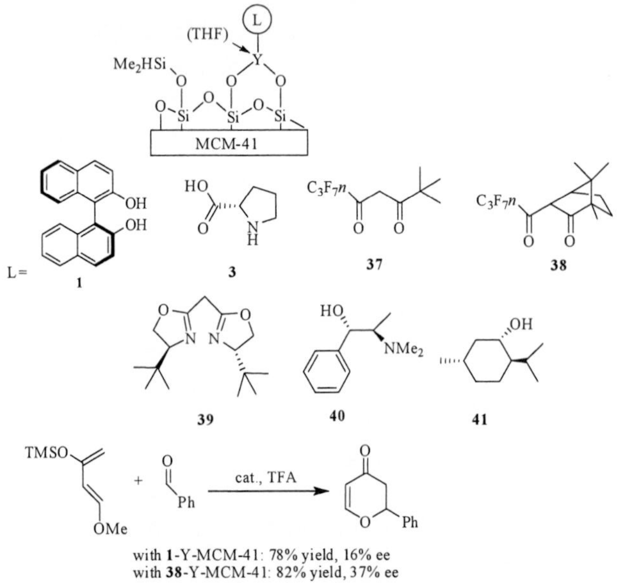

Scheme 24.Inorganic-organic hybrid materials of Yttrium centeres with different chiral ligands on MCM-41.

Seebach et al. reported the immobilization of Cr(salen) complexes on a controlled pore glass (CPG) 42 (Figure 12) which was used as heterogeneous catalyst in hetero-Diels-Alder additions of aldehydes to Danishefskys diene [116]. The Cr complex had shown a linear relationship between the enantiomeric purities of ligand and product under homogeneous conditions. Under heterogeneous condition however, the enantioselecitivity of the hetero-Diels- Alder reaction increased for initial five runs (from an average of 76 to ~83% ee) and thereafter remained constant for another five runs.

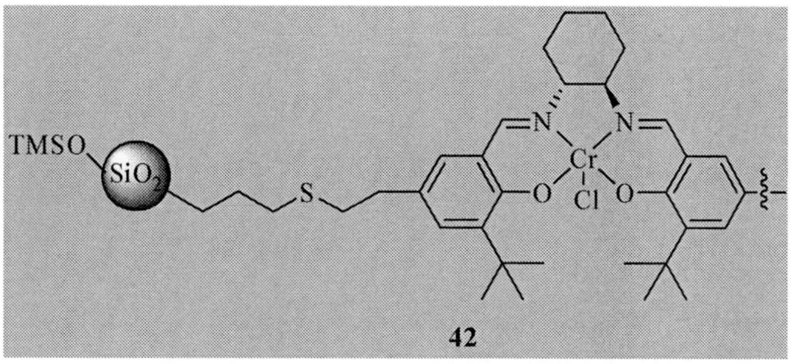

42

Figure 12. Chiral Cr(salen) complex immobilized on silica

Hutchings et al. reported the encapsulation of PhBox-Cu complex in different zeolites and crystalline mesoporous solids by cation exchange method [117]. The immobilized catalysts were a lot less active than the homogeneous counterpart. Regarding enantioselectivity, all the catalysts supported on mesoporous solids gave inferior results (ee; <20%) while the immobilization of these catalysts on zeolite Y showed improved enantioseletivities (ee, ~ 41%) but with a reversal in enantioselectivity with respect to the homogeneous reaction (Scheme 25). This reversal of enantioselection in the all cases is intriguing, but no explanation for it was offered.

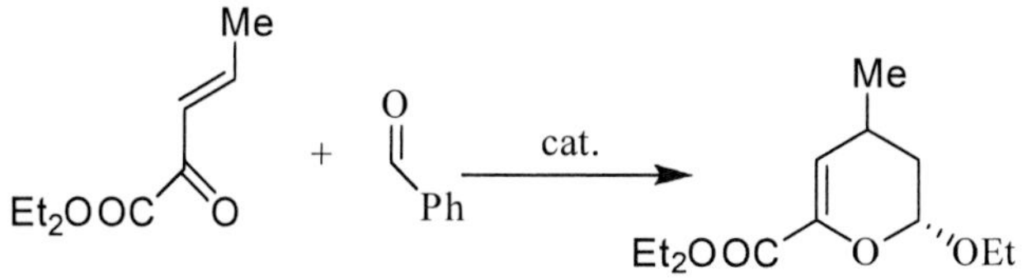

with 7-Al-SBA-15-Cu: 66%yield, 11% ee
with 7-MCM-42-Cu: 81%yield, 7% ee
with 7-HY-Cu: 16%yield, 41% ee

Scheme 25. Hetero-Diels- Alder reaction catalyzed by PhBox-Cu complex immoblized on mesoporous silica and zeolite.

CYCLOPROPANATION

Chiral cyclopropanes are an important class of strained small ring compounds. They can undergo a variety of chemical transformations to give many useful intermediates. In addition, cyclopropyl group is also found as a basic structural unit in a wide range of naturally occurring compounds in plants and microorganisms [118-120]. The development of an important class of photodegradable and low mammalian-toxic insecticides, namely the pyrethroids, made the stereocontrolled synthesis of substituted cyclopropanes an important goal [121]. This directly led to the emergence of new catalysts for asymmetric cyclopropanation reactions. The complexes of Cu and Rh are the most widely used catalysts for asymmetric cyclopropanantion reactions [122,123], which in recent times are being studied for their immoblization to suit their application at a large scale. The first heterogeneous asymmetric cyclopropanation was reported by Matlin et al. [124], where chiral ligand 10-methylene-3-trifluoroacetyl-(+)-camphor, was immobilized on Hypersil 5 mm silica to give heterogenenous catalyst (Scheme 26).

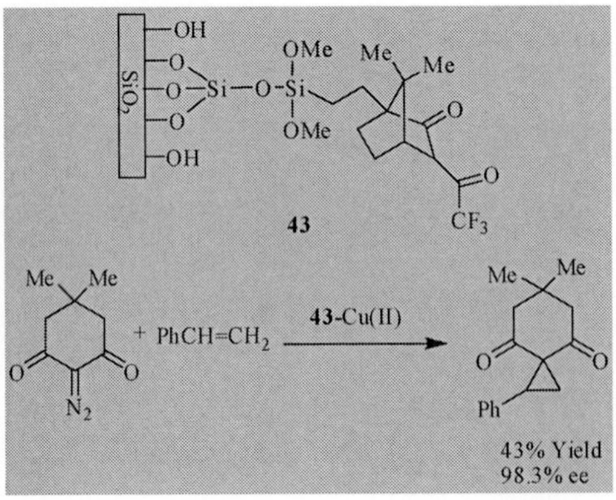

Scheme 26. Immoblized chiral Cu-β-diketonate used in cyclopropanation.

In this system, residual silanol sites on the silica were capped using trimethylsilyl chloride to block the formation of non-enantioselective catalytic sites. The reaction of styrene with 2-diazodimedone with the immobilized chiral α-diketone-Cu complex exhibited 2.4 times higher activity but comparable enantioselectivity (98.3% ee) with its homogeneous analogue. However, recycling of this catalyst failed due to the coating of polystyrene around the silica. The authors claimed that catalytic activity was retained for recycling experiments when a non-polymerisable olefins, such as, indene was used as a substrate. Later on Iglesias et al. [125,126] reported the heterogenization of Cu and Rh complexes of triethoxysilyl modified (S)-proline or (2S,4R)-hydroxyproline ligands by covalent grafting onto ultrastable Y-zeolite (Figure 13).

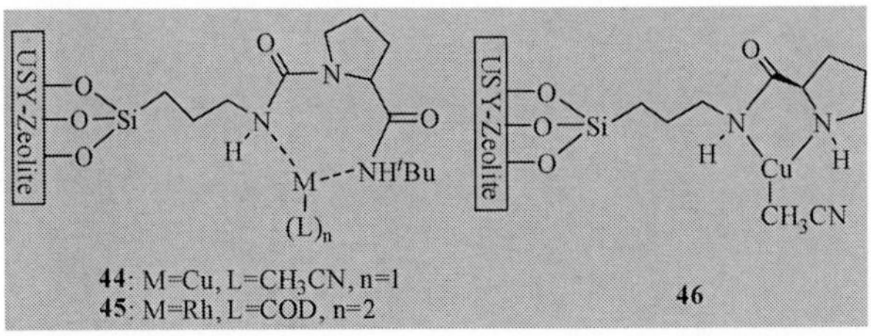

Figure 13. Cu and Rh complex of substituted-pyrrolidine-ligands on USY-Zeolite.

The catalytic properties of the anchored complexes 44, 45 and 46 in the cyclopropanation of styrene and dihydropyrane with alkyl diazoacetate were compared to those of their homogeneous counterparts. The chemical yields obtained with unsupported and zeolite-supported complexes were almost identical (35-100%). However, both homogeneous and heterogeneous catalysts exhibited extremely low enantioselectivities (0-18% ee). Among other chiral ligands, oxazoline-derived ligands have been extensively used in enantioselective catalytic cyclopropanation reactions [118, 127]. Immobilized Cu and Ru complexes of oxazolines like bis(oxazolines) (Box), aza-bis(oxazolines) (azaBox), and pyridine-bis(oxazolines) (Pybox) have been more frequently used as heterogeneous catalysts for this reaction. Mayoral and co-workers for first time reported the non-covalent immobilization of Box-Cu complexes on anionic supports through ionic exchange and thoroughly explored this methodology in their subsequent studies [128-131]. The PhBox-CuCl$_2$ and BnBox-CuCl$_2$ complexes were held strongly by electrostatic interaction on several anionic supports such as clays (Laponite, bentonite, K10 montmorillonite). The supported catalysts were used in the benchmark reaction for cyclopropanation reaction (Scheme 27).

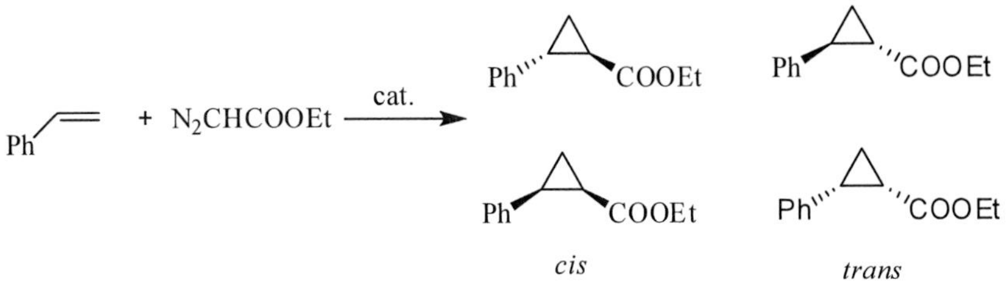

with **7**-laponite: 40% conversion, 42% ee *trans* & 6% ee *cis*
with **7**-nafionsilica:38% conversion, 57% ee *trans* & 46% ee *cis*
with **8**-laponite:37% conversion, 69% ee *trans* & 64% ee *cis*
with **8**-nafionsilica:41% conversion, 23% ee *trans* & 19% ee *cis*

Scheme 27. cyclopropanation reaction using clay and nafion-silica supported bis(oxazoline)-Cu complexes.

In their first study, three different types of clay such as Laponite, bentonite, and K10 montmorillonite were tested as the catalyst support for the asymmetric cyclopropanation of styrene with ethyl diazoacetate. The results depended on the nature of the support, the chiral ligand, and the catalyst precursor and solvent used to carry out the ion exchange. Catalysts supported on Laponite gave better results over the others. The exchanged catalysts were generally recoverable with leaching of complex resulting in the loss of catalytic activity and enantioselectivity of the recoverd catalysts. As compared to Laponite, Nafion was a better support due to its weaker electrostatic interaction with bis(oxazoline)-Cu complex. ^{t}BuBox and Cu(OTf)$_2$ were also used for catalyst immobilization. In general, lower enantioselectivities as well as lower *trans*/*cis* ratios with regard to the homogeneous catalytic systems were observed. Using the PhBox ligand, results were virtually identical to those obtained in homogeneous phase (~60% ee in *trans*-cyclopropanes). Furthermore, the catalyst was easily recoverable and reusable with the same results.

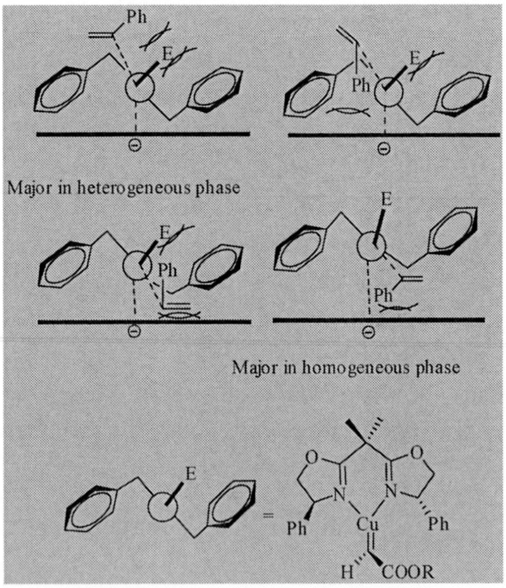

Scheme 28. Steric interactions in the possible approaches of styrene to the copper-carbene intermediate in the laponite-immobilized 8-Cu complex

None of the anionic supports tested above, apart from clays, displayed the "pro-*cis*" behavior generally observed with these lamellar solids. The *trans/cis* selectivity was also in the same 66:34 range as was observed in homogeneous phase. This particular effect has been ascribed to the two-dimensional nature of the clay surface, and can be magnified if a solvent with a dielectric constant lower than dichloromethane is used. Under these conditions, the cationic complex forms a tighter ion pair with the anionic surface, and the steric effect of the latter increases. Thus, when the cyclopropanation reactions were catalyzed by the laponite-exchanged PhBox-Cu complex, in hexane or styrene, a complete reversal of the *trans/cis* diastereoselectivity (31:69) was observed. Even more interestingly, the major *cis*-cyclopropane obtained had the opposite absolute configuration, as compared to results obtained under homogeneous phase. Pybox were used as ligands in the same systems, the same *cis* preference was obtained, but no reversal in the configuration of the major *cis*-cyclopropane was observed. The mechanism of reaction was proposed by authors, (Scheme 28), based on the steric effect of the support surface on the incoming alkene. The transition state lacking this steric interaction turned out to be the most stable in the heterogeneously catalyzed reactions, leading to the major product observed. Thus, when the ^{t}BuBox-Cu(OTf)$_2$ complex was immobilized by cationic exchange on a Nafion-silica nanocomposite (SAC-13), only about 20% ee for *trans*- and *cis*-cyclopropanes was obtained as compared to over 90% ee in homogeneous phase in the reaction of styrene with ethyl diazoacetate. This indicates a probable leaching of the ligand from the immobilized catalysts because of steric interactions with the support. Azabis(oxazoline) (azaBox) ligands possess a higher coordinating ability than their analogous Box ligands, as shown by theoretical calculations and competitive catalytic experiments [132]. The immobilization of Box ligands on silica supports by covalent method was described by Mayoral et al. [133] (Figure 14). The different allyl and vinylbenzyl-functionalized Box ligands were grafted on mercaptopropylsilica to give the immobilized catalyst precursors. Some good results were obtained in the benchmark cyclopropanation reaction (up to 86% ee with the IndaBox-derived catalyst), however, no recovery experiments were described.

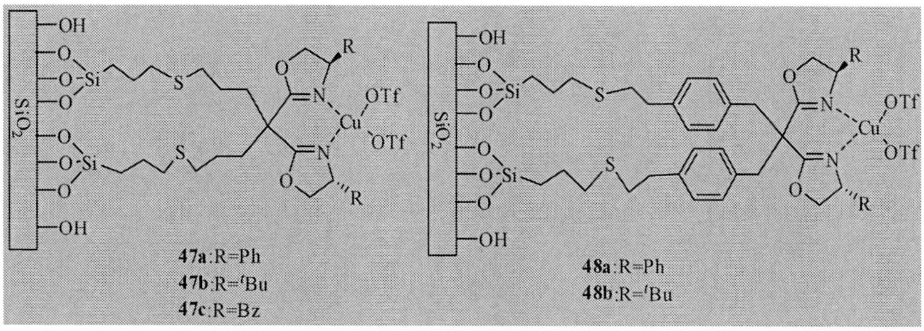

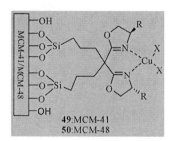

Figure 14. Bis(oxazoline)-Cu complexes supported on silica.

A similar strategy to immobilize a bis(trimethoxysilylpropyl)-modified PhBox on two different mesoporous materials, namely MCM-41 and MCM-48 for cyclopropanation was reported by Shannon et al. [134]. Complexes 49 and 50 were prepared by the addition of Cu(OTf)₂ (Figure 15). The catalytic results in the benchmark cyclopropanation reaction were similar to those obtained in homogeneous phase with the PhBox-Cu(OTf)₂ catalysts but all the immobilized catalysts could be reused only once.

Figure 15. Bis(oxazoline)-Cu complexes supported on MCM-41 and MCM-48.

More recently, the group of Ying and co-workers reported the use of mesocellular foam (MCF), a silica support with a high surface area (>800m²g⁻¹) and ultra-large open pores (25 nm) to support IndaBox ligand through differently modified methylene bridge [101]. The free silanol groups on the silica support were then silanized with trimethylsilyl groups, and the corresponding Cu(OTf) and Cu(OTf)₂ complexes were tested in the benchmark cyclopropanation reaction (Scheme 29). The resulting catalysts showed good but slightly lower activity and enantioselectivities than the corresponding homogenous catalyst. Furthermore, the solids could be recovered and reused two more times with the same results. In a subsequent and more extensive study by the same authors [135], different silica supports (MCM-48, SBA-15 and amorphous silica) were compared with the MCF. The effect of the silanization of the support and that of the linker structure was also studied for benchmark cyclopropanation reaction. Silanization of the free silanol groups of the supports was shown to be important for the catalytic activity and selectivity of the reactions. In general, good enantioselectivities (over 84% ee) were obtained with all the immobilized and silanized catalysts, although better yields were obtained in general with the MCF, due to the better chemoselectivity towards the cyclopropanation reaction. The nature of the linker was also important to obtain good enantioselectivities, the propyl spacer being the best among those tested. Up to eight runs with similar reaction results could be carried out with the best catalyst indicating a good recyclability of these immobilized catalysts.

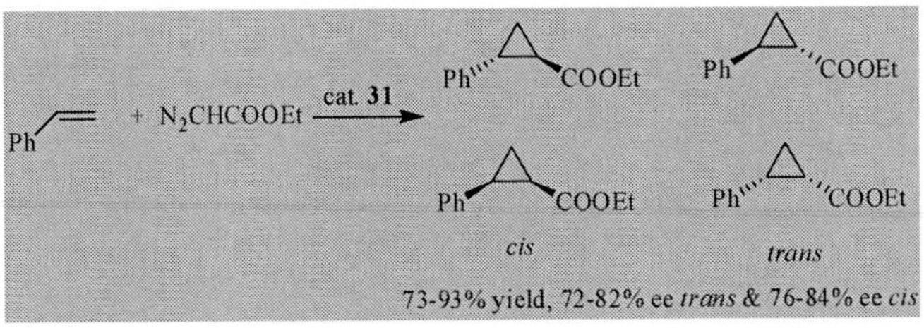

Scheme 29. cyclopropanation using MCF supported bis(oxazoline)-Cu catalyst.

More recently, a modification of this immobilization strategy, where the Box ligand was grafted by only one linker has been described [136]. This modification resulted into significant improvement in the enantioselectivity and recoverability of the catalytic system. In this system, the benchmark cyclopropanation reaction gave up to 95% ee (enantioselectivity at par with homogeneous phase obtained with the use of 'BuBox ligand) with high chemoselectivities (yield of cyclopropanes, 70-80%). The chemo-, diastereo- (trans/cis) and enantioselectivities was retained for eleven recycle experiments and thus can be regarded as one of the best recoverable catalytic systems for this reaction. Mayoral et al, described the synthesis of organic-inorganic hybrid materials containing chiral Box ligands using long alkyl amine as template with $Si(OEt)_4$ as silica source in ethanol water mixture [137].

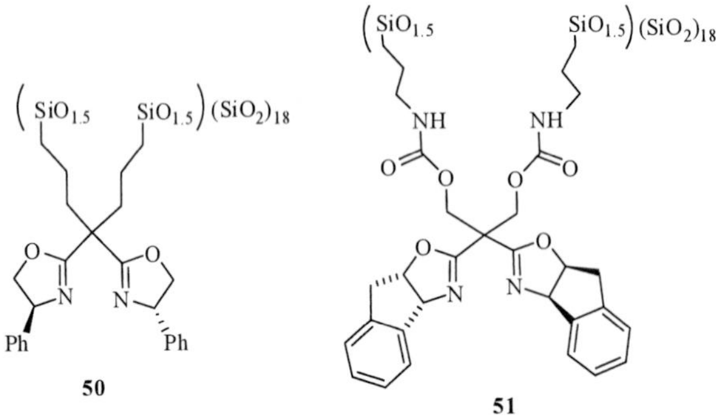

Figure 16. Organic–inorganic hybrid materials with chiral bis(oxazoline) ligands.

A conventional acid hydrolysis method was avoided for the synthesis of this material as Box ligands are very sensitive towards Brønsted acids. Using derivatized PhBox and IndaBox ligands (Figure 16), two hybrid materials with surface area of 240 and 40 $m^2 g^{-1}$, respectively, and a mean pore diameter of 22 Å were prepared. In the case of PhBox with Cu, the catalytic tests in the benchmark cyclopropanation reaction led to selectivities almost identical to those found in homogeneous conditions. This catalyst was easily recovered and used for another catalytic run without any loss in activity and selectivity. On the other hand, enantioselectivity obtained with IndaBox-Cu 51 was far less (ee, 51%) than the corresponding homogeneous system (ee, 83%). In addition, this heterogenized systems showed very little (conversion, 3%) catalytic activity on reuse.

HYDROFORMYLATION

The hydroformylation reaction is one of the most importnet reaction. The attractive features are that a new C-C bond is formed and a new functionality is introduced at the same time, using inexpensive alkenes as starting material, syngas and a small amount of catalyst, making this transformation a highly efficient and atom-economical carbonylation reaction. Depending on the substrate used, a new stereogenic carbon may be obtained making the asymmetric hydroformylation in principle a very attractive tool for the preparation of enantiomerically enriched aldehydes [138,139]. Chiral aldehydes can be precursors for a variety of pharmaceuticals and biologically active compounds (e.g. ibuprofen and naproxen), biodegradable polymers and liquid crystals [140-142] (Scheme 30). Although asymmetric hydroformylation has been widely investigated as an important strategy in chiral synthesis, few catalytic processes are currently applied on a commercial scale [143]. This may be due primarily to the difficulties in controlling the selectivities of hydroformylation and the separation and recycling of the catalysts from the reaction system. In recent years, interest has been gained in heterginization of chiral catalyst for asymmetric hydrofomylation reaction.

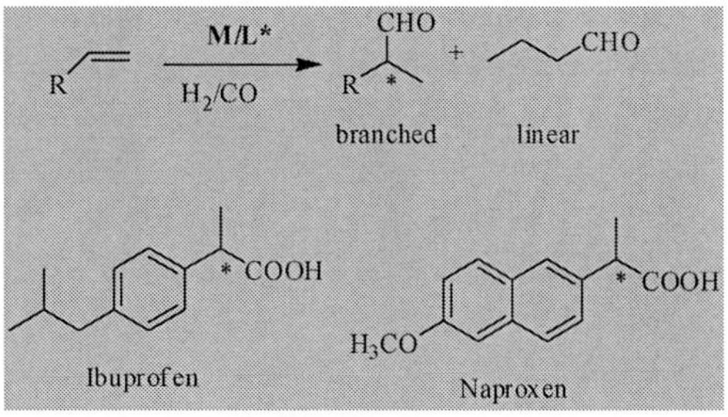

Scheme 30. Asymmetric metal-catalyzed hydroformylation reaction.

M. Iglesias et al. [144] reported the synthesis of diphosphinite ligands derived from (2S,4R), (2S,4S)-1-benzyl-4-hydroxy-4-phenyl-2-(1,1-diphenylmethyl)-pyrrolidinylmethanol and (2S,4R), (2S,4S)-1-(3-triethoxysilyl)propylaminocarbonyl-4-hydroxy-4-phenyl-2-(1,1-diphenylmethyl) pyrrolidinylmethanol in high yields (60–80%) which upon reaction with [RhCl(cod)]₂ yielded cationic [Rh(diphosphinite)(cod)] complexes. The metal complexes bearing a triethoxysilyl group were covalently bonded to USY-zeolite by controlled hydrolysis to get Rh-heterogenised complexes (Figure 17). A comparative study (homogeneous versus supported) for the catalytic activity and selectivity in hydroformylation reactions revealed an enhanced performance for the heterogenized catalysts with the additional advantage of catalysts recycle in successive runs by a simple filtration, without a significant loss of activity. The increase in activity of homogeneous catalyst on heterogenisation was attributed to the unique environment of supermicropores of USY-zeolite containing a large quantity of silanol groups. This protected environment significantly improves the stability of the catalyst for a range of substrates. Despite so many advantages the enantioselectivity of this heterogenized catalyst never reached above 10%.

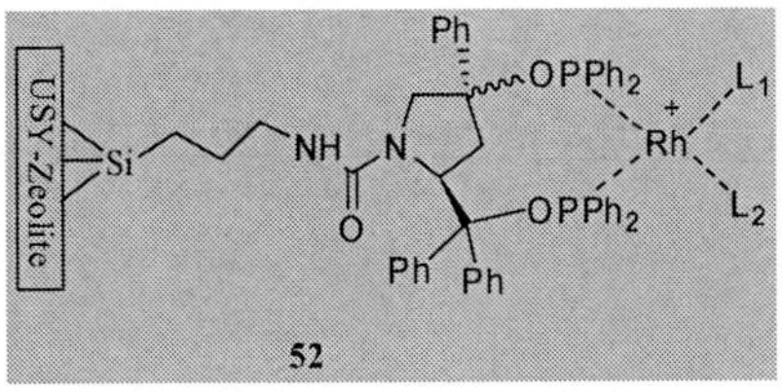

Figure 17. Chiral diphosphinite-Rh complex supported on USY zeolite.

Recently Li et al. [145,146] reported an interesting study on heterogeneous asymmetric hydrofomylation of olefins using nanoparticles of Rh/SiO$_2$ chirally modified with nine chiral O-P, N-P and P-P type ligands. Notable among them are chiral phosphorus ligands *vis.*, (*S,S*)-(+)-2,3-O-isopropylidene-2,3-dihydroxy-1,4-bis(diphenylphosphino) butane ((*S,S*)-DIOP) 53, (*R*)-(+)-2,2'-bis(diphenylphosphino)-1,1'-binaphthyl ((*R*)-BINAP) 54, and (*S*)-6,6'-dimethoxy-2,2'-bis(diphenylphosphino)-1,1'-biphenyl ((*S*)-MeO-BIPHEP) 55 (Scheme 31). The chirally modified Rh/SiO$_2$ catalysts particularly catalyst 54 and 55 exhibited very high activity and regioselectivity (branched:normal ratio, 92:8) for the asymmetric hydroformylation of styrene, however, enantioselectivity was obtained upto 30% only, whereas, with vinyl acetate as substrate, 100% selectivity of branched aldehyde with ee upto 72% was achieved with the use of (*R*)-BINAP–Rh/SiO$_2$ as catalysts. It is noteworthy that the modification of Rh/SiO$_2$ with (*S,S*)-DIOP 53 resulted in increased activity for the hydroformylation of vinyl acetate and gave a TOF of 128 h^{-1}, which was even higher than that of the unmodified Rh/SiO$_2$ catalyst (90 h^{-1}).

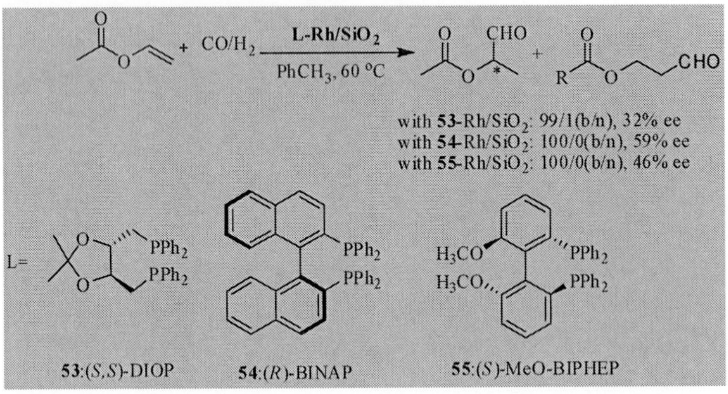

Scheme 31. Asymmetric hydroformylation using chirally modified Rh/SiO$_2$ catalysts.

A catalytic mechanism involving multiple surface sites on the chirally modified Rh/SiO$_2$ catalyst was proposed (Scheme 32). Accordingly rhodium sites actived the olefins while chiral ligand guided the attack of CO through the path that provided high regioselectivity and enantioselectivity in the product. There is a lot of potential for the improvement of enantioselectivity of this system that can accommodate a range of pro-chiral olefins.

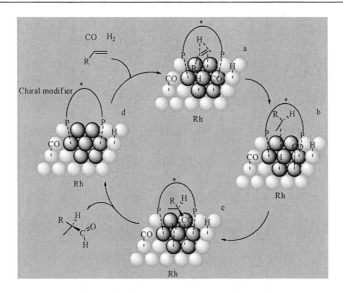

Scheme 32. The proposed mechanism of asymmetric hydroformylation on chirally modified Rh/SiO$_2$ catalyst.

Li et al. recently reported a system similar to the one mentioned in preceeding report where authors synthesized novel rhodium nanoparticles stabilized with chiral phosphorus ligands and corresponding supported catalysts [146]. The resulting highly dispersed rhodium nanoparticles showed small and narrowly distributed core sizes (1.5–2.0 nm). The supported rhodium nanocatalysts showed increased activities compared to the unsupported catalysts (e.g. from 12 to 22 h^{-1} for the hydroformylation of styrene). This nano catalyst gave high regioselectivity (branch/normal: 100/0) and chiral induction (up to 59% ee) in the asymmetric hydroformylation of vinyl acetate under mild conditions. The catalyst was recycled four times without any loss in activity and selectivity.

CARBONYL-ENE AND IMINE-ENE REACTIONS

The ene-reaction, a powerful carbon-carbon bond forming reaction, was discovered by Alder in 1943 [147,148]. It occurs between an alkene having an allylic hydrogen (the "ene") and an electron-deficient double bond (the "enophile"). The reaction is concerted, through a six membered cyclic transition state. The first catalytic enantioselective ene-reaction was promoted by an aluminum complex of a binaphthol derivative [149]. Since then, other chiral complexes, notably of binaphthol, BINAP, and bis(oxazoline) ligands, were shown to catalyze the ene-reaction in an enantioselective manner [150]. Enantiomerically pure α-hydroxy and α-amino carbonyl compounds are in high demand for the production numerous pharmaceuticals. Homogeneously catalyzed carbonyl-ene reactions have been reported using chiral aluminum/titanium–binaphthol complexes, [151] but the scope of these catalysts was very limited due to the difficulty in their recovery and reusability. Nevertheless, there are few reports on heterogenization of carbonyl-ene and imine-ene reactions using inorganic supports. Hutchings et al. [152] have demonstrated the first heterogeneously catalyzed, enantioselective carbonyl-ene and imine-ene reactions using Cu(II)-bisoxazoline supported on zeolite Y

(Scheme 33). The active catalyst was prepared by impregnation of Cu(II) iom in the zeolite cages followed by diffusion of chiral bisoxazoline ligand to give heterogeneous catalyst which was stable under aerobic and ambient conditions. The enantioselectivities obtained with the supported catalysts were similar or even better than those obtained with its homogeneous version.

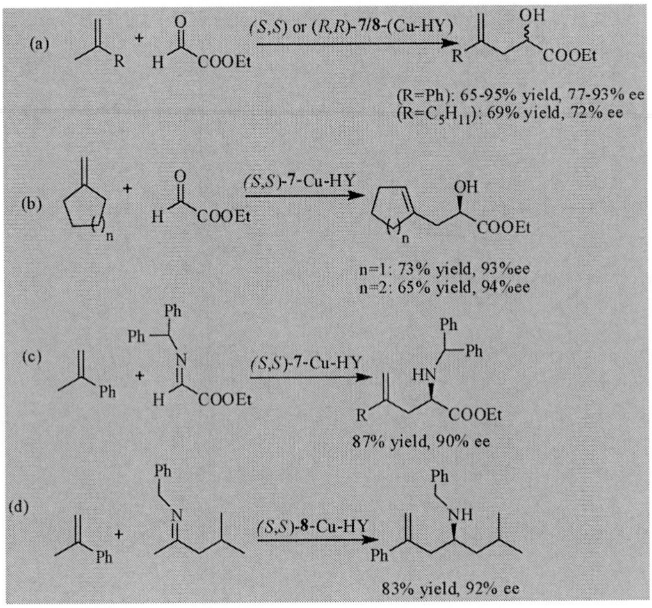

Scheme 33. Carbonyl- and Imino-ene reactions using Cu- bis(oxazoline)- zeolite Y.

For example, enantioselectivity obtained with the substrate α-methylstyrene increased from 66% ee under homogeneous condition to 80% ee under heterogeneous condition. Similarly for methylenecyclopentane it increased from 57 to 93% ee. Less than 1% copper leaching from the supported catalyst was detected under the reaction conditions and the system worked truly under heterogeneous condition. Authors extended the use of Cu-bis(Oxazoline) zeolite Y catalysts to imino-ene reactions as well (Scheme 47, (c) & (d)). The results demonstrated that both α-carboxyl imine (ethyl-N-benzhydryliminoethanoate) and alkylimine (N-benzyl isovalerimine) compounds readily react with α-methylstyrene giving high yields and enantioselectivity with the immobilized copper catalyst. Furthermore, the catalyst can be efficiently recovered by filtration and reused after washing up to three times with the reagent alternatively changed from ethyl glyoxylate to methyl piruvate. F. ono et al. reported gold nanoparticles for the immobilization of Cu-bis(oxazoline) which was used as hetergoneous catalyst in the ene reaction between 2-phenylpropene and ethyl glyoxylate [153]. The catalyst was preparted by functionalizing bis-oxazoline ligands with thiol groups and immobilized onto Au nanoparticles stabilized by hexanethiol (Scheme 34). The corresponding copper complexes were tested in the reaction of α-methylstyrene with ethyl glyoxylate. The dispersion degree in CH_2Cl_2 was highly dependent on the length of the spacer (n), which conditioned the catalyst recoverability by hexane addition and centrifugation. The best results were obtained with a tetramethylene spacer ($n = 1$), leading to five consecutive runs with the same enantioselectivity (84–86% ee) and some loss of activity.

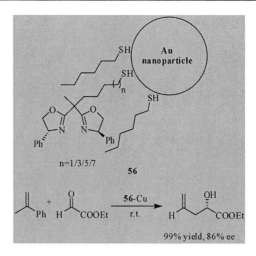

Scheme 34. Ene reaction between 2-phenylpropene and ethyl glyoxylate using nano-sized chiral bisoxazoline catalyst.

ASYMMETRIC ADDITION OF DIALKYLZINCS TO ALDEHYDES

Nucleophilic addition of organometallic reagents to carbonyl substrates constitutes one of the most fundamental operations in organic synthesis [38]. The use of organozinc chemistry, in place of conventional organolithium or organomagnesium chemistry has been developed into an ideal protocol for the catalytic enantioselective alkylation of aldehydes, leading to a diverse array of optically pure secondary alcohols [154,155]. Optically active secondary alcohols are important intermediates for the synthesis of many naturally occurring compounds, biologically active intermediates and materials such as liquid crystals [156]. Various chiral auxiliaries have been widely explored in this important asymmetric C-C bond forming reaction. Chiral β-amino alcohols derived from natural products such as ephedrine, dimethylaminoisoborneol, and proline often exhibit high asymmetric induction ability. Titanium complexes with TADDOLs, BINOLs, and 1,2-disulfonamidocyclohexane ligands show excellent enantioselectivity [157]. It is very attractive to immobilize the homogeneous catalysts on supports for the easy separation of catalysts from products and reaction mixtures, and the recycling of catalysts. In this regard the first report on immobilization of chiral N-alkylnorephedrines on functionalized alumina, silica gel, and silica gel coated with chloromethylated polystyrene for enantioselective addition of dialkylzincs to aldehydes was published by Soai et al. [158]. The alumina supported catalyst gave the highest enantiomeric excess (ee, 59%) (Scheme 35). However, this result was inferior to that of the homogeneous analogue (ee, 67%). Recycling experiment was carried out using heterogeneous catalyst immobilized on silica gel coated with chloromethylated polystyrene, and no loss of activity and enantioselectivity was found.

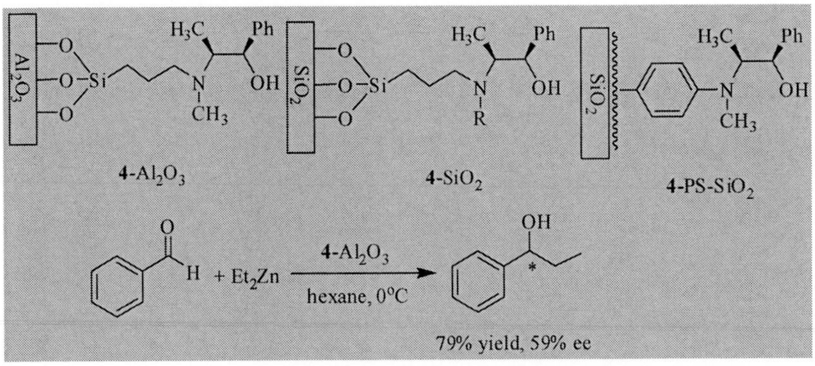

Scheme 35. Asymmetric diethyzinc addition to benzyldehyde using N-alkylnorephedrines on Al$_2$O$_3$, SiO$_2$ and SiO$_2$ coated with chloromethylated polystyrene.

Another interesting work was carried out by Seebach et al [105], where the TADDOL ligand was grafted covalently onto controlled-pore glasses (CPG) with different pore sizes. The supported TADDOL **33** with excess of Ti(O^{i}Pr)$_4$ promoted the addition of Et$_2$Zn to benzaldehyde to give chiral 1-phenylpropanol with high yield 98% and excellent enantioselectivity (up to 96%) comparable to that of the homogeneous catalyst (Scheme 36). Even a high catalyst loading (up to 0.32 mmol TADDOL per g of support) did not have a negative effect on selectivity. The immobilized TADDOL could be recycled for 20 catalytic runs without loss of enantioselectivity by washing with HCl/H$_2$O/acetone after reaction. Possibly, the residual products on the pores, which made the accessibility to some active sites difficult, may account for the minute decrease in the activity.

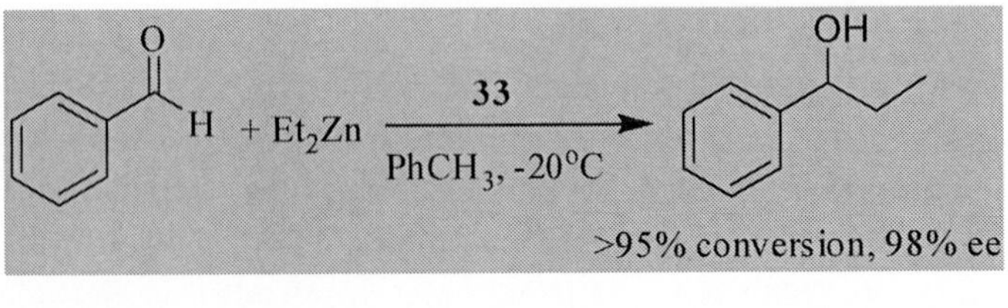

Scheme 36. Asymmetric diethyzinc addition to benzyldehyde using TADDOL-Ti on CPG **33**.

Chiral hybrid organic–inorganic materials were prepared using optically active-aminoalcohols such as (−)-ephedrine anchoring on MCM-41 type silica as a member of micelle-templated silicas (MTS) by Lasperas and coworkers [159, 160]. After this, (-)-ephedrine was anchored through halogen substitution and silylation of the surface was performed with halogenopropyltrimethoxysilane (Figure 18). These materials were used as chiral solid auxiliaries in the enantioselective addition of catalysis diethylzinc to benzaldehyde. Good conversion (61-87%) with low enantioselctivity (up to 38%) was achieved with the heterogenenous catalysts.

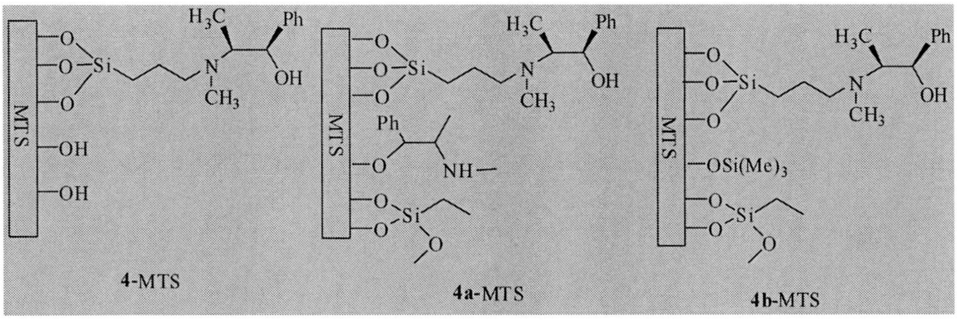

Figure 18. (−)-ephedrine anchored on micelle-templated silicas (MTS).

Authors also studied various factors such as accessibility to the catalytic sites and coverage of the inorganic surface, which affected their efficiency (activity, selectivity and enantioselectivity) either by changing the support (pore diameter, passivation with hexamethyldisilazane) or by dilution of the catalytic sites. Whatever the auxiliary, the enantioselectivities remained moderate. These results are explained by the activity of the naked surface towards the formation of racemic alcohols. They show that the accessibility to the surface was increased when dilution was performed, leading to the grafting of (−)-ephedrine directly on the surface and that passivation by trimethylsilylation was insufficient to prevent the formation of racemic alcohols. Later the same authors published work on covalent immobilization of (1R,2S)-(-)-ephedrine , used as a model molecule of β-aminoalcohols, on the surface of MCM-41-type mesoporous aluminosilicates **4c-Al-MTS**, performed by a new sol–gel method, led to chiral auxiliaries (Figure 19), which showed greatly enhanced rates and ee's in the enantioselective addition of diethylzinc to benzaldehyde [161,162].

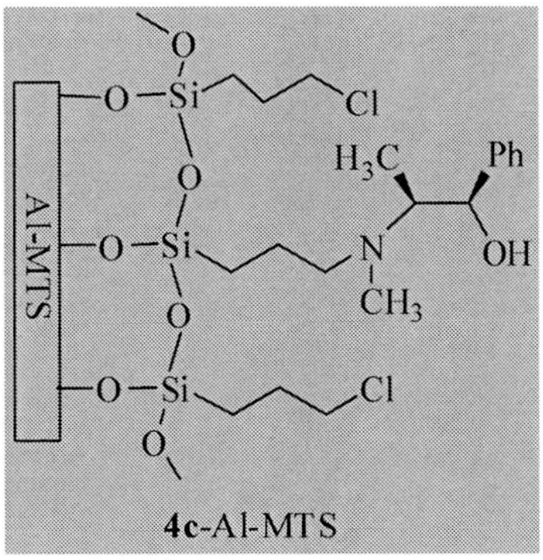

Figure 19. (−)-ephedrine anchored on aluminosilicates.

The accessibility to the catalytic sites was studied by taking Al-MTS supports of 3.6 and 8.3 nm mean pore diameter and dilution of catalytic sites and rigidification of the anchoring arms were also studied in order to determine the role of site–site and site–surface interactions. The coupling alkyl halide moiety was then substituted with (−)-ephedrine and in the heterogeneous catalyst, thus prepared, was evaluated for asymmetric addition of diehtylzinc to benzaldehyde. These materials showed properties which depended mainly on the grafting method. The best results (activity, enantioselectivity) were obtained (98% conversion with

64% ee) with catalysts prepared from supports featuring high initial pore diameter (8.3 nm). Moreover, these catalysts could be reused three times without loss of enantioselectivity. Authors have concluded from their study that two factors that affect the activity and enantioselctivity in this particular reaction are, firstly, the coverage of the surface by organics in order to suppress the activity towards the racemic reaction and secondly, large mesopore diameters ensure the accessibility to the catalytic sites. Later these authors studied factors which influenced the catalytic activity of Al-MTS supported (-)-ephedrine [163]. Various factors including the activity of the naked surface towards the formation of racemic alcohols and the accessibility of the chiral catalytic sites or rigidification of the grafting arms were examined. Bae et al. have immobilized the pyrrolidine methanol derived from proline on mesoporous silicas MCM-41 and SBA-15 with different pore size [164]. These new heterogeneous catalysts have been used in asymmetric addition of diethylzinc to benzaldehyde. High yield (up to 98%) with good enantioselectivity (52-75%) was achieved with SBA-15 based catalyst **59** than MCM-41 and amorphous silica based catalysts (Scheme 37). The enantioselectivity was found to be largely dependent on the pore size of the mesoporous silicas, capping of free silanol moieties with trimethylsilyl (-Si(Me)$_3$) groups and employment of nBuLi. The role of nBuLi in this reaction remained unclear. Among various catalytic systems, the best result (ee, 75%) was obtained with the TMS-capped SBA-15 based catalyst treated with nBuLi.

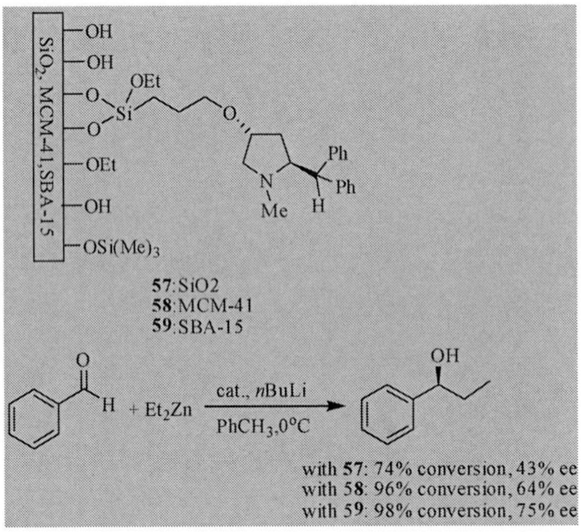

Scheme 37. chiral pyrrolidinemethanol immobilized on different silica and used in asymmetric addition of diethylzinc to benzyldehyde.

Another interesting work was reported by Fraile and coworkers [165] on immobilization of a readily available chiral amino alcohol (R)-1,1,2-triphenyl-2-(piperazin-1-yl)ethanol on silica by sol-gel synthesis and grafting (Figure 20). The solid prepared according to the latter method led to high yield (99%) and enantioselectivity (ee, 77%) in the asymmetric addition of diethylzinc to benzaldehyde at -25°C. Authors gave two explainations on the role of nBuLi, one is the formation of lithium silyloxide species, less active than the analogous zinc species, and the second is the formation of a lithium amino alkoxide species, more active than the analogous zinc derivative. Important improvements in enantioselectivity have been reported by using the lithium salts of amino alcohols, although this behavior has shown to be highly dependent on the structure of the amino alcohol and even on the amount of catalyst [166]. However, the lack of experimental studies regarding the relative activity of lithium and zinc species, both on amino alcohol and surface silanol, precludes any convincing explanation.

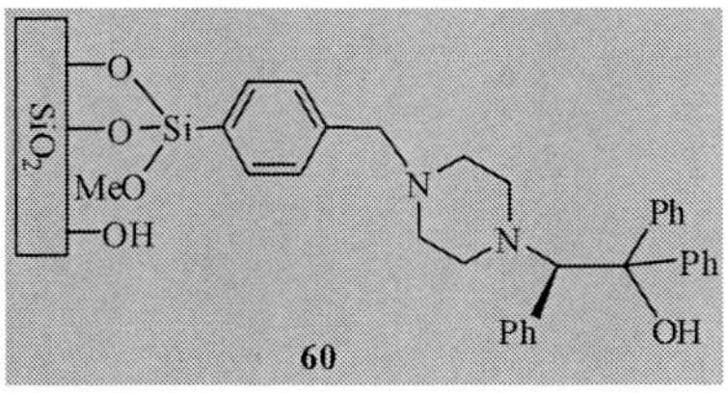

Figure 20. chiral amino alcohol immobilization on silica.

Rhee at al. [167] reported the first work on the use of silica supported dendritic chiral auxiliaries for the enantioselective addition of diethylzinc to benzaldehyde. Dendrimers are highly branched macromolecules and they are generally described to have a structure of spherical shape with a high degree of symmetry. By combining the chirality or asymmetry with their highly symmetrical nature, dendrimers render themselves attractive for the design of asymmetric catalysts. Here, authors have anchored polyamidoamine dendrimer on silica (Figure 21). The control of dendrimer propagation on silica surface is of prime importance to obtain enhanced conversion, selectivity, and enantioselectivity. High yield (up to 92%) with moderate enantioselctivty (ee, 59%) was achieved. Furthermore, the heterogeneous dendritic chiral catalyst can also be recycled and reused upto three cycles without a significant loss of catalytic activity.

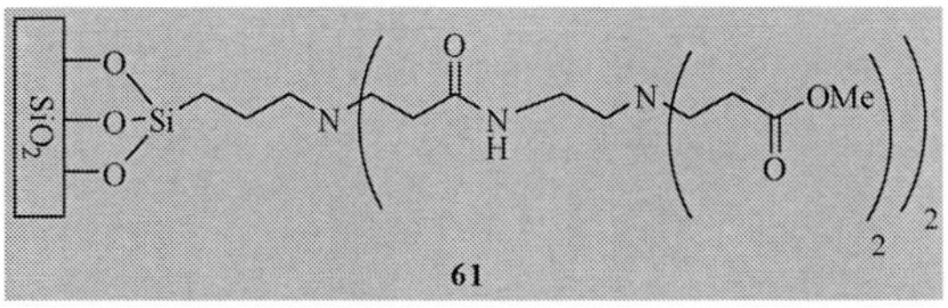

Figure 21. polyamidoamine dendrimer on silica.

Soai et al. [168] reported some very interesting work on helical silica which was prepared by using chiral organogel derived from 1,2-diaminocyclohexane (Figure 22). These artificially designed right- and left-handed helical silicas were used as enantioselective media during the addition reaction of diisopropylzinc to pyrimidine-5-carbaldehyde. Highly enantiomerically enriched (up to 96–97% ee) 5-pyrimidyl alkanol was obtained with both right and left handed helical silica (Scheme 38). Authors also claimed that right- and left-handed helical silica, even after standing for 14 months at room temperature, worked as a chiral initiator. So, the chirality of inorganic architecture is considered to contribute to the asymmetric induction.

Soai et al. reported further work on chiral organic–inorganic hybrid materials like silsesquioxane hybrid derived from silylated trans-1,2-diaminocyclohexane, as a chiral initiator in asymmetric autocatalysis in the enantioselective addition of diisopropylzinc to pyrimidine-5-carbaldehyde [169]. In the presence of chiral hybrid silsesquioxane 63 (Scheme 39), excellent enantioselctivity (ee, 96%) of 5-pyrimidyl alkanol was achieved. The hybrid silica is recoverable, and can be used successively many tines without any loss of enantioselectivity (ee, 90–94%) of product.

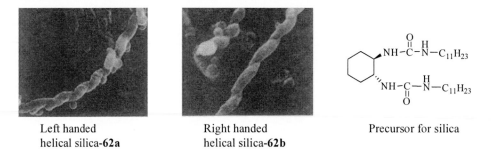

| Left handed
helical silica-**62a** | Right handed
helical silica-**62b** | Precursor for silica |

Figure 22. Helical silica.

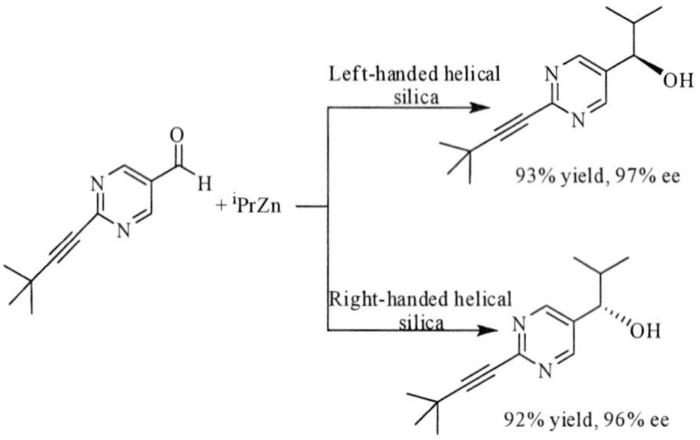

Scheme-38. Enantioselective addition of diisopropylzinc to pyrimidine-5-carbaldehyde using helical silica.

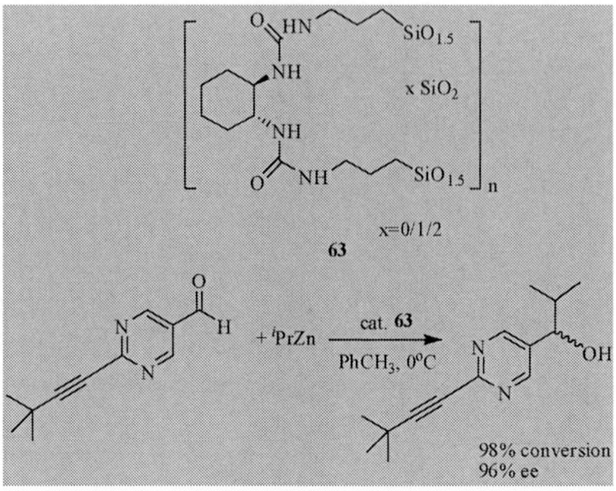

Scheme-39. Enantioselective addition of diisopropylzinc to pyrimidine-5-carbaldehyde using chiral hybrid silsesquioxane.

Chiral *N*-sulfonylated amino alcohols ((1*R*,2*S*)-2-(*p*-toluenesulfonylamino)-1,3-diphenyl-1-propanol) derived catalyst grafted onto amorphous silica gel and mesoporous silica (MCM-41) [170] were also used as solid chiral auxiliaries in Ti catalyzed asymmetric addition to diethylzinc addition to benzyldehyde (Scheme 40). Silica immobilized catalyst imparted good yield (98%) with high enantioselectivity (ee, 80%) as compared to MCM-41 based catalyst (conversion, 86%; ee, 55%). The silica-immobilized catalyst was reused in multiple catalytic runs (up to 10 times) without loss of enantioselectivity.

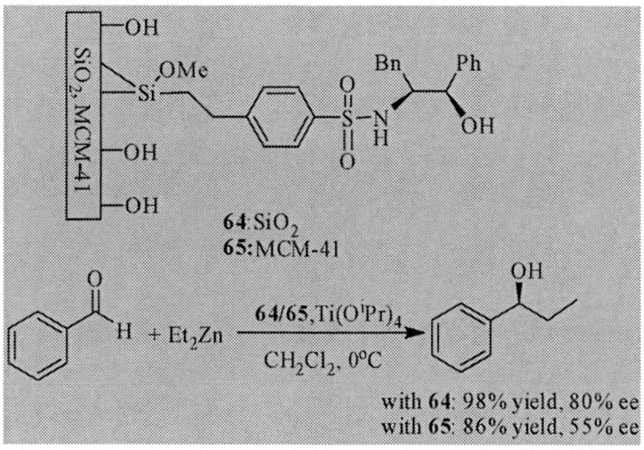

Scheme-40. Silica-immobilized Ti(IV) complexes of *N*-sulfonylated amino alcohol.

Abdi and coworkers, for the first time, reported [171], the grafting of chiral (*S*)-BINOL ligand on ordered mesoporous silicas-MCM-41 and SBA-15 through covalently bonded linkers (Scheme 41). The resulting inorganic–organic hybrid materials were used as chiral auxiliaries in Ti-promoted enantioselective addition of diethyl zinc to aldehydes under heterogeneous reaction conditions. These heterogeneous catalysts showed promising activity and enantioselectivity. The catalyst **67** (Scheme 41) having a large pore diameter (6.8 nm) and with capping of free silanol moiety by trimethylsilyl groups was found to be more active and enantioselective (ee up to 81%) than uncapped catalyst. The reuse of expensive chiral BINOL based catalyst was effectively worked out by washing the used catalyst with 10% HCl in methanol while the regenerated catalyst was used for further catalytic runs with the retention of enantioselectivity.

Recently same authors have extended their work by [172], synthesizing silica-supported chiral BINOL on two different relatively large pore sized mesoporous silicas, SBA-15 (7.5 nm) and MCF (14 nm), by covalent grafting method using *N*-methyl-3-aminopropyltriethoxysilane (MAPTES) as reactive surface modifier. These heterogenized ligands 68 and 69 (Scheme 41) were used in Ti-catalyzed asymmetric addition of diethylzinc to various aldehydes. High catalytic activity with excellent enantioselectivity (up to 94% ee) for secondary alcohols was reported for MCF supported chiral BINOL 69 under heterogeneous reaction conditions. Good to excellent enantioselectivity (ee, 68-91%) was also reported with various small to bulkier aldehydes. The MCF supported catalyst was reused in multiple catalytic runs without loss of enantioselectivity. Authors suggested that the pore size of silica supports and capping of free silanol groups with TMS groups on the silica surface were vital parameters towards achieving high enantioselectivities.

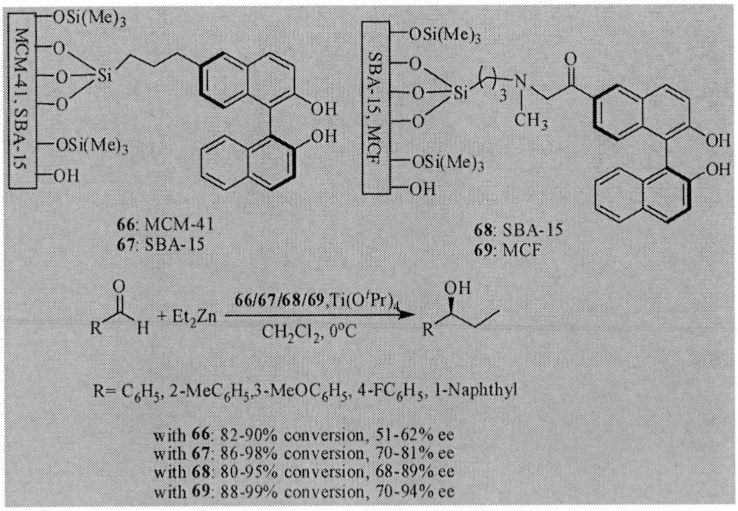

Scheme-41. (*S*)-BINOL covalently grafted to different pore size silica and used in Ti-catalyzed asymmetric diethylzinc addition to aldehydes.

Kiraly et al. recently reported the immobilization of optically active cationic surfactant (*1R,2S*)-(-)-N-dodecyl-N-methylephedrinium bromide (DMEB) in the channel of montmorillonite clay *via* cation exchange [173]. Immobilization of the chiral cationic surfactant DMEB on montmorillonite resulted in the formation of the organophilic clay DMEB-M (Figure 23), which contains optically active surfactant molecules between the clay lamellae. It was found that swelling of montmorillonite was accompanied by delamination in organic solvents. The delaminated material dispersed in the reaction medium to work as efficient catalyst for the addition of dialkyl zinc to aldehydes. The selectivity towards the formation of 1-phenyl-1-propanol surpassed 80% under both homogeneous and heterogeneous conditions irrespective of the reaction time. The conversion of benzaldehyde and the selectivity towards the formation of 1-phenyl-1-propanol were found to increase at higher catalyst loading and temperature.

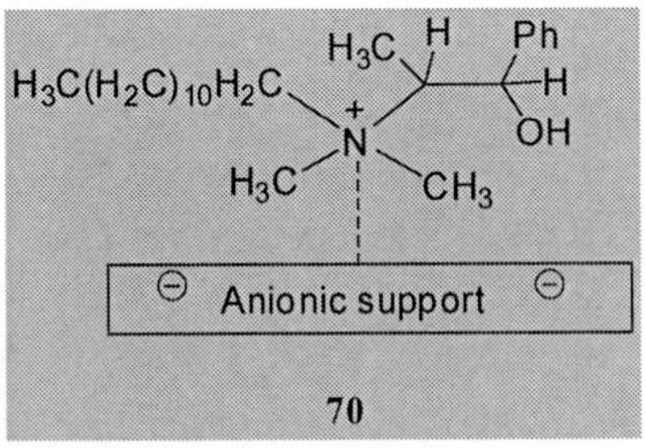

Figure 23. DMEB immobilized on montmorillonite clay.

Recently there were two reports on immobilization of chiral 2-amino-1,2-diphenylethanol into the layers of phosphonates of zirconium [174] and titanium [175] (Figure 24). Both phosphonates are semi-crystalline layered materials with an interlayer space of ~20.11 Å. Zirconium phosphonates enantioselectively catalysized the addition of Et$_2$Zn to benzaldehyde to afford optically active secondary alcohol in >90% yield and 51% e.e., which was 6% lower than the corresponding chiral ligand in homogeneous system. Catalyst was reused more than ten times by washing it with HCl (36%, 5 ml), to remove zinc oxide formed during the catalytic reaction, then with Na_2CO_3 solution and finally with water till near neutral pH. The wet material was finally dried under vacuum at 60–70 °C for 24 h. In the case of chiral 2-amino-1,2-diphenylethanol immobilized on the layered titanium phosphonates the catalytic activity for enantioseletive addition of diethylzinc to benzaldehyde was relatively inferior (yield, 88.5%; ee, 40.8%) to the corresponding zirconium catalyst. The catalyst was similarly treated, as in the case of Z-based catalyst, for its regeneration and reuse. The catalyst worked seven times without any loss in its efficiency.

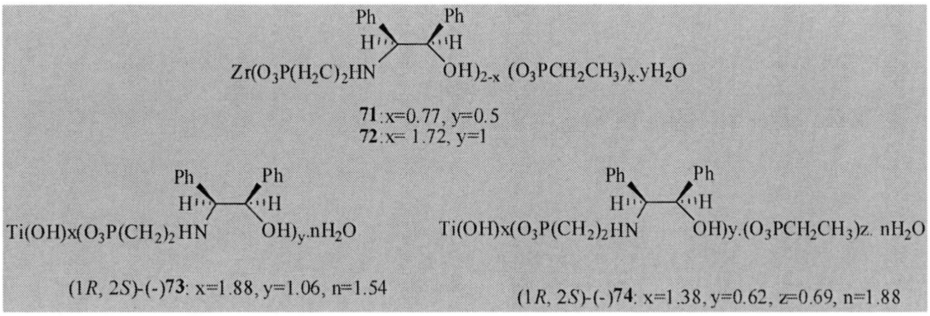

Figure 24. Chiral hybrid zirconium phosphonates and titanium phosphonates.

ENANTIOSELECTIVE ADDITION
OF PHENYLACETYLENE TO ALDEHYDES

The steroselective addition of alkynyl-metal reagents to carbonyls is one of the important nucleophilic reactions. This nucleophilic addition of alkynes to carbonyls produces a new C-C bond with concomitant creation of a stereogenic center in a single transformation. The resulting *secondary/tertiary* propargylic alcohols are versatile precursors to many organic molecules including natural products and pharmaceutical compounds [176-178]. The acetylene and hydroxyl functions of the propargylic alcohol products can be used to construct diverse molecular structures. There are various chiral nitrogen-containing ligands like amino alcohols, alkaloids, chiral BINOLs and SALEN based ligands that have been extensively studied in the enantioselective addition of phenylacetylene to aldehydes, in homogenenous reaction conditions [177]. However, there is only one report published on heterogenization of chiral β-hydroxyamide on silica support for asymmetric addition of phenylacetylene to aldehydes [179]. The authors reported the first use of silica-immobilized titanium(IV) complex of β-hydroxyamide as a catalyst for the asymmetric addition of phenylacetylene to aromatic aldehydes to give products with high yield (up to 95%) and good enantioselectivity (up to 81% ee) (Scheme 42). The catalyst could be reused up to five times without serious loss of enantioselectivity.

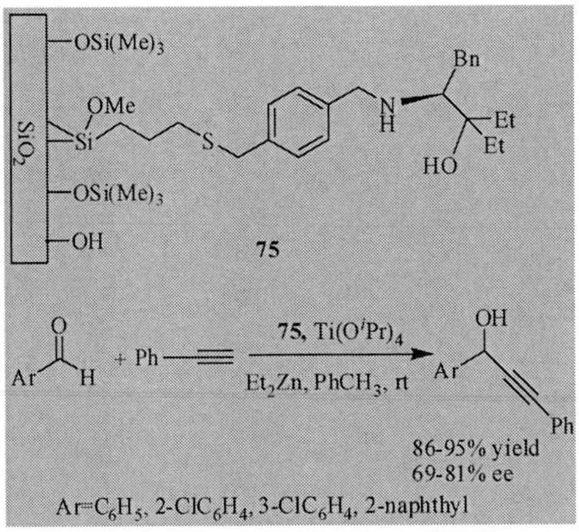

Scheme-42. asymmetric phenylacetylene addition to aldehydes.

CYNOSILYLATION OF ALDEHYDES

Enantioselective addition of cyanide to prochiral ketones and aldehydes constitutes an important organic reaction, since the resulting chiral cyanohydrins are versatile building blocks [180,181]. Except for their use as synthetic intermediates, cyanohydrins also play an important role as pesticides in agriculture. Several enantioselective reactions have been developed based on the use of enzyme, Lewis acid, or Lewis base catalysis. Today, the most important class of reactions are based on Lewis acid catalysis [182]. Several metal complexes have been employed for the enantioselective addition of trimethylsilyl cyanide (TMSCN) to carbonyl compounds [183]. Chiral catalysts, being expensive, are desired to be recycled for their commercial viability. Surprisingly, only few reports have been published on immobilization of chiral metal complexes for their use in asymmetric cynosilylation of aldehydes.

Corma and collaborators have published some very interesting studies on the reutilization of VOsalen catalysts. In 2003, the authors reported the synthesis of supported catalyst by a sol-gel strategy [184]. In this method, a salen-vanadyl complexe substituted with two trimethoxysilyl termini and a structure-directing agent-cetyltrimethylammonium were co-crystallized to develop periodic mesoporous organosilica (PMO) catalysts for their use in cyanosilylation of carbonyl compounds The resultant material (0.25 mmol % to benzaldehyde) showed a high turn over number (320) for the room temperature silylcyanation of benzaldehyde with TMSCN (3 eq.) with an ee of 30 % (Scheme 43). The result is far from the 94 % ee achieved with the non-supported homogeneous catalyst. These low ee values could not be increased upon silylation of the external surface to reduce the population of undesirable silanol groups. Interestingly, the chirality of the resulting solid could be assessed directly by measuring the optical activity of a suspension of the ChiMO in 1,2-dichloroethane, a solvent with a refraction index similar to that of the solid [185].

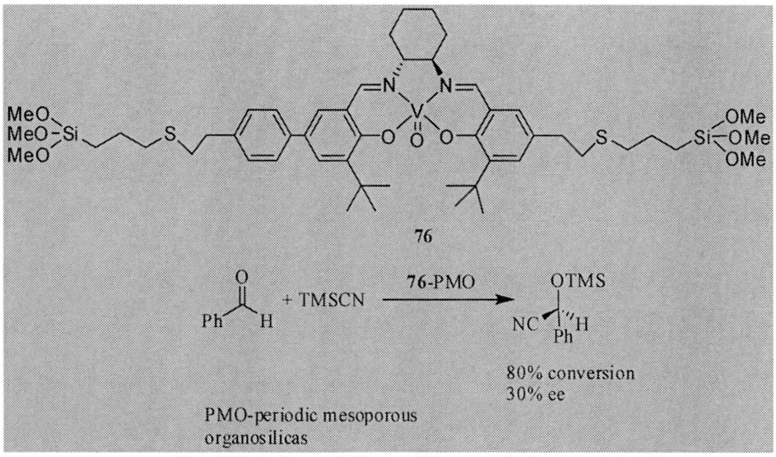

76

PMO-periodic mesoporous organosilicas

Scheme-43. cyanosilylation of benzyldehyde using VO(salen) anchored on PMO.

In their continued efforts to develop recyclable supported catalyst, Corma et al. further synthesized oxo-vanadium–salen complexes anchored on amorphous silica 77, ITQ-2 (a novel delaminated zeolite) 78, MCM-41 79 and single wall carbon nanotube 80 (Scheme 44) through mercaptopropylsilyl groups [186,187]. The resulting solids (catalysts 77, 78 & 79) were tested for their catalytic efficiency in enantioselective addition reaction of trimethylsilyl cyanide on various aldehydes only to get low ee values (ee, 49-63%) in the product. In an attempt to improve the enantioselectivity of the solid catalysts, free silanol groups on the catalyst surface were silylated with trimethylsilyl and the length of linker attaching the salen complex with amorphous silica was increase to 11 C. As a result high enantioselectivity (up to 85%) was achieved using CHCl$_3$ as the solvent at 0°C and the results were close to those achieved with homogeneous complex (ee, 90%). These catalysts were recycled for 5 repeated catalytic runs with gradual loss in activity, though enantioselectivity was somewhat retained. From their study, authors have suggested that the tether linking the complex and the solid should be long enough to permit the complex to have a large conformational freedom and the solid surface has to be modified to reduce the presence of residual silanol groups.

Single-wall carbon nanotubes (SWNT) have a special structure and morphology consisting of long tubes in μm scale and with less than 2 nm in diameter. Taking advantage of these structural and geometrical features of SWNT, authors covalently anchored a styryl-functionalized vanadyl Schiff base on a mercapto-modified SWNT. The salen-vanadyl covalently attached to SWNT 80 (Scheme 44) oexhibited high catalytic activity for the cyanosilylation of aromatic aldehydes with an almost complete conversion and selectivity towards the silylated cyanohydrin. This high catalytic activity is similar to that of an analogous complex in solution, but the solid was reused four times with the same conversion (>95%). Comparison of the performance of the salenvanadyl complex anchored to SWNT with an analogous sample in which the vanadyl complex was attached to activated carbon (Norit, surface area) 1400 m^{2}g^{-1}) [188] was made where the latter gave only a conversion of 83%, indicating that the former was better. The complex anchored to SWNT is truly heterogeneous (no leaching observed) and reusable (no decrease in activity) in five consecutive runs for the asymmetric cyanosilylation of benzaldehyde. Given that the chemistry of SWNT constitutes an emerging research front, it can be predicted that more studies will soon appear using SWNT as support for chiral salen-metal complexes. In the present examples, the salen complexes were anchored to the tips of SWNT by functionalization of the carboxylic acids present at the termini of the tubes. Authors have suggested that it will be more interesting to develop a side-wall functionalization strategy to anchor the complexes, because in this way, the potential loading of catalytically active sites will increase by several orders of magnitude. The immobilization of salen-titanium complexes, either monomeric or dimeric, on MCM-41 was reported by Kim and Kim [189]. The authors have anchored complexes by multigrafting of salicylaldehyde derivatives and diaminocyclohexane on 3-mercaptopropyl-functionalized MCM-41 (81a and 81b, Figure 25) and apical coordinative bond through the metal by means of the silanoxy groups present on the solid surface (82 and 83, Figure 25).

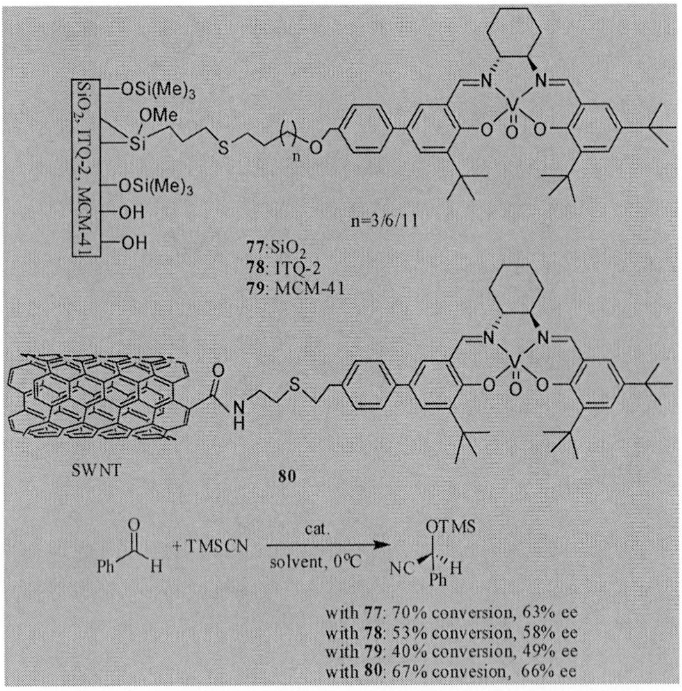

Scheme-44. V–salen complexes anchored on SWNT, SiO_2, ITQ-2 and MCM-41 for cyanosilylation reaction.

These heterogeneous catalysts were tested for the asymmetric cyanohydrin formation of benzaldehydes using cyanotrimethylsilane as reagent, with contrasting results in terms of enantioselectivity. A high level of enantioselectivity was attained using chiral TI(IV) salen complexes immobilized onto MCM-41. The chiral salen TI(IV) complexes immobilized over MCM-41 using the multistep synthesis were stable during these reactions and exhibited relatively high enantioselectivity for reactions (up to 93% ee) as compared to homogeneous ones (ee, 87%) at room temperature. When using direct anchoring method (catalyst 82 and 83), the monomeric Ti(IV) catalyst 82 may exhibit higher enantiomeric excess (ee, 64%) than the supported dimer 83 (ee, 59%), but in both methods only moderate conversions (23–40%) has been achieved. Although no leaching was detected for the supported dimeric salen titanium complex that was simply separated from the reaction mixture by filtration and was recycled two times with a decrease in the reaction rate.

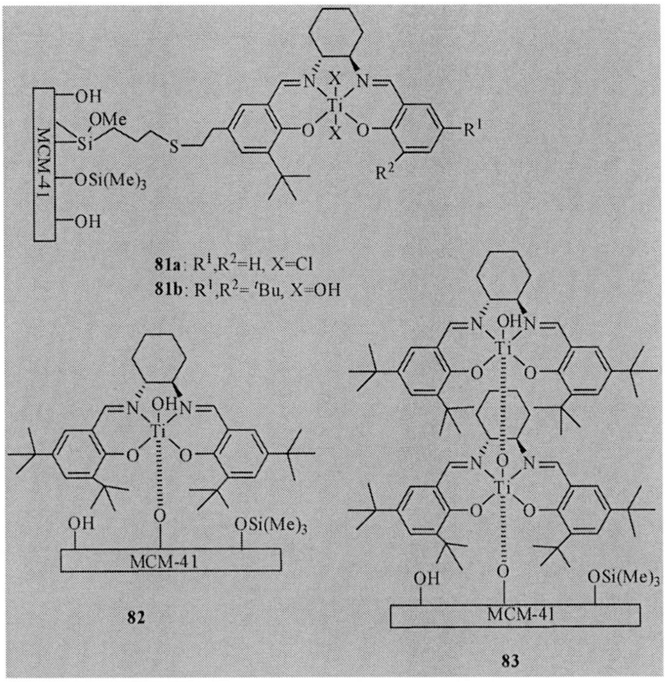

Figure 25. Chiral Ti(iV) salen complexes immobilized on MCMC-41.

FRIDEL-CRAFTS ALKYLATION

The Friedel-Crafts alkylation of aromatic compounds is a classical reaction in synthetic chemistry, in which Lewis acids are required to form a new C-C bond [190, 191]. Nevertheless, performing the Friedel-Crafts reaction in a catalytic asymmetric fashion is still a great challenge for chemists. There are few reports on immobilized chiral catalysts for Friedel-Crafts alkylation in literature. Corma et al. first succefully anchored chiral Cu(II) bis(oxazoline) covalently bonded on silica and MCM-41 and used them as heterogeneous catalysts for Friedel–Crafts hydroxyalkylation of 1,3-dimethoxybenzene with 3,3,3-trifluoropyruvate [192]. The homogeneous pre-formed, well-characterized Cu(II) bisoxazoline complex was anchoring to 3-mercaptopropyl silanized silicas (amorphous silica and MCM-41) using azoisobutyronitrile (AIBN) as radical generator. The strategy of exhaustive silylation with hexamethyldisilazane reduced the presence of free OH on the solid surface which minimized the presence of adventitious uncomplexed copper ions on the surface of the solid that could act as non-enantioselective catalytic sites. The solid catalysts were used in Friedel–Crafts hydroxyalkylation of 1,3-dimethoxybenzene with 3,3,3-trifluoropyruvate in acetonitrile at room temperature (Scheme 45). Good conversions (72-77%) with excellent mass balances (>95%) were obtained in the presence of the solid catalysts. High enantioselctivity (up to 92% ee) was achieved in case of amorphous silica grafted catalyst 84. The reactions were truly heterogeneous, and the enantioselectivity was always (ee, 92%) with amorphous silica and ee, 82% with MCM-41) higher than that obtained with the analogous ligand in solution (ee, 72%). Solid catalysts were reused without any loss of catalytic activity.

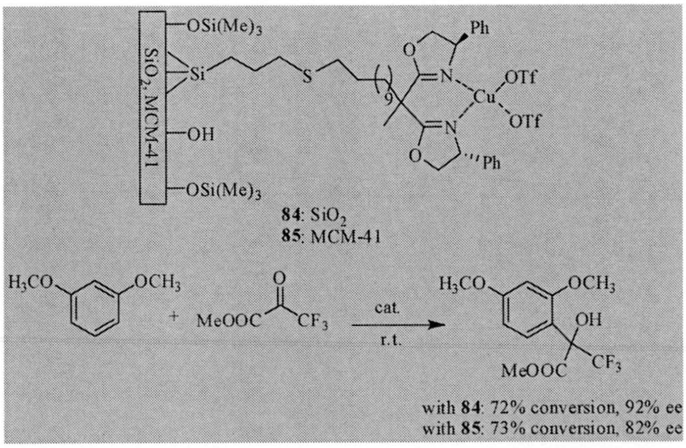

with **84**: 72% conversion, 92% ee
with **85**: 73% conversion, 82% ee

Scheme-45. Friedel–Crafts alkylation using immobilized Cu-bis(oxazoline) on silica,

A highly efficient and enantioselective organocatalyst, imidazolidin-4-one, was immobilized on siliceous MCF and polymercoated MCF by Ying and coworkers [107]. The resulting heterogenized catalyst demonstrated excellent catalytic performance and recyclability for Friedel–Crafts alkylation. The authors observed that partially pre-capping the MCF with TMS groups enabled the attainment of well-dispersed catalysts on siliceous support with optimal performance. They have also developed a polymer-coated MCF, which retained the porous structure of MCF. An ee value as high as 92% was achieved with MCF supported catalyst which was partially pre-capped with trimethylsilyl (TMS) groups when the reaction was run at -30 °C. MCF-supported catalyst and the polymer-coated MCF-supported catalyst provided good conversion (up to 72%) and enantioselectivities (ee, 74%) at room temperature (Scheme 46). Catalysts were recycled with retention of activity.

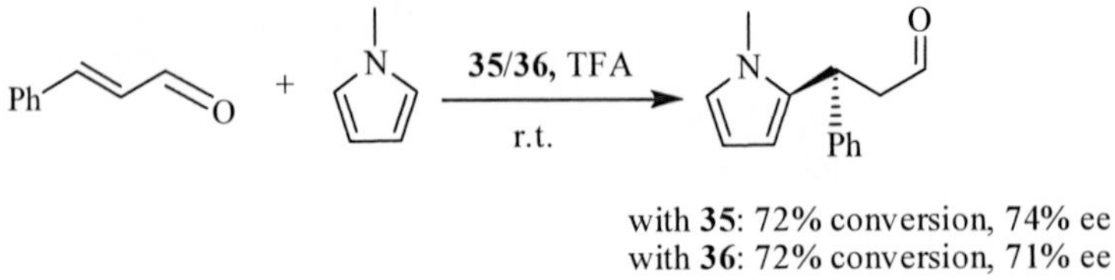

with **35**: 72% conversion, 74% ee
with **36**: 72% conversion, 71% ee

Scheme-46. Friedel–Crafts alkylation using supported catalysts.

Recently He et al. reported the immobilization of 9-thiourea *epi*-quinine on mesoporous silica SBA-15 by covalent linkage through a mercaptopropyl linker to produce a highly enantioselective heterogeneous catalyst for asymmetric Friedel–Crafts reaction of imines with indoles [193]. 9-Thiourea *epi*-quinine supported on mesoporous silica exhibited enhanced enantioselectivity (ee, up to 99.2%) in the asymmetric Friedel–Crafts reaction of imines with indoles (Scheme 47). Interestingly, the reaction was found to be insensitive to the electronic properties of the indole ring. Excellent enantioselectivities (ee, 89->99%) of the corresponding benzenesulfonamides were obtained using supported chiral catalyst. Heterogeneous catalyst could be readily recycled to up to five cycles with retention of enantioselctivity by simple filtration and regenerated by washing with organic solvents. They have observed that the confinement effect as well as the accessibility of the nano-sized pores of the SBA-15 support may play roles in directing the electrophile reaction and efficiently increasing the enantiomeric excess.

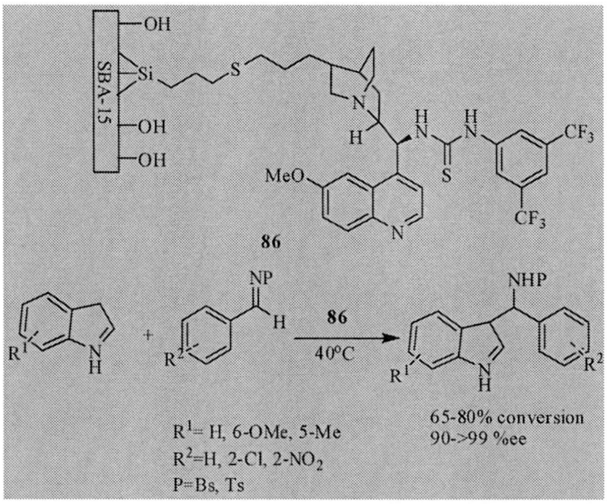

Scheme-47. Friedel–Crafts reaction of imines with indoles using 9-thiourea *epi*-quinine supported on SBA-15.

CONCLUSION

The development of heterogeneous catalysis is, in some ways, getting closer to the enzyme catalyze reactions that mother nature does very efficiently in a continuous flow reactor system -that is the living "Cell". Therefore, the goal before scientists is to develop a catalyst which works the way a natural enzyme works in a biological system i.e. a protein working as either a self-supported catalyst or as a support to a catalytically active metal complex. Virtually all the reactions in a biological system are enantio-controlled reactions. Developing enantioselective reactions out side the cell is a big challenge as high throughput synthesis is necessary in order to make the catalytic system economical. In addition, it is highly desirable for the process to be environmentally sound. To achieve all the virtues in a single system is a Herculean task. Nevertheless, synthetic chemists are approaching this goal at a tremendous speed since more and more novel synthetic and characterization tools are now at their disposal. The past decade has seen some very interesting works in the area of asymmetric catalysis, which could well lead to a break through towards the final goal. Asymmetric catalysis got recognition from the Nobel Prize committee in the year 2000. Our next goal should be to develop an efficient heterogeneous asymmetric catalysis for various commercial organic transformations. One such transformation is enantioselective C-C bond formation.

Good progress has been made in metal catalyzed C-C bond formation reactions under homogeneous condition in the last two decades. At the same time the efforts are on towards immobilization of homogeneous catalysts on organic solid phase, organic liquids phase, self-support organic/organometallic solids, inorganic solid phase and chiral inorganic structures like helical silica. The heterogenization of the potential homogeneous catalytic system on to inorganic solid supports is one of the most vigorously pursued area of research as it is evident from the literature presented in this chapter. Various inorganic supports like silica, alumina, clays and mesoporous silica have been applied as support materials for immobilization of chiral catalysts for C-C bond formations. The figure 26 represents general trend for the use of different inorganic supports for the immobilization of chiral catalysts.

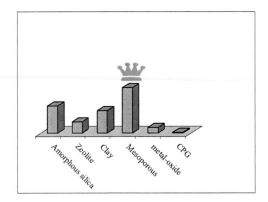

Figure 26. Different inorganic supports for the immobilization of chiral catalysts.

Among various inorganic materials, mesoporous silicas are "crown materials" for heterogenization because of its unique features. Some of the results obtained with the use of mesoporous ordered silica materials as support suggest that this material is not mere spectator to asymmetric catalysis performed by its guest metal complex. This material improves the stability of the active catalyst, create environment which enhances enantioselective capacity of the chiral catalyst and often improves activity as well. Needless to say that they make the homogeneous chiral catalyst recyclable and suitable for desired goal of making it work in a continuous flow reactor. This is not to say that other inorganic solids are in any way inferior. Some of the truly remarkable results were achieved by the use of amorphous silica as support. With the advent of theoretical molecular calculations and simulation studies in the area of catalysis and continued efforts of synthetic chemists it would be possible in the near future to choose the right kind of support for specific chiral catalyst required to catalyze a particular organic transformation.

ACKNOWLEDGEMENTS

Authors are thankful to DST and CSIR Net Work Project on Catalysis, K. Pathak to CSIR for providing Senior Research Fellowship. We are also thankful to Dr. P.K. Ghosh, the Director of the Institute, for providing the necessary facilities.

REFERENCES

[1]   Noyori, R., (1994) *Asymmetric Catalysis in Organic Synthesis*, New York, Wiley-Interscience.

[2]   Jacobsen, E. N.; Pfaltz, A.; Yamamoto, H., (1999) *Comprehensive Asymmetric Catalysis*, Berlin, Springer, Vols. I–III.

[3]   Ojima, I., (2000) *Catalytic Asymmetric Synthesis*, New York, Wiley-VCH, 2nd Ed.

[4]   Stinson, S. C., (2001) Chiral chemistry *Chem. Eng. News*, *79*, 45-57.

[5]   Noyori, R., (2003) Asymmetric catalysis: Science and opportunities (Nobel Lecture 2001) *Adv. Synth. Catal. 345*, 15-32.

[6]   Blaser, H. U.; Pugin, B.; Spindler, F., (2005) Progress in enantioselective catalysis assessed from an industrial point of view *J. Mol. Catal. A: Chem. 231*, 1-20.

[7]   Blaser, H. U.; Spindler, F.; Studer, M. (2001) Enantioselective catalysis in fine chemicals production, *Appl. Catal. A: 221*, 119-143.

[8]   De Vos, D. E.; Vankelecom, I. F.; Jacobs, P. A., (2000) *Chiral catalyst immobilization and recycling*, Weinheim, Wiley-VCH.

[9]   Fan, Q.; Li, Y.-M.; Chan, A. S. C., (2002) Recoverable catalysts for asymmetric organic synthesis, *Chem. Rev. 102,* 3385 – 3466.

[10]   Song, C. E.; Lee, S.-G., (2002) Supported chiral catalysts on inorganic materials, *Chem. Rev. 102*, 3495 –3524.

[11]   De Vos, D. E.; Dams, M.; Sels, B. F.; Jacobs, P. A.; (2002) Ordered mesoporous and microporous molecular sieves functionalized with transition metal complexes as catalysts for selective organic transformations, *Chem. Rev. 102,* 3615 –3640.

[12]   Li, C. (2004) Chiral synthesis on catalysts immobilized in microporous and mesoporous materials, *Chem. Soc. Rev. 46,* 419-492.

[13]   Baleizao, C.; Garcia, H.; (2006) Chiral salen complexes: An overview to recoverable and reusable homogeneous and heterogeneous catalysts, *Chem. Rev. 106*, 3987-4043.

[14]   Heitbaum, M.; Glorius, F.; Escher, I.; (2006) Asymmetric heterogeneous catalysis, *Angew. Chem. Int. Ed. 45,* 4732 – 4762.

[15]   Noyori, R.; Kitamura, M., (1989) *Functional groups/miscellaneous/O=C-C-X/Cl/O=C-C-X/Br/haloketones/metals/reduction/O=C/ketones.* Sheffold, R. (Ed.), In *Modern Synthetic Methods,* (Vol. 5, pp, 115-198), Berlin, Springer.

[16]   Dai, L.-X. (2004) Chiral metal–organic assemblies—A new approach to immobilizing homogeneous asymmetric catalysts, *Angew. Chem. Int. Ed., 43*, 5726 –5729.

[17]   Wu, C.-D.; Hu, A.; Zhang, L.; Lin, W., (2005) A homochiral porous metal-organic framework for highly enantioselective heterogeneous asymmetric catalysis, *J. Am. Chem. Soc. 127*, 8940-8941.

[18]   Cho, S.-H.; Ma, B.; Nguyen, S. T.; Hupp, J. T.; Albrecht-Schmitt, T. E.; (2006) A metal–organic framework material that functions as an enantioselective catalyst for olefin epoxidation, *Chem. Commun.* 2563–2565.

[19]   Saluzzo, C.; Halle, R. te.; Touchard, F.; Fache, F.; Schulz, E.; Lemaire, M., (2000) Recent progress in asymmetric heterogeneous catalysis: use of polymer-supported catalysts, *J. Organomet. Chem. 603,* 30-39.

[20]   Bergbreiter, D. E., (2002) Using soluble polymers to recover catalysts and ligands, *Chem. Rev. 102,* 3345-3384.

[21]   Dickerson, T. J.; Reed, N. N.; Janda, K. D., (2002) Soluble polymers as scaffolds for recoverable catalysts and reagents, *Chem. Rev. 102,* 3325-3344.

[22]   Brinker, C. J.; Scherer, G. W., (1990) *Sol-gel Science, the Physics and Chemistry of Sol-gel Processing*, London, Academic Press.

[23] Thomas, J. M.; Ramdas, S.; Millward, G. R.; Klinowski, J.; Audier, M.; Conzalec-Calbet, J.; Fyfe, C. A., (1982) Surprises in the structural chemistry of zeolites, *J. Solid State Chem. 45*, 368-380.

[24] Corma, A., (1997) From microporous to mesoporous molecular sieve materials and their use in catalysis, *Chem. Rev. 97*, 2373-2420.

[25] Garcia, H.; Roth, H. D., (2002) Generation and Reactions of Organic Radical Cations in Zeolites, *Chem. Rev. 102*, 3947-4008.

[26] Corma, A.; Garcia, H., (2006) Silica-bound homogenous catalysts as recoverable and reusable catalysts in organic synthesis, *Adv. Synth. Catal. 348*, 1391 – 1412.

[27] Yoon, T. P.; Jacobsen, E. N., (2003) Privileged chiral catalysts, *Science 299*, 1691-1693.

[28] Brunel, J. M., (2005) BINOL: A versatile chiral reagent, *Chem. Rev. 105*, 857-898.

[29] Arai, T.; Sasai, H.; Aoe, K.; Okamura, K.; Date, T.; Shibasaki, M., (1996) A new multifunctional heterobimetallic asymmetric catalyst for Michael additions and tandem Michael-aldol reactions, *Angew. Chem. Int. Ed. 35*, 104-106.

[30] Matsunaga, S.; Ohshima, T.; Shibasaki, M., (2002) Linked-BINOL: An approach towards practical asymmetric multifunctional catalysis, *Adv. Synth. Catal. 344*, 3- 15.

[31] Jacobsen, E. N.; (2000) Asymmetric catalysis of epoxide ring-opening reactions, *Acc. Chem. Res. 33*, 421-431.

[32] Jacobsen, E. N.; Zhang, W.; Guler, M. L., (1991) Electronic tuning of asymmetric catalysts, *J. Am. Chem. Soc. 113*, 6703-6704.

[33] Seebach, D.; Beck, A. K.; Heckel, A., (2001) TADDOLs, their derivatives, and TADDOL analogues: versatile chiral auxiliaries, *Angew. Chem. Int. Ed. 40*, 92 – 138.

[34] Seebach, D.; Hayakawa, M.; Sakaki, J.; Schweizer, W. B., (1993) Derivatives of α,α,α',α'-Tetraaryl-2,2-dimethyl-1,3-dioxolan-4,5-dimethanol(TADDOL) conntaining nitrogen, sulfur, and phosphorus atoms. New ligands and auxiliaries for enantioselective reactions, *Tetrahedron 49*, 1711-1724.

[35] Ghosh, A. K.; Mathivanan, P.; Capiello, J., (1998) C_2-symmetric chiral bis(oxazoline)-metal complexes in catalytic asymmetric synthesis, *Tetrahedron Asymmetry 9*, 1-45.

[36] Fache, F.; Schulz, E.; Tommasino, M. L.; Lemaire,M., (2000) Nitrogen-containing ligands for asymmetric homogeneous and heterogeneous catalysis, *Chem. Rev. 100*, 2159-2231.

[37] Evans, D. A.; Johnson, J. S.; Olhava, E. J., (2000) Enantioselective synthesis of dihydropyrans. Catalysis of hetero Diels-Alder reactions by bis(oxazoline) copper(II) complexes, *J. Am. Chem. Soc. 122*, 1635-1649.

[38] Noyori, R.; Kitamura, M. (1991) Enantioselective addition of organometallic reagents to carbonyl compounds: Chirality transfer, multiplication, and amplification, *Angew. Chem. Int. Ed. Engl. 30*, 49-69.

[39] Ager, D. J.; Prahash, I.; Schaad, D. R., (1996) 1,2-Amino alcohols and their heterocyclic derivatives as chiral auxiliaries in asymmetric synthesis, *Chem. Rev. 96*, 835-876.

[40] Carreira, E. M., (1999) *Mukaiyama aldol reaction*. In *Comprehensive Asymmetric Catalysis*, Jacobsen, E. N.; Pfaltz, A.; Yamamoto, H. (Eds.) (Vol III. Pp. 997-1066) Berlin, Springer-Verlag Press.

[41] Machajewski, T. D.; Wong, C. H., (2000) The catalytic asymmetric aldol reaction, *Angew. Chem. Int. Ed. 39*, 1352-1374.

[42] Palomo, C.; Oiarbide, M.; Garcia, J. M., (2004) Current progress in the asymmetric aldol addition reaction, *Chem. Soc. Rev. 33*, 65-75.

[43] Dalko, P. I.; Moisan, L., (2004) In the golden age of organocatalysis, *Angew. Chem. Int. Ed. 43*, 5138 – 5175.

[44] List, B.; Lerner, R.; Barbas III.; C. F., (2000) Proline-catalyzed direct asymmetric Aldol reactions, *J. Am. Chem. Soc. 122*, 2395 – 2396.

[45] Nortrup, A. B.; MacMillan, D. W. C.; (2004) Two-step synthesis of carbohydrates by selective aldol reactions, *Science* 1752 – 1755.

[46] Choudary, B. M.; Kavita, B.; Chowdari, N. S.; Sreedhar, B.; Kantam, M. L.; (2002) Layered double hydroxides containing chiral organic guests: Synthesis,characterization and application for asymmetric C-C bond forming reactions, *Catal. Lett. 78*, 373-377.

[47] Beadham, I.; Dhar, D.; Chandrasekaran, S.; (2003) Proline and benzylpenicillin derivatives grafted into mesoporous MCM-41: Novel organic-inorganic hybrid catalysts for direct aldol reaction, *Proc. Indian Acad. Sci.-Chem. Sci. 115*, 365-372.

[48] Calderòn, F.; Fernandez, R.; Sànchez, F.; Mayoralas, A. F., (2005) Asymmetric aldol reaction using immobilized proline on mesoporous support, *Adv. Synth. Catal. 347*, 1395 –1403.

[49] Doyagüez, E. G.; Calderòn, F.; Sànchez, F, Mayoralas, A. F., (2007) Asymmetric aldol reaction catalyzed by a heterogenized proline on a mesoporous support. The role of the nature of solvents, *J. Org. Chem. 72*, 9353-9356.

[50] An, Z.; Zhang, W.; Shi, H.; He, J.; (2006) An effective heterogeneous L-proline catalyst for the asymmetric aldol reaction using anionic clays as intercalated support, *J. Catal. 241*, 319–327.

[51] Lin, Z.; Jianliang, X.; Li C.; (2007) Direct asymmetric aldol reactions on heterogeneous bifunctional catalyst, *Chinese J. Catal. 28*, 673–675.

[52] Gruttadauria, M.; Riela, S.; Aprile, C.; Meo P. L.; D'Anna F.; Noto, R.; (2006) Supported ionic liquids. New recyclable materials for the L-proline-catalyzed Aldol reaction, *Adv. Synth. Catal. 348*, 82–92.

[53] Aprile, C.; Giacalone, F.; Gruttadauria, M.; Marculescu, A. M.; Noto R.; Revellc, J. D.; Wennemers, H.; (2007) New ionic liquid-modified silica gels as recyclable materials for L-proline or H–Pro–Pro–Asp–NH$_2$-catalyzed aldol reaction, *Green Chem. 9*, 1328–1334.

[54] Luo, S.; Li, J.; Xu, H.; Zhang, L.; Cheng, J.-P.; (2007) Chiral amine-polyoxometalate hybrids as highly efficient and recoverable asymmetric enamine catalysts, *Org. lett. 9*, 3675-3678.

[55] Kozhevnikov, I. V., (2002) *Catalysis by polyoxometalates*, England, Wiley: Chichester.

[56] Giménez, R.; Swager, T. M.; (2001) Dinuclear pincer-palladium(II) complexes and their use as homogeneous or heterogeneous catalyst for the aldol reaction of methyl isocyanoacetate, *J. Mol. Catal. A: Chem. 166*, 265–273.

[57] Yamamoto, H., (1999) *Lewis acids in organic synthesis*, Wiley-VCH: Weinheim.

[58] Nelson, S. G.; (1998) Catalyzed enantioselective aldol additions of latent enolate equivalents *Tetrahedron: Asymmetry 9*, 357-389.

[59] Denmark, S. E.; Stavenger, R.; (2000) Asymmetric catalysis of aldol reactions with chiral Lewis bases, *Acc. Chem. Res. 33*, 432-440.

[60] Sodeoka, M.; Ohrai, K.; Shibasaki, M.; (1995) Catalytic asymmetric aldol reaction *via* chiral Pd(II) enolate in wet DMF, *J. Org. Chem. 60*, 2648-2649.

[61] Fraile, J. M.; Pèrez, I.; Mayoral, J. A.; (2007) Comparison of immobilized Box and azaBox–Cu(II) complexes as catalysts for enantioselective Mukaiyama aldol reactions, *J. Catal. 252*, 303–311.

[62] Doherty, S.; Goodrich, P.; Hardacre, C.; Pârvulescu, V.; Paun, C.; (2008) Efficient heterogeneous asymmetric catalysis of the Mukaiyama Aldol reaction by silica- and ionic liquid-supported Lewis acid Copper(II) complexes of bis(oxazolines), *Adv. Synth. Catal, 350*, 295 – 302.

[63] Yamada, Y. M. A.; Yoshikawa, N.; Sasai, H.; Shibasaki, M., (1997) Direct catalytic asymmetric Aldol reactions of aldehydes with unmodified ketones, *Angew. Chem. Int. Ed. 36,* 1871-1873.

[64] Boruwa, J.; Gogoi, N.; Saikia, P. P.; Barua, N. C., (2006) Catalytic asymmetric Henry reaction, *Tetrahedron: Asymmetry*, *17,* 3315–3326.

[65] Watanabe, M.; Murata, K.; Ikariya, T., (2002) Practical synthesis of optically active amino alcohols *via* asymmetric transfer hydrogenation of functionalized aromatic ketones *J. Org. Chem. 67*, 1712-1715.

[66] Narayana, C.; Reddy, N. K.; Kabalka, G. W., (1992) Sodium percarbonate: A mild reagent for conversion of tosylhydrazones and nitroalkanes to carbonyl compounds, *Synth. Commun. 22*, 2587–2592.

[67] Matt, C.; Wagner, A.; Mioskowski, C., (1997) Novel transformation of primary nitroalkanes and primary alkyl bromides to the corresponding carboxylic acids, *J. Org. Chem. 62*, 234–235.

[68] Berner, O. M.; Tedeschi, L.; Enders, D., (2002) Asymmetric Michael additions to nitroalkenes, *Eur. J. Org. Chem.* 1877–1894.

[69] Tamura, R.; Kamimura, A.; Ono, N., (1991) Displacement of aliphatic nitro groups by carbon and heteroatom nucleophiles, *Synthesis*, 423–434.

[70] Sasai, H.; Suzuki, T.; Arai, S.; Arai, T.; Shibasaki, M., (1992) Basic character of rare earth metal alkoxides. Utilization in catalytic carbon-carbon bond-forming reactions and catalytic asymmetric nitroaldol reactions, *J. Am. Chem. Soc. 114,* 4418–4420.

[71] Shibasaki, M.; Yoshikawa, N., (2002) Lanthanide complexes in multifunctional asymmetric catalysis, *Chem. Rev. 102*, 2187–2209.

[72] Palomo, C.; Oiarbide, M.; Laso, A., (2007) Recent advances in the catalytic asymmetric nitroaldol (Henry) reaction, *Eur. J. Org. Chem.* 2561–2574.

[73] Choudary, B. M.; Ranganath, K. V. S.; Pal, U.; Kantam, M. L.; Sreedhar, B. (2005) Nanocrystalline MgO for asymmetric henry and michael reactions, *J. Am. Chem. Soc, 127*, 13167-13171.

[74] Bhatt, A. P.; Pathak, K.; Jasra, R. V.; Kureshy, R. I.; Khan, N. H.; Abdi, S. H. R., (2006) Chiral lanthanum–lithium–binaphthol complex covalently bonded to silica and MCM-41 for enantioselective nitroaldol (Henry) reaction, *J. Mol. Catal. A: Chem. 244*, 110–117.

[75] Kobayashi, S.; Ishitani, H., (1999) Catalytic enantioselective addition to imines, *Chem. Rev. 99,* 1069-1094.

[76] Anderson, J. C.; Peace, S.; Pih, S., (2000) The Lewis acid catalysed addition of 1-trimethylsilyl nitropropanate to imines, *Synlett.* 850-852.

[77] Adams, H.; Anderson, J. C.; Peace, S.; Pennell, A. M. K., (1998) The nitro-mannich reaction and its application to the stereoselective synthesis of 1,2-diamines, *J. Org. Chem. 63*, 9932 – 9934.

[78] Westermann, B., (2003) Asymmetric catalytic Aza-Henry reactions leading to 1,2-diamines and 1,2-diaminocarboxylic acids, *Angew. Chem. Int. Ed. 42*, 151-153.

[79] Ballinim, R.; Petrini, M., (2004) Recent synthetic developments in the nitro to carbonyl conversion (Nef reaction), *Tetrahedron 60*, 1017 – 1047.

[80] Yoon, T. P.; Jacobsen, E. N., (2005) Highly enantioselective thiourea-catalyzed Nitro-Mannich reactions, *Angew. Chem. Int. Ed. 44*, 466 –468.

[81] Nishiwaki, N.; Kundsen, K. R.; Gothelf, K.; Jørgensen, K. A. (2001) Catalytic enantioselective addition of nitro compounds to imines - A simple approach for the synthesis of optically active β-nitro-α-amino esters, *Angew. Chem. Int. Ed. 40*, 2992-2995.

[82] Handa, S.; Gnanadesikan, V.; Matsunaga, S.; Shibasaki, M., (2007) *syn*-Selective catalytic asymmetric Nitro-Mannich reactions using a heterobimetallic Cu-Sm-schiff base complex, *J. Am. Chem. Soc. 129*, 4900-4901.

[83] Lee, A.; Kim, W.; Lee, J.; Hyeon, T.; Kim, B. M., (2004) Heterogeneous asymmetric nitro-Mannich reaction using a bis(oxazoline) ligand grafted on mesoporous silica, *Tetrahedron: Asymmetry 15*, 2595–2598.

[84] List, B., (2004) Enamine catalysis is a powerful strategy for the catalytic generation and use of carbanion equivalents, *Acc. Chem. Res. 37*, 548-557.

[85] Dalko, P. I.; Moisan, L., (2004) In the golden age of organocatalysis, *Angew. Chem. Int. Ed. 43*, 5138 – 5175.

[86] Notz, W.; Tanaka, F.; Barbas, C. F. III., (2004) Enamine-based organocatalysis with proline and diamines: The development of direct catalytic asymmetric Aldol, Mannich, Michael, and Diels-Alder reactions, *Acc. Chem. Res. 37*, 580-591.

[87] Corma, A.; Iborra, S.; Rodrìguez, I.; Iglesias, M.; Sànchez, F., (2002) Heterogenized chiral amines as base catalysts for enantioselctive Michael reaction, *Catal. lett. 82*, 237-242.

[88] Vijaikumar, S.; Dhakshinamoorthy, A.; Pitchumani, K., (2008) L-proline anchored hydrotalcite clays: An efficient catalyst for asymmetric Michael addition, *Applied Catal. A: 340*, 25-32.

[89] Kagan, H. B.; Riant, O., (1992) Catalytic asymmetric Diels-Alder reactions, *Chem. Rev. 92*, 1007-1019.

[90] Nicolaou, K. C.; Snyder, S. A.; Montagnon, T.; Vassilikogiannakis, G., (2002) The Diels-Alder reaction in total synthesis, *Angew. Chem. Int. Ed. 41*, 1668-1698.

[91] Takao, K.; Munakata, R.; Tadano, K.; (2005) Recent advances in natural product synthesis by using intramolecular Diels-Alder reactions, *Chem. Rev. 105*, 4779-4807.

[92] Evans, D. A.; Miller, S. J.; Lectka, T.; von Matt, P., (1999) Chiral bis(oxazoline)copper(II) complexes as Lewis acid catalysts for the enantioselective Diels-Alder reaction, *J. Am. Chem. Soc. 121*, 7559-7573.

[93] Corey, E. J.; Imai, N.; Zhang, H.-Y., (1991) Designed catalyst for enantioselective Diels-Alder addition from a C2-symmetric chiral bis(oxazoline)-iron(III) complex *J. Am. Chem. Soc. 113*, 728-729.

[94] Rechavi, D.; Lemaire, M., (2002) Enantioselective catalysis using heterogeneous bis(oxazoline) ligands: Which factors influence the enantioselectivity?, *Chem. Rev. 102*, 3467-3494.

[95] Fraile, J. M.; Garcia, J. I.; Mayoral, J. A., (2008) Recent advances in the immobilization of chiral catalysts containing bis(oxazolines) and related ligands, *Coord. Chem. Rev. 252*, 624–646.

[96] Fraile, J. M.; Garcia, J. I.; Mayoral, J. A.; Royo, A. J., (1996) Chiral Lewis acids supported on silica gel and alumina, and their use as catalysts in Diels-Alder reactions of methacrolein and bromoaerolein, *Tetrahedron: Asymmetry 7*, 2263-2276.

[97] Rechavi, D.; Lemaire, M., (2001) Heterogenization of a chiral bis(oxazoline) catalyst by grafting onto silica, *Org. Lett. 3*, 2493-2496.

[98] Rechavi, D.; Lemaire, M., (2002) Enantioselective catalysis of Diels–Alder reactions by heterogeneous chiral bis(oxazoline) catalysts, *J. Mol. Catal. A: 182*, 239-247.

[99] Rechavi, D.; Albela, B.; Bonneviot, L.; Lemaire, M., (2005) Understanding the enantioselectivity of a heterogeneous catalyst: the influence of ligand loading and of silica passivation, *Tetrahedron 61*, 6976-6981.

[100] Park, J. K.; Kim, S. W.; Hyeon, T.; Kim, B. M., (2001) Heterogeneous asymmetric Diels–Alder reactions using a copper–chiral bis(oxazoline) complex immobilized on mesoporous silica, *Tetrahedron: Asymmetry 12*, 2931-2935.

[101] Lancaster, T. M.; Lee, S. S.; Ying, J. Y., (2005) Effect of surface modification on the reactivity of MCF-supported IndaBOX, *Chem. Commun.*, 3577-3579.

[102] Tada, M.; Tanaka, S.; Iwasawa, Y., (2005) Enantioselective Diels–Alder reaction Promoted by achiral functionalization of a SiO$_2$-supported Cu–BOX [bis(oxazoline)] catalyst, *Chem. Lett. 34*, 1362-1363.

[103] Tada, M.; Iwasawa, Y., (2006) Advanced chemical design with supported metal complexes for selective Catalysis, *Chem. Commun.*, 2833–2844.

[104] Tanaka, S.; Tada, M.; Iwasawa, Y., (2007) Enantioselectivity promotion by achiral surface functionalization on SiO$_2$-supported Cu-bis(oxazoline) catalysts for asymmetric Diels–Alder reactions, *J. Catal. 245*, 173-183.

[105] Heckel, A.; Seebach, D., (2000) Immobilization of TADDOL with a high degree of loading on porous silica gel and first applications in enantioselective catalysis, *Angew. Chem. Int. Ed. 39*, 163-165.

[106] Selkälä, S. A.; Tois, J.; Pihko, P. M.; Koskinen, A. M. P., (2002) Asymmetric organocatalytic Diels-Alder reactions on solid support, *Adv. Synth. Catal. 344,* 941-945.

[107] Zhang, Y.; Zhao, L.; Lee, S. S.; Ying, J. Y., (2006) Enantioselective catalysis over chiral imidazolidin-4-one immobilized on siliceous and polymer-coated mesocellular foams, *Adv. Synth. Catal. 348*, 2027-2032.

[108] Fraile, J. M.; Garcìa, J. I.; Harmer, M. A.; Herrerìas, C. I.; Mayoral, J. A. (2001) Bis(oxazoline)-metal complexes immobilised by electrostatic interactions as heterogeneous catalysts for enantioselective Diels–Alder reactions, *J. Mol. Catal. A: Chem. 165*, 211-218.

[109] O'leary, P.; Krosveld, N. P.; De Jong, K. P.; van Koten, G.; Klein Gebbink, R. J. M., (2004) Facile and rapid immobilization of copper(II) bis(oxazoline) catalysts on silica: application to Diels–Alder reactions, recycling, and unexpected effects on enantioselectivity, *Tetrahedron Lett. 45,* 3177-3180.

[110] Wang, H.; Liu, X.; Xia, H.; Liu, P.; Gao, J.; Ying, P.; Xiao, J.; Li, C., (2006) Asymmetric Diels–Alder reactions with hydrogen bonding heterogeneous catalysts and mechanistic studies on the reversal of enantioselectivity, *Tetrahedron 62,* 1025-1032.

[111] Hong, W.; Jian, L.; Peng, L.; Qihua, Y.; Jianliang, X.; Li, C., (2006) Asymmetric Diels-Alder reactions on supported bis(oxazoline) catalysts, *Chinese J. Catal. 27,* 946–948.

[112] Tietze, L. F.; Kettschau, G.; Gewart, J. A.; Schuffenhauer, A. (1998) Hetero-Diels-Alder reactions of 1-oxa-1,3-butadienes, *Curr. Org. Chem. 2,* 19-62.

[113] Tietze, L. F.; Kettschau, G., (1997) Hetero Diels-Alder reactions in organic chemistry, *Top. Curr. Chem.189,* 1-120.

[114] Westley, J. W.; Evans, R. H.; Williams, T.; Stempel, A., (1970) Biosynthesis of antibiotic X- 537A, *Chem. Commun.* 1467-1468.

[115] Gerstberger, G.; Anwander, R., (2001) Screening of rare earth metal grafted MCM-41 silica for asymmetric catalysis, *Microporous Mesoporous Mater. 44-45,* 303-310.

[116] Heckel, A.; Seebach, D., (2002) Enantioselective heterogeneous epoxidation and Hetero-Diels-Alder reaction with Mn- and Cr-salen complexes immobilized on silica gel by radical grafting, *Helvetica chimica acta 85,* 913-926.

[117] Wan, Y.; McMorn, P.; Hancock, F. E.; Hutchings, G. J.; (2003) Heterogeneous enantioselective synthesis of a dihydropyran using Cu-exchanged microporous and mesoporous materials modified by bis(oxazoline), *Catal. Lett. 91,* 145-148.

[118] Lebel, H.; Marcoux, J. F.; Molinaro, C.; Charette, A. B., (2003) Stereoselective cyclopropanation reactions, *Chem. Rev.* 103, 977-1050.

[119] Reissig, H. U., (1996) Recent developments in the enantioselective syntheses of cyclopropanes, *Angew. Chem., Int. Ed. Engl.* 35, 971-973.

[120] Doyle, M. P., (1993) *Catalytic asymmetric cyclopropanations.* Ojima, I., (Ed.) In *Catalytic Asymmetric Synthesis* (pp 63-99). New York, VCH.

[121] Salaun, J., (1989) Optically active cyclopropanes, *Chem. Rev.* 89, 1247-1270.

[122] Pfaltz, A., (1999) *Cyclopropanation and C-H insertion with Cu.* In Jacobsen, E. N., Pfaltz, A., Yamamoto, H., (Eds.) *Comprehensive Asymmetric Catalysis.* (Vol. 2, pp. 514-538), Berlin, Springer-Verlag.

[123] Lydon, K. M., McKervey, M. A., (1999) *Cyclopropanation and C-H insertion with Rh.* In Jacobsen, E. N., Pfaltz, A., Yamamoto, H., (Eds.) *Comprehensive Asymmetric Catalysis.* (Vol. 2, pp. 539-580), Berlin, Springer- Verlag.

[124] Matlin, S. A.; Lough, W. J.; Chan, L.; Abram, D. M. H.; Zhou, Z., (1984) Asymmetric induction in cyclopropanation with homogeneous and immobilized chiral metal β-diketonate catalysts, *J. Chem. Soc. Chem. Commun.* 1038-1039.

[125] Carmona, A.; Corma, A.; Iglesias, M.; Sànchez, F.; (1996) Synthesis and characterisation of chiral Cu(I) complexes with substituted-pyrrolidine-ligands bearing a triethoxysilyl group and preparation of heterogenised catalysts on USY-zeolites, *Inorg. Chim. Acta* 244, 79-85.

[126] Alcon, M. J.; Corma, A.; Iglesias, M.; Sànchez, F., (1999) Cyclopropanation reactions catalysed by copper and rhodium complexes homogeneous and heterogenised on a modified USY-zeolite. Influence of the catalyst on the catalytic profile, *J. Mol. Catal. A: Chem. 144,* 337-346.

[127] Desimoni, G.; Faita, G.; Jørgensen, K. A., (2006) C_2-symmetric chiral bis(oxazoline) ligands in asymmetric catalysis, *Chem. Rev. 106,* 3561-3651.

[128] Fraile, J. M.; Garcìa, J. I.; Mayoral, J. A.; Tarnai, T., (1997) Asymmetric cyclopropanation catalysed by cationic bis(oxazoline)-Cu[II] complexes exchanged into clays, *Tetrahedron: Asymmetry 8,* 2089-2092.

[129] Fraile, J. M.; Garcìa, J. I..; Mayoral, J. A.; Tarnai, T.; (1998) Clay-supported bis(Oxazoline)-copper complexes as heterogeneous catalysts of enantioselective cyclopropanation reactions, *Tetrahedron: Asymmetry 9*, 3997-4008.

[130] Alonso, P. J.; Fraile, J. M.; Garcìa, J.; Garcìa, J. I.; Martinez, J. I.; Mayoral, J. A.; Sànchez, M. C., (2000) Spectroscopic study of the structure of bis(oxazoline)copper complexes in solution and immobilized on Laponite clay. Influence of the structure on the catalytic performance, *Langmuir 16*, 5607-5612.

[131] Fraile, J. M.; Garcìa, J. I.; Mayoral, J. A.; Tarnai, T.; Harmer, M. A., (1999) Bis(oxazoline)–copper complexes, supported by electrostatic interactions, as heterogeneous catalysts for enantioselective cyclopropanation reactions: Influence of the anionic support, *J. Catal.*, *186*, 214-221.

[132] Fraile, J. M.; Garcìa, J. I.; Herrerìas,C. I.; Mayoral, J. A.; Reiser, O.; Socuellamos, A.; Werner, H. (2004) The role of binding constants in the efficiency of chiral catalysts immobilized by electrostatic interactions: The case of azabis(oxazoline)-copper complexes, *Chem. Eur. J. 10*, 2997-3005.

[133] Burguete, M. I.; Fraile, J. M.; Garcìa, J. I.; Garcìa-Verdugo, E.; Herrerìas, C. I.; Luis, S.V.; Mayoral, J.A., (2001) Bis(oxazoline)copper complexes covalently bonded to insoluble support as catalysts in cyclopropanation reactions, *J. Org. Chem. 66*, 8893-8901.

[134] Clarke, R. J.; Shannon, I. J.; (2001) Mesopore immobilised copper bis(oxazoline) complexes for enantioselective catalysis, *Chem. Commun.* 1936-1937.

[135] Lee, S. S.; Ying, J. Y., (2006) Siliceous mesocellular foam-supported chiral bisoxazoline: Application to asymmetric cyclopropanation, *J. Mol. Catal. A: Chem. 256*, 219-214.

[136] Lee, S. S.; Hadinoto, S.; Ying, J. Y., (2006) Improved enantioselectivity of immobilized chiral bisoxazolines by partial precapping of the siliceous mesocellular foam support with trimethylsilyl groups, *Adv. Synth. Catal. 348*, 1248-1254.

[137] Fraile, J. M.; Garcìa, J. I.; Herrerìas, C. I.; Mayoral, J. A., (2005) The first synthesis of organic–inorganic hybrid materials with chiral bis(oxazoline) ligands, *Chem. Commun.* 4669-4671.

[138] Van Leeuwen, P. W. N. M.; Claver, C. (2000) *Rhodium Catalyzed Hydroformylation*, Dordrecht, Kluwer Academic.

[139] Beller, M.; Cornils, B.; Frohning, C. D.; Kohlpainter, C. W. (1995) Progress in hydroformylation and carbonylation, *J. Mol. Catal.A: Chem. 104*, 17-85.

[140] Botteghi, C.; Paganelli, S.; Schinato, A.; Marchetti, M. (1991) The asymmetric hydroformylation in the synthesis of pharmaceuticals, *Chirality 3*, 355-369.

[141] Gladiali, S.; Bayon, J. C.; Claver, C., (1995) Recent advances in enantioselective hydroformylation, *Tetrahedron: Asymmetry 6*, 1453-1474.

[142] Beller, M.; Kumar, K., (2004) *Hydroformylation: Applications in the synthesis of pharmaceuticals and fine chemicals.* In Beller, M; Bolm, C. (Eds.) *Transition metals for organic synthesis* (2nd Ed.,Vol. 1, Chapter 2.1, pp. 29-51). Weinheim, Wiley-VCH.

[143] Rouhi, A. M., (2004) Chiral chemistry, *Chem. Eng. News 82*, 47-62.

[144] Fuerte, A.; Iglesias, M.; Sànchez, F., (1999) New chiral diphosphinites: synthesis of Rh complexes. Heterogenisation on zeolites, *J. Organomet. Chem. 588*, 186–194.

[145] Han, D.; Li, X.; Zhang, H.; Liu, Z.; Li, J.; Li, C., (2006) Heterogeneous asymmetric hydroformylation of olefins on chirally modified Rh/SiO$_2$ catalysts, *J. Catal.* 243, 318–328.

[146] Han, D.; Li, X.; Zhang, H.; Liu, Z.; Hu, G.; Li, C., (2008) Asymmetric hydroformylation of olefins catalyzed by rhodium nanoparticles chirally stabilized with (*R*)-BINAP ligand, *J. Mol. Cata. A: Chem. 283*, 15–22.

[147] Dias, L. C., (2000) Chiral Lewis acid catalyzed Ene-reactions, *Curr. Org. Chem. 4*, 305-342.

[148] Mikami, K.; Terada, K. (1999) In Jacobsen, E. N.; Pfaltz, A.; Yamamoto, H. (Eds.) *Comprehensive Asymmetric Catalysis III*, (Vol. III, pp. 1143-1174.) Springer, Berlin.

[149] Maruoka, K.; Hoshino, Y.; Shirasaka, T.; Yamamoto, H., (1988) Asymmetric ene reaction catalyzed by chiral organoaluminum reagent, *TetrahedronLett. 29*, 3967-3970.

[150] Mikami, K.; Shimizu, M., (1992) Asymmetric ene reactions in organic synthesis, *Chem. Rev. 92*, 1021-1050.

[151] Mikami, K.; Terada, M.; Nakai, T., (1990) Catalytic asymmetric glyoxylate-ene reaction: a practical access to alpha.-hydroxy esters in high enantiomeric purities, *J. Am. Chem. Soc. 112*, 3949-3954.

[152] Caplan, N.A.; Hancock, F. E.; Bulman, P. P. C.; Hutchings, G. J., (2004) Heterogeneous enantioselective catalyzed carbonyl- and imino-ene reactions using copper bis(Oxazoline) zeolite Y, *Angew. Chem. Int. Ed. 43*, 1685-1688.

[153] Ono, F.; Kanemasa, S.; Tanaka, J., (2005) Reusable nano-sized chiral bisoxazoline catalysts, *Tetrahedron Lett. 46*, 7623-7626.

[154] Knochel, P.; Jones, P., (1999) *Organozinc reagents: A practical approach*, New York, Oxford Press.

[155] Edrick, E., (1996) *Organozinc reagents in organic synthesis*, New York, CRC Press.

[156] Laumen, K.; Breitgoff, D.; Schneider, M. P., (1988) Enzymic preparation of enantiomerically pure secondary alcohols. Ester synthesis by irreversible acyl transfer using a highly selective ester hydrolase from *Pseudomonas sp.*; an attractive alternative to ester hydrolysis, *J. Chem. Soc. Chem. Commun.* 1459-1461.

[157] Pu, L.; Ye, H. B., (2001) Catalytic asymmetric organozinc additions to carbonyl compounds, *Chem. Rev. 101*, 757-824.

[158] Soai, K.; Watanabe, M.; Yammoto, A., (1990) Enantioselective addition of dialkylzincs to aldehydes using heterogeneous chiral catalysts immobilized on alumina and silica-gel, *J. Org. Chem. 55*, 4832-4835.

[159] Laspèras, M.; Bellocq, N.; Brunel, D.; Moreau, P., (1998) Chiral mesoporous templated silicas as heterogeneous inorganic-organic catalysts in the enantioselective alkylation of benzaldehydes, *Tetrahedron: Asymmetry 9*, 3053-3064.

[160] Bellocq, N.; Abramson, S.; Laspèras, M.; Brunel, D.; Moreau, P., (1999) Factors affecting the efficiency of hybrid chiral mesoporous silicas used as heterogeneous inorganic–organic catalysts in the enantioselective alkylation of benzaldehyde, *Tetrahedron: Asymmetry 10*, 3229-3241.

[161] Abramson, S.; Laspèras, M.; Galamear, A.; Desplantier-Giscard, D.; Brun, D., (2000) Best design of heterogenized b-aminoalcohols for improvement of enantioselective addition of diethylzinc to benzaldehyde, *Chem. Commun.* 1773-1774.

[162] Abramson, S.; Laspèras, M.; Chiche, B., (2001) Direct immobilization of chiral auxiliaries on mineral supports and heterogeneous enantioselective alkylation of benzaldehyde with diethylzinc, *J. Mol. Catal. A: Chem. 165*, 231–242.

[163] Abramson, S.; Laspèras, M.; Brunel, D., (2002) Design of mesoporous aluminosilicates supported (1*R*,2*S*)-(−): evidence for the main factors influencing catalytic activity in the enantioselective alkylation of benzaldehyde with diethylzinc, *Tetrahedron: Asymmetry 13*, 357–367.

[164] Bae, S. J.; Kim, S. W.; Hyeon, T.; Kim, B., (2000) New chiral heterogeneous catalysts based on mesoporous silica: asymmetric diethylzinc addition to benzaldehyde, *Chem. Commun.* 31-32.

[165] Fraile, J. M., Mayoral, J. A., Serrano, J.; Pericas, M. A.; Sola, L.; Castellnou, D., (2003) New silica-immobilized chiral amino alcohol for the enantioselective addition of diethylzinc to benzaldehyde, *Org. Lett. 5,* 4333-4335.

[166] Mehler, T.; Martens, J., (1994) New thioether derivatives as catalysts for the enantioselective addition of diethylzinc to benzaldehyde, *Tetrahedron: Asymmetry 5,* 207-210.

[167] Chung, Y.-M.; Rhee, H.-K., (2002) Dendritic chiral auxiliaries on silica: a new heterogeneous catalyst for enantioselective addition of diethylzinc to benzaldehyde, *Chem. Commun.* 238–239.

[168] Sato, I.; Kadowaki, K.; Urabe, H.; Jung, J. H.; Ono, Y.; Shinkai, S.; Soai, K., (2003) Highly enantioselective synthesis of organic compound using right- and left-handed helical silica, *Tetrahedron Lett. 44*, 721–724.

[169] Kawasaki, T.; Ishikawa, K.; Sekibata, H.; Sato, I.; Soai, K., (2004) Enantioselective synthesis induced by chiral organic–inorganic hybrid silsesquioxane in conjunction with asymmetric autocatalysis, *Tetrahedron Lett. 45*, 7939–7941.

[170] Huang, L-N.; Hui, X.-P.; Xu, P.-F., (2006) Asymmetric addition of diethylzinc to benzaldehyde catalyzed by silica-immobilized titanium(IV) complexes of *N*-sulfonylated amino alcohols, *J. Mol. Catal. A: Chem. 258*, 216–220.

[171] Pathak, K.; Bhatt, A. P.; Abdi, S. H. R.; Kureshy, R. I.; Khan, N. H.; Ahmad, I.; Jasra, R. V., (2006) Enantioselective addition of diethylzinc to aldehydes using immobilized chiral BINOL–Ti complex on ordered mesoporous silicas, *Tetrahedron: Asymmetry 17*, 1506–1513.

[172] Pathak, K.; Ahmad, I.; Abdi, S. H. R.; Kureshy, R. I.; Khan, N. H.; Jasra, R. V., (2008) The synthesis of silica-supported chiral BINOL: Application in Ti-catalyzed asymmetric addition of diethylzinc to aldehydes *J. Mol. Catal. A: Chem. 280*, 106–114.

[173] Mastalir, À.; Kiràly, Z., (2008) Catalytic investigation of (1*R*,2*S*)-(-)-N-dodecyl-N-methylephedrinium bromide immobilized on montmorillonite in the enantioselective alkylation of benzaldehyde with diethylzinc, *Catal. Commun. 9,* 1404–1409.

[174] Wu, X.; Ma, X.; Ji, Y.; Wang, Q.; Jia, X.; Fu, X., (2007) Synthesis and characterization of a novel type of self-assembled chiral zirconium phosphonates and its application for heterogeneous asymmetric catalysis, *J. Mol. Catal. A: Chem. 265*, 316–322.

[175] Ji, Y.; Ma, X.; Wu, X.; Wang, N.; Wang, Q.; Zhou, X., (2007) Novel type of chiral titanium phosphonates and hybrid titanium phosphonates for heterogeneous asymmetric catalysis, *Catal. Lett. 118,* 187-194.

[176] Pu, L., (2003) Asymmetric alkynylzinc additions to aldehydes and ketones, *Tetrahedron 58*, 9873-9886.

[177] Cozzi, P. G.; Hilgraf, R.; Zimmermann, N., (2004) Acetylenes in catalysis: Enantioselective additions to carbonyl groups and imines and applications beyond, *Eur. J. Org. Chem.* 4095-4105.

[178] Lu, G.; Li, Y. M.; Li, X. S.; Chan, A. S. C., (2005) Synthesis and application of new chiral catalysts for asymmetric alkynylation reactions, *Coord. Chem. Rev. 249,* 1736-1744.

[179] Huang, L. N.; Hui, X.-P.; Chen, Z.-Ce.; Yin, C.; Xu, P.-F.; Yu, X.-X.; Cheng, S.-Y., (2007) Enantioselective addition of phenylacetylene to aldehydes catalyzed by silica-immobilized titanium(IV) complex of β-hydroxyamide, *J. Mol. Catal. A: Chem. 275,* 9–13.

[180] Brunel, J. M.; Holmes, I. P., (2004) Chemically catalyzed asymmetric cyanohydrin syntheses, *Angew. Chem. Int. Ed. 43,* 2752-2778.

[181] North, M., (2003) Synthesis and applications of non-racemic cyanohydrins, *Tetrahedron: Asymmetry 14,* 147-176.

[182] Gregory, R. J. H., (1999) Cyanohydrins in nature and the laboratory: Biology, preparations, and synthetic applications, *Chem. Rev.* 99, 3649-3682.

[183] Khan, N. H.; Kureshy, R. I.; Abdi, S. H.R.; Agrawal, S.; Jasra, R. V., (2008) Metal catalyzed asymmetric cyanation reactions, *Corrd. Chem. Rev. 252,* 593–623.

[184] Baleizão, C.; Gigante, B.; Das, D.; Alvaro, M.; Garcìa, H.; Corma, A., (2003) Synthesis and catalytic activity of a chiral periodic mesoporous organosilica (ChiMO) *Chem. Commun.* 1860-1861.

[185] Baleizão, C.; Gigante, B.; Das, D.; Alvaro, M.; Garcìa, H.; Corma, A. (2004) Periodic mesoporous organosilica incorporating a catalytically active vanadyl schiff base complex in the framework, *J. Catal. 223,* 106-113.

[186] Baleizão, C.; Gigante, B.; Garcìa, H.; Corma, A., (2003) Chiral vanadyl Schiff base complex anchored on silicas as solid enantioselective catalysts for formation of cyanohydrins: optimization of the asymmetric induction by support modification, *J. Catal. 215,* 199-207.

[187] Baleizão, C.; Gigante, B.; Garcìa, H.; Corma, A., (2004) Vanadyl salen complexes covalently anchored to single-wall carbon nanotubes as heterogeneous catalysts for the cyanosilylation of aldehydes, *J. Catal, 221,* 77-84.

[188] Baleizão, C.; Gigante, B.; Garcìa, H.; Corma, A., (2004) Chiral vanadyl salen complex anchored on supports as recoverable catalysts for the enantioselective cyanosilylation of aldehydes. Comparison among silica, single wall carbon nanotube, activated carbon and imidazolium ion as support, *Tetrahedron 60,* 10461–10468.

[189] Kim, J.-H.; Kim, G.-J., (2004) Enantioselectivities of chiral Ti(IV) salen complexes immobilized on MCM-41 in asymmetric trimethylsilylcyanation of benzaldehyde *Catal. Lett. 92,* 123-130.

[190] Roberts, R. M.; Khalaf, A. A., (1984) *Friedel-Crafts alkylation chemistry. A century of discovery,* New York, Dekker.

[191] Heaney, H., (1991) *C-C bond formation/aromatics/chain extension.* In Trost, B. M.; Fleming, I., (Eds.), *Comprehensive Organic Synthesis* (Vol. 2, pp. 733-752) Oxford, Pergamon Press.

[192] Corma, A.; Garcìa, H.; Moussaif, A.; Sabater, M. J.; Zniber, R.; Redouane, A., (2002) Chiral copper(II) bisoxazoline covalently anchored to silica and mesoporous MCM-41

as a heterogeneous catalyst for the enantioselective Friedel–Crafts hydroxyalkylation, *Chem. Commun.* 1058-1059.

[193] Yu, P.; He, J.; Guo, C.; (2008) 9-Thiourea Cinchona alkaloid supported on mesoporous silica as a highly enantioselective, recyclable heterogeneous asymmetric catalyst, *Chem. Commun.* DOI: 10.1039/b800640g.

In: Modern Trends in Macromolecular Chemistry
Editor: Jon N. Lee

ISBN: 978-1-60741-252-6
© 2009 Nova Science Publishers, Inc.

Chapter 7

SMART HYBRID NANOPARTICLES BASED ON STIMULI-RESPONSIVE PEGYLATED NANOGELS CONTAINING METAL NANOPARTICLES FOR BIOMEDICAL APPLICATIONS

Motoi Oishi,[1,2,3] *and Yukio Nagasaki*[* 1,2,3,4,5]

[1]Tsukuba Interdisciplinary Materials Science (TIMS),
University of Tsukuba, Tsukuba, Ibaraki, Japan.
[2]Center for Tsukuba Advanced Research Alliance (TARA),
University of Tsukuba, , Tsukuba, Ibaraki, Japan.
[3]Graduate School of Pure and Applied Sciences,
University of Tsukuba, Tsukuba, Ibaraki, Japan.
[4]Master's School of Medical Sciences, Graduate School of Comprehensive Human
Sciences, University of Tsukuba, Tsukuba, Ibaraki, Japan.
[5]Satellite Laboratory, International Center for Materials Nanoarchitectonics (MANA),
National Institute for Materials Science (NIMS), Tsukuba, Ibaraki, Japan.

ABSTRACT

We described here the preparation, characterization, and application of the smart hybrid nanoparticles based on the stimuli-responsive PEGylated nanogels containing metal nanoparticles such as gold nanoparticles (AuNPs) and platinum nanoparticles (PtNPs). The stimuli-responsive PEGylated nanogels composed of cross-linked poly[2-(*N,N*-diethylamino)ethyl methacrylate] (PEAMA) gel core and poly(ethylene glycol) (PEG) tethered chains bearing a functional group (acetal group or carboxylic acid group) as a platform moiety for the installation of bio-tag were synthesized by emulsion polymerization of vinylbenzyl group-ended heterotelechelic PEG, amine monomer, and cross-linker. The resulting PEGylated nanogels showed unique reversible volume phase transition of the polyamine gel core in response to various stimuli, such as pH, ionic strength, and temperature. Note that cross-linked PEAMA gel core of the PEGylated nanogels acts as not only nanoreactor but also stimuli-responsive nanomatrix to produce and immobilize AuNPs and PtNPs. For instance, one-pot synthesis of pH-responsive

PEGylated nanogels containing AuNPs was successfully carried out through the self-reduction of HAuCl$_4$ without any reducing agents. The PEGylated nanogel containing AuNPs (fluorescence quencher) in the core and fluorescence dye-labeled DEVD peptide at the tethered PEG chain end showed the recovery of pronounced fluorescence signals in response to apoptotic cells. Additionally, the PEGylated nanogels containing PtNPs showed the on-off regulation of the catalytic activity for reactive oxygen species in response to pH. Thus, smart hybrid nanoparticles based on stimuli-responsive PEGylated nanogels containing metal nanoparticles described here can be utilized to novel functional biomedical nanomaterials.

INTRODUCTION

Small clusters of metals such as Pt, Au, Ag, and Cu have received considerable attention and interest in the field of biomedical applications, because of their unique mechanical, electronic, magnetic, optical, and chemical properties different from those of bulk materials and single atom.[1] Such unique properties of metal nanoparticles (MNPs) are known to sensitively depend on the size, size distribution, shape, and chemical environments. Nevertheless, MNPs dispersed by the electrostatic repulsion force of the absorbed ions on their surfaces tend to show coagulation themselves in high ionic strength milieu as well as nonspecific absorption of biological components such as proteins and cells, leading to the significant reduced their performances, especially *in vivo*.[2] Accordingly, one of the most fascinating subjects in biomedical applications of MNPs is to control the size, size distribution, shape, and dispersion stability of the MNPs under physiological conditions.

In this regard, a variety of polymers with metal-ion affinities (polyamines) can been used to sequester metal ions into localized domains that can subsequently be converted into MNPs.[3] The polymer matrix usually serves as nanoreactor to both control particle size and stabilize the MNPs against coagulation. Recently, next generation of MNPs with new properties and functions can have a significant impact on biomedical applications. A major key to the construction of next generation of MNPs is the incorporating of the MNPs into smart polymer matrixes that can be triggered to significant change in the characteristics of matrixes in response to the stimuli such as pH and temperature.[4]

Worth noting in this regard is stimuli-responsive PEGylated nanogels constructed from a cross-linked poly[2-(*N,N*-diethylamino)ethyl methacrylate] (PEAMA) core and biocompatible poly(ethylene glycol) (PEG) chains that possess a carboxylic acid group or acetal group as a platform moiety for the installation of bio-tag.[5] Indeed, the stimuli-responsive PEGylated nanogels showed extremely high dispersion stability under physiological conditions as well as reversible volume phase transition (swelling) in response to the pH, temperature, and ionic strength due to the protonation of the tertiary amino groups in the cross-linked PEAMA core surrounded by the tethered PEG chains. Eventually, the cross-linked PEAMA core of the nanogels acts as both nanoreactor and stimuli-sensitive matrix to produce and stabilize the MNPs. In this chapter, we described the preparation, characterization, and biomedical application of the smart hybrid nanoparticles based on the stimuli-responsive PEGylated nanogels containing metal nanoparticles such as gold nanoparticles (AuNPs) and platinum nanoparticles (PtNPs). The resulting PEGylated nanogels containing AuNPs or PtNPs showed unique functions synchronizing with the reversible volume phase transition of the PEAMA gel core in response to pH and enzyme.

Thus, we believe that the use of stimuli-responsive PEGylated nanogels for the preparation of hybrid MNPs represents a promising strategy for creation of novel functional biomedical nanomaterials.

SYNTHESIS AND CHARACTERIZATION OF STIMULI-RESPONSIVE PEGYLATED NANOGELS

A major step forward in the preparation of stimuli-responsive PEGylated nanogels has been the synthesis of a new type of heterobifunctional PEG macromonomer that posses a 4-vinylbenzyl group at the one end and a reactive functional group (acetal group and carboxylic acid group) at the other end of PEG. The synthesis of polymeric particle via an emulsion polymerization technique is a well-established and widely utilized technique in industrial fields.[6] PEG macromonomers are often used as a stabilizer of monomer droplets during the polymerization.[7] In this case, PEG macromonomers act not only as the stabilizers of the monomer droplets but also as comonomers that are incorporated into the polymeric particle. The compound 2-(N,N-diethylamino)ethyl methacrylate (EAMA) has been chosen as a comonomer because it has a hydrophobic nature in the basic region and changes its hydrophilicity with pH changes. The copolymerization of heterobifunctional PEG macromonomer (CH$_2$=CH-Ph-CH$_2$O-PEG-COOK; M_n=2000: 2k; M_n=4000: 4k; and M_n=8000: 8k) and EAMA was conducted in the presence of KPS (potassium persulfate) as the initiator and 1.0 mol% ethyleneglycol dimethacrylate (EGDMA) as the cross-linker, as shown in Scheme 1. The sizes of the obtained nanogels with PEG-2k chains (PEGylated nanogel-2k) were controllable by the feed ratios of the PEG macromonomer to EAMA, while retaining the low-size distribution factors ($\mu_2/\Gamma^2 < 0.15$), as determined by DLS measurement (Figure. 1). With the increasing PEG macromonomers versus EAMA gives smaller particles because of decrease in surface activity of PEG used. Thus, versions of PEGylated nanogels-2k with diameters in the range of 50 to 680 nm were obtained.

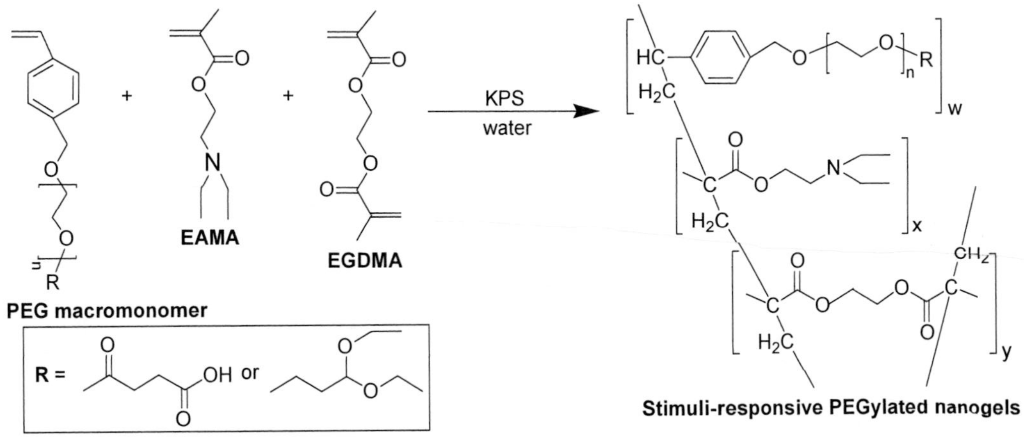

Scheme 1. Synthesis of stimuli-responsive PEGylated nanogels via emulsion polymerization.

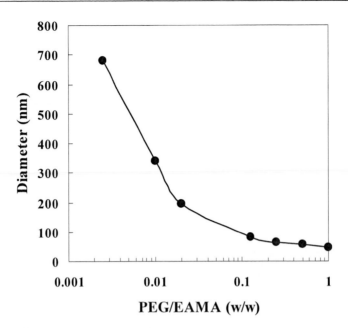

Figure 1. Effects of PEG/EAMA ratio on the size of the stimuli-responsive PEGylated nanogel-2k (DLS measurement: angle 90°; temperature 25 °C).

Figure 2 shows the zeta-potentials of the nanogels with various PEG molecular weights (2k, 4k, and 8k) obtained by changing the environmental pH. The zeta-potential in the acidic region was basically positive due to existence of the protonated amino groups in the core. There was a significant decrease in the zeta-potential at around pH 7, reflecting the pK_a of PEAMA ($pK_a \sim 7.4$).[8] In the acidic region, the zeta-potential of the PEGylated nanogel-2k with short PEG chains was +32 mV. With the increasing molecular weight of PEG on the nanogel surface, the zeta-potential decreased (PEGylated nanogel-4k: +26 mV; PEGylated nanogel-8k: +16 mV), which indicates that the charge-shielding effect of PEG becomes more significant with increasing molecular weight.[9] In contrast, the reverse tendency for the surface charge was observed in the basic region. The cross-linked PEAMA core was expected to be totally deprotonated, and eventually, the zeta-potential of the PEGylated nanogels showed a shift in the negative direction owing to the existence of the caboxylate group at the end of the PEG tethered chains. There was a change in the zeta-potential in the positive direction with the increasing molecular weight of the PEG tethered chain in the basic region. We have reported previously that the average surface area occupied by each PEG molecule is not influenced by the size of the particle obtained by the dispersion polymerization in the presence of PEG macromonomers with the same molecular weight.[10] However, Wu et al.[11] have reported that the density of the tethered PEG chain decreases with increasing molecular weight, which eventually influences the zeta-potentials of the PEGylated nanogels because the carboxyl group locates adjacent to the end of PEG tethered chain.

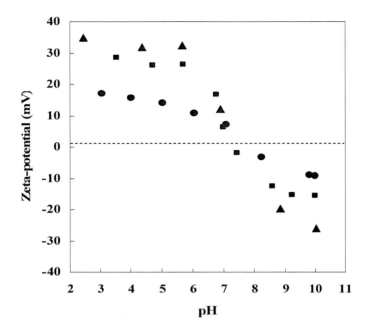

Figure 2. Effects of pH and PEG chain length on the zeta-potentials of the stimuli-responsive PEGylated nanogels with 2k (triangle), 4k (square), or 8k (circle) of PEG chains (laser Doppler electophoresis measurement: temperature 25 ˚C; ionic strength: 0.01 M).

Since the PEAMA gel is known to show volume phase transition as a function of pH,[12] the effects of the environmental pH on the sizes of the PEGylated nanogels with various PEG molecular weights (2k, 4k, and 8k) were assessed (Figure 3). Sharp pH sensitivity was revealed in the pH range of 7.0-7.5 for all PEGylated nanogels with unimodal size distribution ($\mu_2/\Gamma^2 < 0.15$), viz., all of the PEGylated nanogels in the acidic region showed 5.1 ~ 6.8-fold larger hydrodynamic volumes than the PEGylated nanogels in the basic region. This behavior is almost consistent with the α/pH curve of the nanogel. This finding strongly suggests that protonation of the amino groups in the PEAMA core triggers nanogel swelling due to an increase in the ion osmotic pressure together with polymer solvation.[13] Neither coagulation nor precipitation of the PEGylated nanogels was observed over the entire pH change, indicating its high dispersion stability due to existence of PEG tethered chains. The effect of ionic strength on the size of PEGylated nanogel-4k is shown in Figure 4. The size of PEGylated nanogel-4k decreased with increasing ionic strength (I) into the acidic region due to the influence of the osmotic pressure of the PEAMA core of the nanogel. Note that the pH of the volume phase transition point was shifted to the basic region with increasing ionic strength, because the pK_a value of the amino groups in the PEAMA core of the nanogel was also shifted to the basic region (I = 0 M: pK_a = 7.0, I = 0.15 M: pK_a = 7.4). As shown in Figure 5, the size of the nanogel-4k increased proportionally with unimodal size distribution ($\mu_2/\Gamma^2 < 0.2$) with decreasing temperature, reaching a 5.5-fold larger hydrodynamic volume at 20 ˚C (diameter = 136 nm) compared to that at 37 ˚C (diameter = 77 nm). This is most likely due to the change in the pK_a value of the amino groups in the PEAMA core.

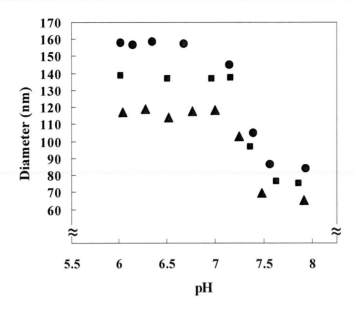

Figure 3. Effects of pH and PEG chain length on the size of the stimuli-responsive PEGylated nanogels with 2k (triangle), 4k (square), or 8k (circle) of PEG chains (DLS measurement: angle 90˚; temperature 30 ˚C; ionic strength: 0.15 M).

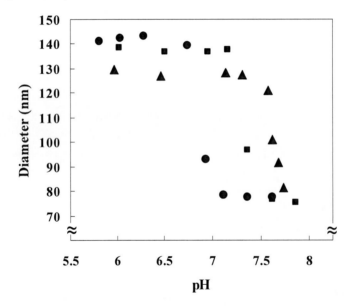

Figure 4. Effect of ionic strength on the size of the stimuli-responsive PEGylated nanogels-4k The ionic strengths are 0.025 M (circle), 0.15 M (square), and 1.5 M (triangle) (DLS measurement: angle 90˚; temperature 30 ˚C; pH: 7.4).

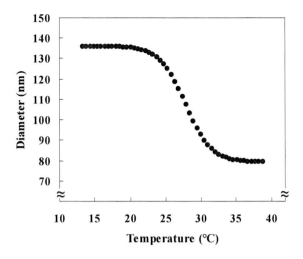

Figure 5. Effects of temperature on the size of the stimuli-responsive PEGylated nanogels-4k (DLS measurement: angle 90°; ionic strength: 0.15 M; pH: 7.4).

PREPARATION OF pH-RESPONSIVE PEGYLATED NANOGELS CONTAINING GOLD NANOPARTICLES

Among MNPs, of particular interest is AuNPs,[14] which show a very intense surface plasmon band (SPB) in the visible spectrum (typically around 520 nm bright-pinkish color). AuNPs are useful for biomedical applications such as biological markers,[15] DNA sensors,[16] immunoassays,[17] drug delivery systems,[18] molecular recognition systems,[19] and efficient fluorescent quenchers in the field of bioimaging.[20] We have recently also reported a novel approach to the preparation of stimuli-responsive PEGylated nanogels containing AuNPs through an ion-exchange process between $AuCl_4^-$ H^+ and the $-N^+Et_2H$ Cl^- in the protonated PEAMA core, followed by the reduction of Au(III) ions to Au(0) particles without any additional reducing agents, viz., tertiary amino groups in the cross-linked PEAMA core play a crucial role in the self-reduction of Au(III) ions as well as in the immobilization of the AuNPs (Figure 6).

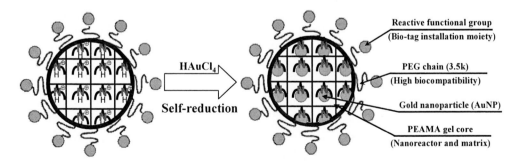

Figure 6. Schematic illustration of the pH-responsive PEGylated nanogel containing AuNPs.

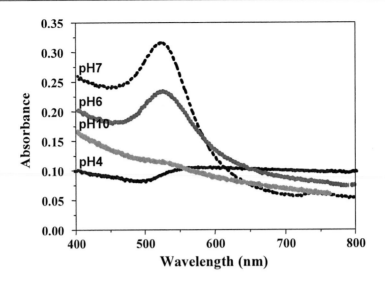

Figure 7. UV-vis spectra of the pH-responsive PEGylated nanogels-3.5k containing AuNPs (N/Au = 8) synthesized at pH 4.0, pH 6.0, pH 7.0, and pH 10.

Figure 7 shows the UV-vis spectra of the reaction solutions of tetrachloroauric acid (HAuCl$_4$) and the pH-responsive PEGylated nanogel-3.5k with 0.1 mol% cross-linking density at an N/Au = 8 (molar ratio of 8 for the tertiary amino group : HAuCl$_4$) under several different pH conditions. The reaction solutions at pH 6.0 and pH 7.0 gradually changed in color from yellow to pinkish-red without any additional reducing agents, and a significant increase in the absorbance at 523 nm attributed to the SPB was also observed, strongly indicating the formation of AuNPs through the self-reduction of HAuCl$_4$.[21] In sharp contrast, neither an increase in the absorbance at SPB nor or color change was not observed in the reaction solution at pH 10.0, whereby the AuCl$_4^-$ complexes changed in AuCl(OH)$_3^-$ and Au(OH)$_4^-$ complexes that are relatively unreducible species. Furthermore, the reaction solution at pH 4.0 showed no absorbance at SPB, while the color of the solution changed to purple, leading to the coagulation and precipitation of the AuNPs. This is presumably due to the low coordination ability of protonated PEAMA core under low pH (high proton concentration) conditions. The transmission electron microscopy (TEM) image of pH-responsive PEGylated nanogel-3.5k containing AuNPs synthesized at pH 6.0 shows that AuNP clusters are clearly evident, whereby the TEM shows not only AuNPs (high contrast) but also the cross-linked PEAMA core of pH-responsive PEGylated nanogel-3.5k (lower contrast; < 100 nm) (Figure 8). The average number of AuNPs in a single nanogel (cluster) and the average diameter of the AuNPs were about 9.6 particles/nanogel and 6 nm, respectively. In addition, the observed SPB of pH-responsive PEGylated nanogel-3.5k containing AuNPs was a somewhat longer wavelength compared to that of commercially available citrate-stabilized AuNPs of the same size (~ 10 nm),[22] which suggests the coordination of the amino groups on the surfaces of the AuNPs. These findings provide direct proof that AuNPs are formed within the PEAMA core of pH-responsive PEGylated nanogel-3.5k through the self-reduction process, i.e., within the "nanoreactors".

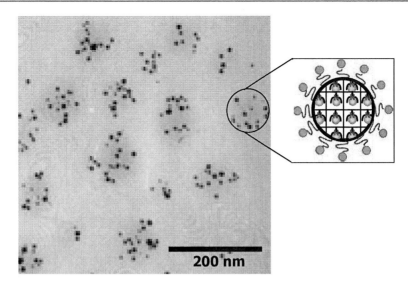

Figure 8. TEM image of pH-responsive PEGylated nanogels-3.5k containing AuNPs synthesized at N/Au = 8 at pH 6.0.

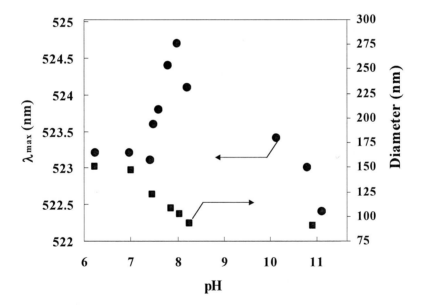

Figure 9. Plots of the variations of the SPB (circle) and size (square) of the pH-responsive PEGylated nanogels-3.5k containing AuNPs (N/Au = 8) at various pH values (DLS measurement: angle, 90˚; temperature, 25 ˚C, ionic strength: 0.15 M).

Figure 9 shows the plots of the position of the SPB and the diameter of the pH-responsive PEGylated nanogels-3.5k containing AuNPs synthesized at N/Au = 8 under 0.15 M ionic strength conditions. The SPB of nanogel-3.5k that contained AuNPs (pH < 7, λ_{max} = 523.2 nm) was slightly red-shifted with increasing pH of the medium from 7.0 to 8.0, reaching a λ_{max} value of 524.8 nm. Note that this behavior is almost consistent with pH region of the volume phase transition of pH-responsive PEGylated nanogel-3.5k. This indicates that coordination between deprotonated amino groups and AuNPs proceeds along with the conversion of the protonated amino groups ($-N^+Et_2HCl^-$) to deprotonated amino groups ($-NEt_2$), which probably leads to the changes in the electronic environments on the surfaces of AuNPs. In contrast, a blue shift of the SPB was observed at pH values above 8.0, since the environments surrounding the AuNPs become less hydrophilic arising from the volume phase transition of the cross-linked PEAMA core. The cross-linked PEAMA core of nanogels-3.5k acts not only as a nanoreactor but also as a pH-responsive matrix to produce new classes of AuNPs, *viz.,* the pH-responsive PEGylated nanogel containing AuNPs can be utilized as pH sensor.

ENZYME-RESPONSIVE PEGYLATED NANOGELS CONTAINING GOLD NANOPARTICLES FOR APOPTOSIS IMAGING

Apoptosis is a normal physiological process that occurs during embryonic development as well as in the maintenance of tissue homeostasis. However, improperly regulated apoptosis can lead to several pathological conditions, including cancer and neurodegenerative diseases. Since the most anticancer drugs generally kill cells by activating apoptosis in several days, apoptosis imaging techniques may be capable of monitoring the cancer response to therapy at an early stage.[23] One of the most frequently activated cysteine proteases during the apoptosis process is caspase-3,[24] which cleaves specifically proteins containing the Asp-Glu-Val-Asp (DEVD) peptide sequence.[25] A variety of fluorescence probes bearing acceptor (quencher) and donor molecules at the terminus of DEVD-peptide has been developed to specifically detect apoptotic cells *in vitro* through fluorescence resonance energy transfer (FRET),[26] however, these probes have been not proven to be effective under *in vitro* and *in vivo* conditions owing to their nonspecific accumulation in normal tissues/cells, preferential renal clearance, and low fluorescence-quenching efficiency, leading to the low fluorescence imaging resolution. A promising approach to imaging the apoptotic cells *in vitro* and *in vivo* is the development of smart nanoprobe capable of complete quenching/dequenching of the signals in response to activated caspase-3 after preferential accumulation in solid tumor tissues through the enhanced permeability and retention (EPR) effect.[27] Recently, we reported a biocompatible, caspase-3-responsive, and fluorescence-quenching smart apoptosis nanoprobe based on PEGylated nanogel containing AuNPs and fluorescein isothiocyanate (FITC)-labeled DEVD peptide at the tethered PEG chain end.[28] The fluorescence signal is quenched in the absence of activated caspase-3 (in normal cells) through FRET process between AuNPs and FITC molecules due to existence of AuNPs close to the FITC-labeled DEVD peptide, whereas recovery of the pronounced fluorescence signal arises from release of the FITC molecules through the cleavage of DEVD peptide by activated caspase-3, allowing high fluorescence imaging resolution (Figure 10).

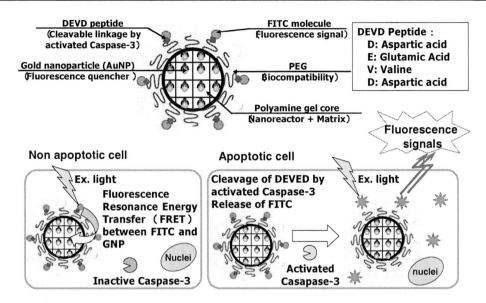

Figure 10. Schematic illustration of the smart fluorescence-quenching apoptosis nanoprobe based on the PEGylated nanogel-2.3k containing AuNPs and FITC-labeled DEVD peptide at the tethered PEG chain end.

A synthetic route to caspase-3-responsive PEGylated nanogel containing AuNPs and FITC-labeled DEVD peptide with 1.0 mol% cross-linking density is shown in Scheme 2. Low molecular weight of PEG (M_n =2360) was used in this study, because FRET efficiency depends on the distance between acceptor and donor molecules. Since the PEG strands immobilized on nanoparticles (micelles and liposomes) are reported to adopt a conformation of slightly stretched random coil,[29] the end-to-end distance of PEG [MW = 2360 g/mol, degree of polymerization (DP) = 54] based on random coil models[30] was calculated to be ca. 2.3 nm, viz., the distance between FTIC molecules located at the PEG chain end and the core of nanogel will be sufficient to efficiently cause the FRET.[31] Conversion of the acetal group into an aldehyde group was carried out by acidic treatment of the PEGylated nanogel-2.3k. The introduction of the FITC-labeled DEVD peptide to the aldehyde group was performed using 0.5 equivalents of H_2N-Cys-Gly-Gly-DEVED-Gly-Gly-Lys(FITC) through the formation of a thiazolidine ring between the aldehyde group and N-terminal Cys moiety,[32] followed by the addition of $NaBH_4$ to reduce the unreacted aldehyde groups to hydroxyl groups. The degree of functionality of the FITC-labeled DEVD peptide (FITC-DEVD-nanogel-2.3k) was determined to be 53% based on the standard curve corresponding to the fluorescence intensity of the FITC.

The synthesis of the FITC-DEVD-nanogel-2.3k containing AuNPs (FITC-DEVD-AuNP-nanogel-2.3k) was carried out at N/Au ratio = 1.5 (molar ratio of amino groups in the nanogel :$HAuCl_4$ = 1.5) by the self-reduction of $HAuCl_4$ (III) with FITC-DEVD-nanogel-2.3k at pH 6.0 without any additional reducing agents. The resulting FITC-DEVD-AuNP-nanogel-2.3k was well-dispersed under physiological conditions (in PBS at 37°C), although slight increase in the size (78.8 nm) with unimodal size distribution (μ_2/Γ^2 = 0.072) was observed probably due to the formation of the AuNPs into the PEAMA gel core. A transmission electron microscopy (TEM) image of the FITC-DEVD-AuNP-nanogel-2.3k shows that AuNP clusters were observed, and the average number of AuNPs in a single nanogel (cluster) and the average diameter of the AuNPs were about 6.7 ± 3.2 particles/nanogel and 8.5 ± 2.2 nm, respectively. Figure 11 shows the fluorescence spectra of the FITC-DEVD-nanogel-2.3k and FITC-DEVD-AuNP-nanogel-2.3k. The FITC-DEVD-nanogel-2.3k showed pronounced fluorescence at 525 nm attributed to FITC molecules, whereas almost no fluorescence was observed for the FITC-DEVD-AuNP-nanogel-2.3k. Note that fluorescence spectrum of the FITC-DEVD-AuNP-nanogel-2.3k after the addition of sodium cyanide (NaCN) to etch the AuNPs[33] was almost the same as that of the FITC-DEVD-nanogel-2.3k, which indicates that fluorescence-quenching of the FITC-DEVD-AuNP-nanogel-2.3k is due to the FRET between the AuNPs and FITC molecules. The quenching efficiency of the FITC-DEVD-AuNP-nanogel-2.3k was found to be 98% as calculated from the ratio of fluorescence intensity between the FITC-DEVD-AuNP-nanogel-2.3k and the FITC-DEVD-AuNP-nanogel-2.3k treated with cyanide etching.

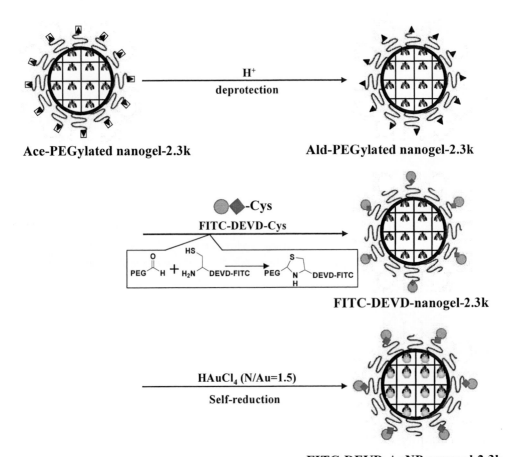

Scheme 2. A synthetic route to the caspase-3-responsive PEGylated nanogel-2.3k containing GNPs and FITC-labeled DEVD peptide.

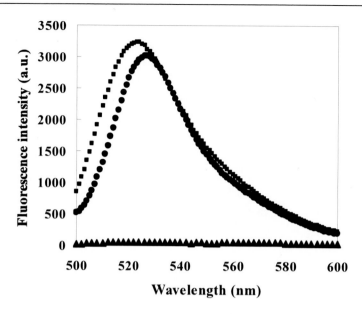

Figure 11. Fluorescence spectra of the FITC-DEVD-nanogel-2.3k (circle), FITC-DEVD-AuNP-nanogel-2.3k (triangle), and the FITC-DEVD-AuNP-nanogel-2.3k after treatment of cyanide etching (square).

Figure 12 shows the relative fluorescence intensity of the FITC-DEVD-AuNP-nanogel-2.3k in the presence of activated caspase-3 as a function of time. An approximately 4.8–fold increase in fluorescence intensity of the FITC-DEVD-AuNP-nanogel-2.3k was immediately observed. The fluorescence intensity plateaus at 90 min, thus, 23.5 ± 0.33 (%) of the fluorescence intensity (FL intensity = 368) were recovered through the cleavage of DEVD peptide by activated caspase-3, as calculated from the fluorescence intensity (FL intensity = 1565) of the FITC-DEVD-AuNP-nanogel-2.3k after cyanide etching treatment. This is most likely due to the susceptibility of the cleavable DEVD peptide being decreased by the steric hindrance of tethered PEG chains of the FITC-DEVD-AuNP-nanogel-2.3k, decreasing DEVD peptide accessibility by activated caspase-3. In sharp contrast, the fluorescence intensity did not recover in the presence of activated caspase-9. Note that no change in fluorescence intensity of the FITC-DEVD-AuNP-nanogel-2.3k was observed in the presence of both activated caspase-3 and caspase-3 inhibitor, indicating that the increase in fluorescence intensity of the FITC-DEVD-AuNP-nanogel-2.3k is indeed a caspase-3-specific event.

To examine whether the FITC-DEVD-AuNP-nanogel-2.3k acts as an apoptosis nanoprobe for monitoring the cancer response, monolayer-cultured HuH-7 cells were visualized under a confocal fluorescence microscope after 4 h incubation in the presence or absence of staurosporine[34] as an apoptosis inducer agent. Before the treatment of the staurosporine for 4 h, HuH-7 cells were incubated with the FITC-DEVD-AuNP-nanogel-2.3k (20 µg/mL) for 24 h. As seen in Figure 13, almost no or small fluorescence signal was observed for HuH-7 cells incubated with the FITC-DEVD-AuNP-nanogel-2.3k alone (Figure 13a)), yielding approximately 4% of fluorescence positive cells (Figure 13c)). In sharp contrast, pronounced fluorescence signals were observed for all apoptotic HuH-7 cells induced by staurosporine in the presence of the FITC-DEVD-AuNP-nanogel-2.3k (Figure

13b)), thus, approximately 95% of fluorescence positive cells were observed, confirming high specificity of the FITC-DEVD-AuNP-nanogel-2.3k nanoprobe for apoptotic cells.

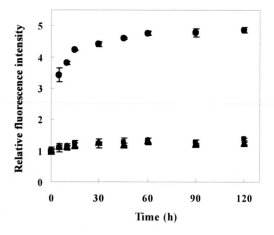

Figure 12. Relative fluorescence intensity of the FITC-DEVD-AuNP-nanogel-2.3k in the presence of caspase-3 (circle), caspase-3 with caspase-3 inhibitor (square), and caspase-9 (triangle) as function of incubation time at 37°C. The plotted data are the average of three experiments ± SD.

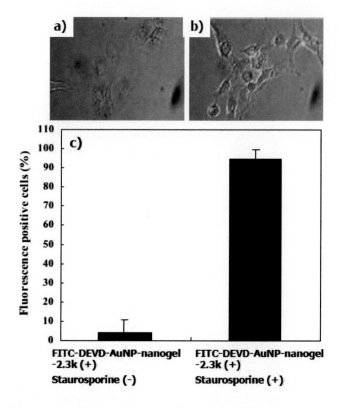

Figure 13. Confocal fluorescence images of HuH-7 cells incubated with the FITC-DEVD-AuNP-nanogel-2.3k in the a) absence and b) presence of staurosporine after 4h incubation. c) Relative fluorescence positive HuH-7 cells in the absence or presence of staurosporine after 4h incubation.

PH-RESPONSIVE PEGYLATED NANOGELS CONTAINING PLATINUM NANOPARTICLES AS SKIN-SPECIFIC NANOZYME

Reactive oxygen species (ROS), such as superoxide (O_2^-) and hydrogen peroxide (HO·), are produced in the body and play several important roles in the self-defense system.[35] Excessive production of ROS, however, has a harmful effect on cellular structures if the normal cells can not scavenge efficiently the generated ROS. In particular, the skin is sensitive to the effects of ROS, since the skin is always in contact with oxygen and is continuously exposed to sun radiation, both of which facilitate the formation of ROS.[36] The consequences of the generated ROS are accelerated skin aging and cell carcinoma. The first lines of skin-defense are the superoxide dismutase and catalase, which protect enzymes that catalyze the conversion of ROS into H_2O and O_2. However, the skin levels of protective enzymes are known to decrease with aging. To circumvent this limitation, there is considerable interest in developing synthetic ROS-scavengers for use in the skin. Recently, platinum nanoparticles (PtNP) have emerged as a new class of synthetic ROS scavengers, since they show the nontoxicity, and catalytic activities for the various radical species, including HO·.[37] In this regard, we have recently reported that pH-responsive PEGylated nanogels containing PtNPs can be utilized as skin-specific catalysts (nanozyme) for ROS. The environment of the skin has been known to be slightly acidic pH (4.5 ~ 7.0) compared with that of other tissues (pH ~ 7.4).[38] Thus, pH-responsive PEGylated nanogels containing PtNPs showed on-off regulation of catalytic activity for ROS, which synchronizes with the reversible volume phase transition of cross-linked PEAMA core in response to an acidic skin pH (Figure 14).

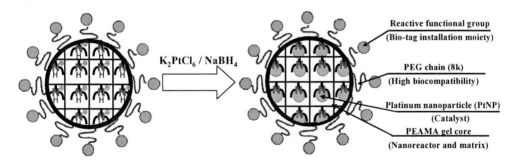

pH-Responsive PEGylated nanogel containing PtNPs

Figure 14. Schematic illustration of pH-responsive PEGylated nanogels containing PtNPs as the skin-specific nanozyme of ROS.

A strategy for the preparation of the pH-responsive PEGylated nanogels containing PtNPs is the formation of PtNPs in the cross-linked PEAMA core through an ion-exchange process between $PtCl_6^{2-}$ $2K^+$ and the $-N^+Et_2H$ Cl^- in the protonated PEAMA core, followed by the reduction of Pt(IV) ions to Pt(0) particles with reducing agents. Immobilization of the $PtCl_6^{2-}$ ions into the PEGylated nanogels-8k with fully protonated amino groups in the cross-linked PEAMA core was carried out with 15 mM (5 μmol) potassium chloroplatinate (K_2PtCl_6) at pH 4 at several N/Pt ratios (2, 3, and 5), followed by three round of ultrafiltaration to remove the free Pt ions. The amount of the Pt ions immobilized into the

PEGylated nanogels-8k, as determined by ICP-AES analysis, was found to be almost equivalent to quantitative immobilization of Pt ions into the cross-linked PEAMA core. This finding indicates that the ion-exchange process proceeds efficiently, because the introduction of divalent $PtCl_6^{2-}$ ions instead of monovalent Cl^- ions provides a strong driving force for the immobilization of Pt ions in the cross-linked PEAMA core. After reduction of the Pt ions located in PEGylated nanogels-8k with $NaBH_4$, a typical TEM image of the PEGylated nanogels containing PtNPs prepared at N/Pt = 3 shows clearly the PtNP clusters, *viz.*, the PtNPs (high contrast) as well as the cross-linked PEAMA core (lower contrast; ~ 50 nm) were observed (Figure15a)). The average diameter of the PtNPs was found to be 1.4 ± 0.4 nm (n = 30). In addition, the average diameter of the PtNPs for other versions of PEGylated nanogels-8k containing PtNPs was also less than 2 nm. As shown in Figure 15b), TEM-EDS reveals the peaks attributed to the Pt (2.2, 9.4, and 11.2 eV),[39] which strongly suggests the formation of PtNPs in the cross-linked PEAMA core. Note that Pt(IV) ions into the cross-linked PEAMA core quantitatively reduced to Pt(0) nanoparticles based on the disappearance of the absorption band at 260 nm this is attributed to the Pt(IV) ions.

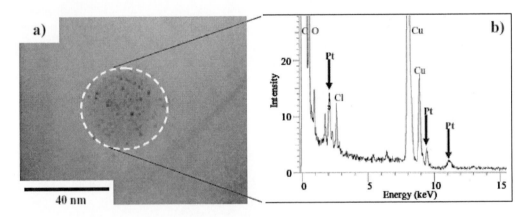

Figure 15. a) TEM image and b) EDS of pH-responsive PEGylated nanogel-8k containing PtNPs prepared at N/Pt = 3.

An ESR spin-trapping method[40] was employed to evaluate the ROS scavenging activities of PEGylated nanogels-8k containing PtNPs (63 nmol) prepared at N/Pt = 5 under acidic (pH 6.9) and physiological (pH 7.8) environments. In this method, 5, 5-dimethyl-1-pyrroline-N-oxide (DMPO) (66 nmol) was used as a spin-trapping agent to quench O_2^- generated by the reaction of hypoxanthine (HPX) (125 nmol) with xanthine oxidase (XOD), since the generated O_2^- (calcd. 250 nmol, 2 eq. of HPX) could react with DMPO to give relatively stable spin-adduct DMPO-OOH (several minutes of half-life time). Thus, ROS-scavenging ability (catalytic activity) of pH-responsive PEGylated nanogel-8k containing PtNPs (Pt: 25 mol % for calcd. O_2^-) could be evaluated based on disturbance of the spectrum of DMPO-OOH (Scheme 3). As shown in Figure 16, almost no ROS-scavenging ability was observed in the presence of PEGylated nanogel-8k containing PtNPs at physiological pH (Figure 16 a) vs. c)) (17% average scavenging efficiency, n = 3), whereby the cross-linked PEAMA core of the PEGylated nanogel-8k containing PtNPs was deprotonated and made hydrophobic through the deprotonation of the amino groups (diameter 97.2 nm); the penetration of water molecules (H_2O) was presumably inhibited. In sharp contrast, swelling PEGylated nanogel-8k containing PtNPs (diameter 129.7 nm) resulted in significant ROS-scavenging ability at acidic pH (Figure 16 b) vs. d)), providing 73% (n = 3) average scavenging efficiency for the generated ROS. This finding suggests that the PEGylated nanogel-8k containing PtNPs has catalytic activity for the O_2^-. The rate constant (K_{Pt}) of the catalytic activity of the PEGylated nanogel-8k containing PtNPs was evaluated using the following equations:

$$F/(1-F) = R[Pt]/[DMPO] \tag{1}$$

$$R = K_{Pt}/K_{DMPO} = F[DMPO]/(1-F)[Pt] \tag{2}$$

where R is the degree of reaction, F is the degree of decrease in ESR signal intensity, K_{Pt} and K_{DMPO} are rate constants for PEGylated nanogel-8k containing PtNPs and DMPO, and [Pt] (250 µM) and [DMPO] (265 µM) are the concentrations of PtNP and DMPO, respectively. In this instace, the value for K_{DMPO} by DMPO is taken as 18 ($M^{-1}s^{-1}$).[41] The K_{Pt} values of the PEGylated nanogel-8k containing PtNPs at pH 6.9 and pH 7.8 were calculated to be K_{Pt} = 53.6 $M^{-1}s^{-1}$ and K_{Pt} = 3.9 $M^{-1}s^{-1}$, respectively. This almost 14-fold increase in the apparent K_{Pt} of the catalytic activity of PEGylated nanogel-8k containing PtNPs in response to pH is remarkable. Note that ROS-scavenging ability of PEGylated nanogel-8k containing PtNPs is almost consistent with the pH range of the volume phase transition of the PEGylated nanogel-8k. We anticipate that the pH-responsive PEGylated nanogel containing PtNPs can be utilized as a skin-specific nanozyme for ROS.

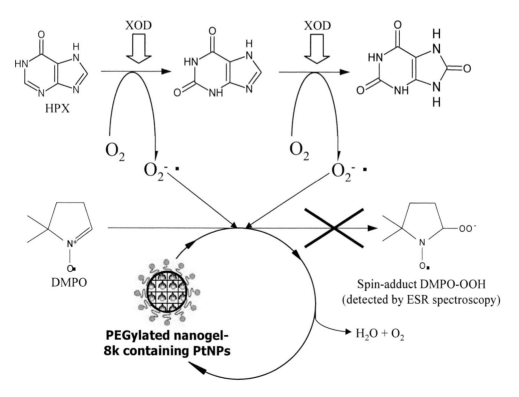

Scheme 3. Schematic illustration of the principle of ESR spin-trapping method.

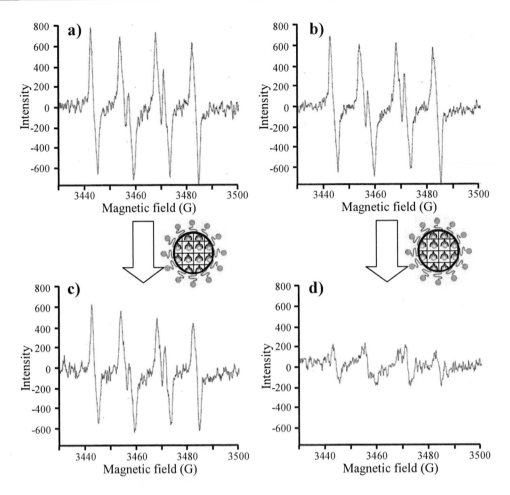

Figure 16. ESR spectra of the stable spin-adduct DMPO-OOH formed by the reaction of DMPO with O_2^- in the absence of pH-responsive PEGylated nanogel-8k containing PtNPs at a) pH 7.8 and b) pH 6.9, and in the presence of pH-responsive PEGylated nanogel-8k containing PtNPs at c) pH 7.8 and d) pH 6.9. These spectra are the typical spectra obtained in triplicate experiments.

CONCLUSION

We describe a novel approach to the synthesis of the smart stimuli-responsive PEGylated nanogels containing MNPs and their applications. In particular, stimuli-responsive PEGylated nanogels composed of cross-linked PEAMA core and PEG tethered chains that bear a caboxylic acid or acetal group as a platform moiety for the installation of a bio-tag showed not only unique properties in response to pH but also acted as nanoreactor to produce and immbilize AuNPs and PtNPs. The pH-responsive PEGylated nanogels containing AuNPs or PtNPs have potential utility in novel nanomaterials such as pH sensor, nanoprobe for apoptosis imaging, and nanocatlyst for reactive oxygen species (nanozyme). We anticipate that the stimuli-responsive PEGylated nanogels can be used to produce other metal nanoparticles, such as Pd nanoparticles, Rh nanoparticles, CdS nanoparticles, and magnetic nanoparticles, and the resulting smart hybrid nanoparticles can be utilized to novel functional biomedical nanomaterials.

ACKNOWLEDGMENT

The authors would like to thank Dr. Hisato Hayashi, Mr. Atsushi Tamura, and Mr. Takahito Nakamura for technical support of this research.

REFERENCES

[1] L. N. Lewis, *Chem. Rev.* 1993, *93,* 2693.

[2] H. Otsuka, Y. Akiyama, Y. Nagasaki, K. Kataoka, *J. Am. Chem. Soc.* 2001, *123,* 8226. b) W. P. Wuelfing, S. M. Gross, D. T. Miles, R. W. Murray, *J. Am. Chem. Soc.* 1998, *120,* 12696.

[3] Y. Nagasaki, *Chem. Lett.* 2008, *37,* 564. b) M-C. Daniel, D. Astruc, *Chem. Rev.* 2004, *104,* 293. c) T. Ishii, H. Otsuka, K. Kataoka, Y. Nagasaki, *Langmuir* 2004, *20,* 561. d) R. M. Crooks, M. Zhao, L. Sun, V. Chechik, L. Yeung, *Acc. Chem. Res.* 2001, *34,* 181.

[4] P. Zheng, X. Jiang, X. Zhang, W. Zhang, L. Shi, *Langmuir* 2006, *22,* 9393. b) R. R. Bhattacharjee, M. Chakraborty, T. K. Mandal, *J. Phys. Chem. B* 2006, *110,* 6768. c) J-H. Kim, T. R. Lee, *Chem. Mater.* 2004, *16,* 3647. d) J. Raula, J. Shan, M. Nuopponen, A. Niskanen, H. Jiang, E. I. Kauppinen, H. Tenhu, *Langmuir* 2003, *19,* 3499. e) J. Du, S. P. Armes, *J. Am. Chem. Soc.* 2005, *127,* 12800. f) S. Liu, J. V. M. Weaver, M. Save, S. P. Armes, *Langmuir* 2002, *18,* 8350.

[5] M. Oishi, Y. Nagasaki *React. Funct. Polym.* 2007, *67,* 1311. b) M. Oishi, S. Sumitani, Y. Nagasaki, *Bioconjugate Chem.* 2007, *18,* 1379. c) M.Oishi, H. Hayashi, M. Iijima, Y. Nagasaki, *J. Mater. Chem.* 2007, *17,* 3720. d) M. Oishi, N. Miyagawa, T. Sakura Y. Nagasaki, *React. Funct. Polym.* 2007 *67,* 662. e) M. Oishi, H. Hayashi, K. Itaka, K. Kataoka, Y. Nagasaki, *Colloid Polym. Sci.* 2007, *285,* 1055.

[6] G.W. Poehlein, *Encyclopedia of Polymer Science and Engineering Second Edition* vol. 6 1.

[7] M. J. Westby, *Colloid Polym. Sci.* 1988, *266.* 46. b) F. Hoshino, M. Sasaki, H. Kawaguchi, Y. Ohtsuka, *Polym. J.* 1987, *19,* 383.

[8] V. P. Wetering E. E. Moret, M. M. E. Schuurmans-Nieuwenbroek, V. M. J. Steenbergen, W. E. Hennink, *Bioconjugate Chem.* 1999, *10,* 589.

[9] S. Liu, J. V. M. Weaver, Y. Tang, N. C. Billingham, S. P. Armes, K. Tribe, *Macromolecules* 2002, *35,* 6121.

[10] R. Ogawa, Y. Nagasaki, K. Kataoka, *Polym. J.* 2002, *34,* 868.

[11] C. Wu, M. Akashi, O.M. Chen, *Macromolecules* 1997, *30,* 2187.

[12] K. Podual, J.F. Doyle III, A.N. Peppas, *Biomaterials* 2000, *21,* 1439.

[13] Y. Hirokawa, T. Tanaka, E. Sato, *Macromolecules* 1985, *18,* 2782. b) I. Ohmine, T. Tanaka, *J. Chem. Phys.* 1982, *77,* 5725.

[14] M-C. Daniel, D. Astruc, *Chem. Rev.* 2004, *104,* 293. b) C. M. Niemeyer, *Angew. Chem. Int. Ed.* 2001, *40,* 4128.

[15] J. W. Slot, H. J. Geuze, *J. Cell Biol.* 1981, *90*, 533.

[16] K. Sato, K. Hosokawa, M. Maeda, *J. Am. Chem. Soc.* 2003, *125*, 8012. b) C. A. Mirkin, R. L. Letsinger, R. C. Mucic, J. J. Storhoff, *Nature* 1996, *382*, 607.

[17] N. T. K. Thanh, Z. Rosenzweig, *Anal. Chem.* 2002, *74*, 1624.

[18] M. Oishi, J. Nakaogami, T. Ishii, Y. Nagasaki, *Chem. Lett.* 2006, *35*, 1046.

[19] S. Morokoshi, K. Ohhori, K. Mizukami, H. Kitanao, *Langmuir* 2004, *20*, 8897.

[20] S. Lee, E.-J. Cha, K. Park, S.-Y. Lee, J.-K. Hong, I.-C. Sun, S. Y. Kim, K. Choi, I. C. Kwon, K. Kim, C.-H. Ahn, *Angwe. Chem. Int. Ed.* 2008, *47*, 2804. b) T. Pons, I. L. Medintz, K. E. Sapsford, S. Higashiya, A. F. Grimes, D. S. English, H. Mattoussi, *Nano Lett.* 2007, *7*, 3157. c) C. S. Yun, A. Javier, T. Jennings, M. Fisher, S. Hira, S. Peterson, B. Hopkins, N. O. Reich, G. F. Strouse, *J. Am. Chem. Soc.* 2005, *127*, 3115. d) B. Dubertret, M. Calame, A. J. Libchaber, *Nat. Biotechnol.* 2001, *19*, 365.

[21] M. Oishi, H. Hayashi, T. Uno, T. Ishii, M. Iijima, Y. Nagasaki, *Macromol. Chem. Phys.* 2007, *208*, 1176.

[22] J. Du, S. P. Armes, *J. Am. Chem. Soc.* 2005, *127*, 12800. b) S. Liu, J. V. M. Weaver, M. Save, S. P. Armes, *Langmuir* 2002, *18*, 8350.

[23] J. Schoenberger, J. Bauer, J. Moosbauer, C. Eilles, D. Grimm, *Curr. Med. Chem.* 2008, *15*, 187.

[24] N. A. Thornberry, Y. Lazebnik, *Science* 1998, *281*, 1312. b) Y.A. Lazebnik, S.H. Kaufmann, S. Desnoyers, G. G. Poirier, W. C. Earnshaw, *Nature* 1994, *371*, 346.

[25] H. Wyllie, *Nature* 1980, *284*, 555.

[26] K. Stefflova, J. Chen, D. Marotta, H. Li, G. Zheng, *J. Med. Chem.* 2006, *49*, 3850. b) Y. Wu, D. Xing, S. Luo, Y. Tang, Q, Chen, *Cancer Lett.* 2006, *235*, 239.

[27] Y. Matsumura, H. Maeda, *Cancer Res.* 1986, *46*, 6387.

[28] M. Oishi, A. Tamura, T. Nakamura, Y. Nagasaki, *Adv. Funct. Mater.* 2008, *19*, 827.

[29] A. Harada, K. Kataoka, *Macromolecules* 1998, *31*, 288. b) A. K. Kenworthy, K. Hristova, D. Needham, T. J. McIntosh, *Biophys. J.* 1995, *68*, 1921.

[30] C. Tanford, Y. Nozaki, M. F. Rohde, *J. Phys. Chem.* 1977, *81*, 1555.

[31] Z. Gueroui, A. Libchaber, *Phys. Rev. Lett.* 2004, *93*, 166108.

[32] L. Zhang, T. R. Torgerson, X.-Y. Liu, S. Timmons, A. D. Colosia, J. Hawiger, J. P. Tam, *Proc. Natl. Acad. Sci. USA* 1998, *95*, 9184.

[33] S. S. Agasti, C.-C. You, P. Arumugam, V. M. Rotello, *J. Mater. Chem.* 2008, *18*, 70. b) S. R. Isaacs, E. C. Cutler, J.-S. Park, T. R. Lee, Y.-S. Shon, *Langmuir* 2005, *21*, 5689. c) A. C. Templeton, M. J. Hostetler, C. T. Kraft, R. W. Murray, *J. Am. Chem. Soc.* 1998, *120*, 1906.

[34] E. Falcieri, A. M. Matelli, R. Bareggi, A. Cataldi, L. Cocco, *Biochem. Biophys. Res. Commun.* 1993, *193*, 19.

[35] N. A. Simonian, J. T. Coyle, *Annu. Rev. Pharmacol. Toxicol.* 1996, 36, 83.

[36] Y. Yamamoto, *J. Dermatological Sci.* 2001, *27 Suppl.1*, S1.

[37] T. Aiuchi, S. Nakajo, K. Nakaya, *Biol. Pharm. Bull.* 2004, *27*, 736. b) H. Tsuji, *U. S. Patent* 2002, 6455594 B1.

[38] J. W. Fluhr, P. M. Elias, *Exogenous Dermatology* 2002, *1*, 163. b) P. Dykes, *Int. J. Cosmetic Sci.* 1998, *20*, 53. c) K. Chikakane, H. Takahashi, *Clinics in Dermatology* 1995, *13*, 299.

[39] M. F. Lengke, M. E. Fleet, G. Southam, *Langmuir* 2006, *22*, 7318.

[40] L. L. Dugan, D. M. Turetsky, C. Du, D. Lobner, M. Wheeler, C. R. Almli, C. K.-F. Shen, T.-Y. Luh, D. W. Choi, T.-S. Lin, *Proc. Natl. Acad. Sci. USA* 1997, *94,* 9434.
[41] E. Finkelstein, G. M. Rosen, E. J. Rauckman, *J. Am. Chem. Soc.* 1980, *102,* 4994.

In: Modern Trends in Macromolecular Chemistry
Editor: Jon N. Lee

ISBN: 978-1-60741-252-6
© 2009 Nova Science Publishers, Inc.

Chapter 8

PREPARATION OF POLYMERS WITH WELL-DEFINED NANOSTRUCTURE IN THE POLYMERIZATION FIELD

*Yoshiro Kaneko and Jun-ichi Kadokawa**

Graduate School of Science and Engineering, Kagoshima University, Kagoshima, Japan

ABSTRACT

In this chapter, we review the preparation methods of the polymers with well-defined nanostructures in the polymerization field. Generally, nanostructures of the synthetic polymers have been controlled by various external forces and intermolecular interactions using previously prepared polymeric materials. On the other hand, the methods to control nanostructures of the polymeric materials during the polymerization reaction have been limited.

As the examples of these methods, we first describe the preparation of amylose-synthetic polymer inclusion complexes by the polymerization of α-D-glucose 1-phosphate catalyzed by phosphorylase in the presence of synthetic polymers, i.e., the formation of inclusion complexes was achieved during amylose-forming polymerization. As the second topic, we describe the investigation of a parallel enzymatic polymerization system to form inclusion complexes. When two enzymatic polymerizations, i.e., the aforementioned amylose-forming polymerization and the lipase-catalyzed polycondensation of dicarboxylic acids and diols leading to the aliphatic polyesters, were simultaneously performed, the inclusion complexes composed of amylose and strongly hydrophobic polyesters were produced. We also describe the preparation of synthetic cellulose with controlled crystalline structure as the third topic, which was achieved by enzymatic polymerization of β-cellobiosyl fluoride monomer using cellulase as a catalyst. Finally, we describe the preparation of crystalline nanofiber of linear polyethylene by the polymerization of ethylene using mesoporous silica fiber-supported titanocene and methylalumoxane as a cocatalyst. These methods allow the control of nanostructures of the polymeric materials during the polymerization reaction.

* Correspondence to: Tel: +81-99-285-7743, Fax: +81-99-285-3253, E-mail: kadokawa@eng.kagoshima-u.ac.jp

1. INTRODUCTION

Biological polymers such as proteins, nucleic acids, and polysaccharides are vital materials for important *in vivo* functions associated with "living". The important biological functions of these polymers appear to be controlled not only by their primary structures but also by higher-ordered structures or so-called nanostructures, which are often constructed during a polymerization reaction *in vivo*, e.g., formation of a double helix during a replication of DNA [1].

On the other hand, nanostructures of the synthetic polymers have been generally controlled by various external forces and intermolecular interactions using previously prepared polymeric materials. For example, control of orientation of the polymers has been achieved by using magnetic field [2], electric field [3], and biaxial drawing [4]. In addition, a number of polymeric assemblies have been constructed by using intermolecular interactions such as hydrogen bond, electrostatic interaction, and hydrophobic interaction, i.e., supramolecular chemistry, using previously prepared polymeric materials [5-7].

In contrast, as shown in aforementioned biological systems, the methods to control nanostructures of the polymeric materials during the polymerization reaction have been limited. Therefore, developments of the preparation methods for the polymeric assemblies with controlled nanostructures by a combination of "polymerization chemistry" and "supramolecular chemistry" would be important from the viewpoint of both academic and application fields.

As one of the examples of these methods, it was reported that the preparation of amylose-synthetic polymer inclusion complexes was performed by the polymerization of α-D-glucose 1-phosphate (G-1-P) using phosphorylase as a catalyst in the presence of synthetic polymers, i.e., the formation of inclusion complexes was achieved during amylose-forming polymerization [8-16]. In addition, the preparation of synthetic cellulose with controlled crystalline structure was reported, which was carried out by enzymatic polymerization of β-cellobiosyl fluoride (β-CF) monomer using cellulase as a catalyst [17,18]. Crystalline structures of celluloses depended on the purification grades of cellulase. As another method to construct controlled a nanostructure, the preparation of crystalline nanofiber of linear polyethylene was reported [19]. This was achieved by the polymerization of ethylene using mesoporous silica fiber-supported titanocene and methylalumoxane as a cocatalyst. These methods allow the control of nanostructures of the polymeric materials during the polymerization reaction. In this chapter, we review the preparation methods for the polymers with well-defined nanostructures in the polymerization field.

2. PREPARATION OF INCLUSION COMPLEXES COMPOSED OF AMYLOSE AND SYNTHETIC POLYMERS IN AMYLOSE-FORMING POLYMERIZATION FIELD: VINE-TWINING POLYMERIZATION

2.1. Amylose-Polymer Inclusion Complexes and Vine-Twining Polymerization

Amylose, a natural linear polysaccharide with helical conformation linked through $(1 \rightarrow 4)$-α-glycosidic linkages (Figure 1), is a well-known host molecule that readily forms inclusion complexes with slender guest molecules having relatively lower molecular weight by hydrophobic interaction between guest molecules and the cavity of amylose [20-26]. However, little had been reported regarding the formation of inclusion complexes composed of amylose and polymeric compounds [27-31]. The principal difficulty for incorporating polymeric materials into the cavity of amylose is that the driving force for the binding is only due to weak hydrophobic interactions. Amylose, therefore, does not have the sufficient ability to include the long chains of polymeric guests into its cavity.

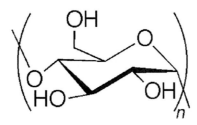

Figure 1. Structure of amylose.

Phosphorylase-catalyzed enzymatic polymerization using G-1-P proceeds with the regio- and stereoselective construction of an α-glycosidic bond under mild conditions, leading to the direct formation of amylose in aqueous media [32]. This polymerization is initiated from a maltooligosaccharide primer such as maltoheptaose (G_7). Then, the propagation proceeds through the following reversible reaction to produce a $(1 \rightarrow 4)$-α-glucan chain, i.e., amylose.

$$[(\alpha, 1 \rightarrow 4)\text{-}G]_n + G\text{-}1\text{-}P \leftrightarrow [(\alpha, 1 \rightarrow 4)\text{-}G]_{n+1} + P$$

In the reaction, a glucose unit is transferred from G-1-P to the nonreducing 4-OH terminus of a $(1 \rightarrow 4)$-α-glucan chain, resulting in inorganic phosphate (P).

By means of the enzymatic polymerization for direct construction of amylose, the method for the preparation of inclusion complexes composed of amylose and synthetic polymers has been developed [8-16]. The representation of this reaction system is similar to the way that vines of plants grow twining around a rod. Accordingly, it has been proposed that this polymerization method for the preparation of amylose-polymer inclusion complexes is named "vine-twining polymerization" (Scheme 1). In this part, we describe the procedure, the principal result, the proposed mechanism, and the application for the vine-twining polymerization.

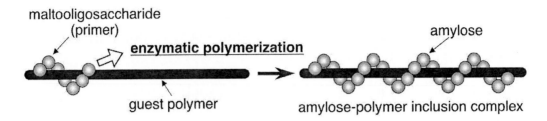

Scheme 1. Concept of vine-twining polymerization.

2.2. Preparation of Amylose-PolyTHF Inclusion Complex

The first example of the method leading to an amylose-polymer inclusion complex was achieved by amylose-forming polymerization in the presence of polyTHF (PTHF) (M_n = 4000) as a guest polymer [8]. The phosphorylase-catalyzed enzymatic polymerization of G-1-P from G_7 as the primer was performed in the presence of a telechelic PTHF with hydroxy end groups in sodium citrate buffer at ca. 40 °C (Scheme 2a). The resulting product was isolated and characterized by means of X-ray diffraction (XRD) and ^{1}H NMR measurements. The XRD pattern of the product was completely different from that of amylose and PTHF, and was similar to that of the inclusion complexes of amylose with monomeric compounds, as shown in previous study [33].

The ^{1}H NMR spectrum in DMSO-d_6 of the product contained signals not only due to amylose but also due to PTHF, in spite of washing with methanol, which was a good solvent of PTHF. Furthermore, the methylene peak of PTHF was broadened and shifted upfield compared to that of the original PTHF. This is because each methylene group of PTHF is immobile and interacts with the protons inside the amylose cavity. When PTHF was added to the NMR sample of the product in DMSO-d_6, two different signals due to methylene protons of PTHF were observed. This result suggested that the PTHF of the product was in an environment different from an original PTHF unit. To further confirm the structure of the product, spin-lattice relaxation time (T_1) measurements were investigated, as T_1 measurements of inclusion complexes have often been used for identification of their structures. In general, the T_1 values of the inclusion complexes are shorter than those of the corresponding individual molecules. Actually, the T_1 value of the methylene peak of the PTHF in the product was shorter than that of the original PTHF. The shorter T_1 in the product indicated restriction of the methylene movement due to the included conditions.

In addition, a similar NMR pattern was also observed in NaOD/D_2O solvent. The original PTHF was insoluble in NaOD/D_2O, and no peak due to PTHF appeared in the ^{1}H NMR spectrum of the suspension of PTHF in NaOD/D_2O. The PTHF in the product was probably solubilized in the alkaline solution by its inclusion in the cavity of amylose. These XRD and NMR data supported the conclusion that helical inclusion complexes were synthesized.

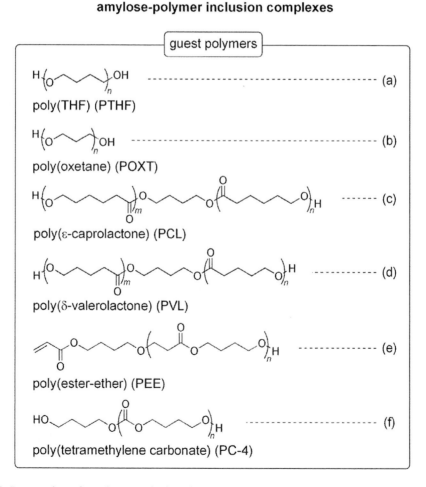

Scheme 2. Preparation of amylose-synthetic polymer inclusion complexes by vine-twining polymerization.

Generally, one helical turn of amylose is composed of ca. 6 repeating glucose units when linear molecules of small cross-sectional area, e.g., fatty acids, are included [34]. The repeat distance of the helix of amylose was reported as 0.795 nm [34], whereas the length of one unit of PTHF was calculated as ca. 0.60 nm, as shown in Figure 2. Therefore, 4.5 repeating glucose units in amylose correspond to the length of one PTHF unit (Figure 2). On the basis of the above calculations, the integrated ratio of the ^{1}H NMR between the signal due to amylose and the signal due to PTHF in the ideal inclusion complex can be calculated. The integrated ratio of these two signals in the ^{1}H NMR spectrum of the product was in good agreement with the calculated value [8]. This also supported the structure of the inclusion complex.

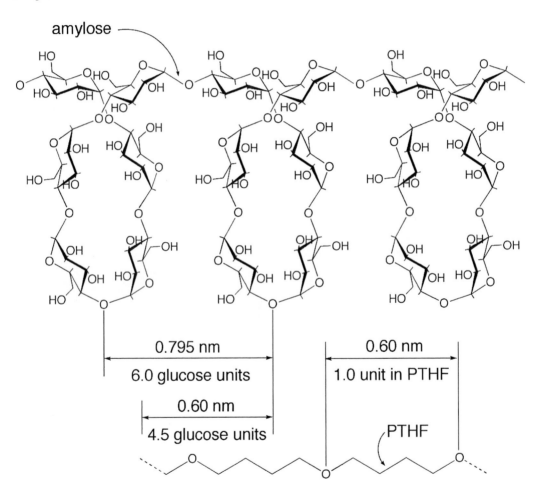

Figure 2. Illustration of repeat distance of amylose helix and unit lengths of PTHF.

The molecular weights of the amylose and PTHF in the inclusion complex were evaluated by means of ^{1}H NMR measurement, respectively, when the molecular weight of PTHF = 4000 was used. The molecular weight of the amylose was 12200-14600, corresponding to 9.9-11.9 nm of molecular length in helical form. On the other hand, the molecular weight of the included PTHF was ca. 2800, suggesting that the chain length is ca.

23.0 nm. Therefore, one PTHF molecule is probably included by two amylose molecules on the average.

Mixing amylose and PTHF in a buffer did not form the inclusion complex. This observation suggested that the inclusion complex formed during enzymatic polymerization. To study the relation between the formation of the inclusion complex and the enzymatic polymerization process, the following experiment was performed (Figure 3). When PTHF was added to the reaction solution immediately after the general enzymatic polymerization of G-1-P had started, an identical inclusion complex to that mentioned above was obtained. However, the contents of PTHF in the products decreased as the time between the start of the enzymatic polymerization and the addition of the PTHF to the solution was increased. These observations revealed that the inclusion complexes were not formed after polymerization produced amyloses with relative higher molecular weights. These results indicated that polymerization proceeded with the formation of the inclusion complex.

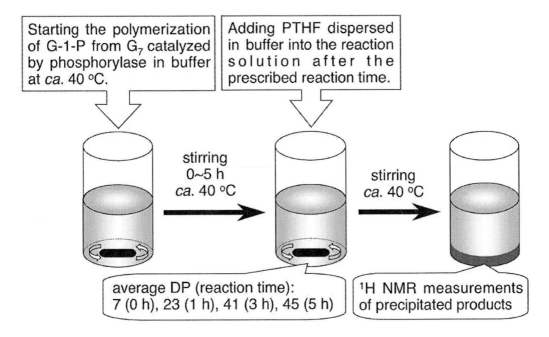

Figure 3. Investigation of relation between the formation of the inclusion complex and the enzymatic polymerization process.

Preparation of inclusion complexes using PTHFs with various M_ns (1000, 2000, 10000, 14000) was also performed. When PTHFs with M_ns = 1000 and 2000 were used as the guest polymers, inclusion complexes were obtained, which were confirmed by means of the XRD and ^{1}H NMR measurements. In contrast, PTHFs with higher M_ns, such as 10000 and 14000, were not dispersed well in the buffer of the polymerization solvent, and accordingly, inclusion complexes were not formed from these PTHFs. To obtain the inclusion complexes using these PTHFs, the polymerization was performed in the following two-phase system. The PTHF was dissolved in diethyl ether and the buffer was added to the solution (1:5, v/v). Enzymatic polymerization of G-1-P then occurred with vigorous stirring to disperse the diethyl ether phase in the buffer. The XRD patterns of the products from the PTHFs with M_ns

= 10000 and 14000 obtained by the two-phase system indicated the formation of the inclusion complexes.

The effect of the end groups of the telechelic PTHFs as the guest polymers was investigated. The end groups employed were the hydroxy (–OH, as mentioned above), methoxy (–OCH$_3$), ethoxy (–OCH$_2$CH$_3$), and benzyloxy (–OCH$_2$Ph) groups. The content of PTHF in the product obtained from the methoxy-terminated PTHF was close to that obtained from the hydroxy-terminated PTHF, indicating the formation of the inclusion complex. On the other hand, the content of PTHF in the product obtained from the ethoxy-terminated PTHF was much lower than that obtained from the aforementioned guest PTHFs. In addition, when the benzyloxy-terminated PTHF was used, no inclusion complex was obtained. These results indicated that the formation of the inclusion complexes was strongly affected by the bulkiness of the end groups of the PTHFs.

2.3. Preparation of Inclusion Complexes Using Other Polyethers as Guest Polymers

To investigate the effect of the alkyl chain lengths of the guest polyethers for the formation of inclusion complexes, the enzymatic polymerization was carried out in the presence of polyethers with different alkyl chain lengths, i.e., PTHF (Scheme 2a), poly(oxetane) (POXT, Scheme 2b), and poly(ethylene glycol) (PEG) [10]. The structures of the products were characterized by means of the XRD and ^{1}H NMR measurements. The XRD profiles of the product from POXT showed the same pattern as that from PTHF. Furthermore, the ^{1}H NMR spectrum of the product in DMSO-d_6 showed the signals due to both amylose and POXT. These observations indicated that the inclusion complex was formed by using POXT as the guest polyether. On the other hand, the XRD pattern of the product obtained with PEG was similar to that of amylose. In addition, the peak due to PEG was not observed in the ^{1}H NMR spectrum of the product. No formation of the inclusion complex with PEG was detected from these analytical data. This was probably attributed to the hydrophilicity of PEG, which caused less hydrophobic interaction between PEG and the cavity of amylose. These results indicated that the hydrophobicity of the guest polymers was an important factor for the formation of the inclusion complexes in this type of the polymerization system.

2.4. Preparation of Amylose-Polyester Inclusion Complexes

Since importance of the hydrophobicity of the guest polymers was found, hydrophobic polyesters as guest polymers such as telechelic poly(ε-caprolactone)s (PCLs) and telechelic poly(δ-valerolactone)s (PVLs) were employed for the preparation of amylose–polyester inclusion complexes [9,11]. The phosphorylase-catalyzed enzymatic polymerization of G-1-P from G$_7$ was performed in the presence of the polyesters in the sodium citrate buffer (Scheme 2c, d). The resulting products were characterized by means of ^{1}H NMR and XRD measurements, which supported the structures of the inclusion complexes. When PCL with higher molecular weight (M_n = 2000) was used as the guest polymer for the experiment same to above, the PCL was not dispersed well in the citrate buffer, and accordingly, the inclusion complex was not formed. To be dispersed in the reaction solvent, the polymerization was conducted in a mixed solvent of citrate buffer with acetone (5:1, v/v). The obtained product was characterized by means of the ^{1}H NMR and XRD measurements to be the inclusion complex. The inclusion complex was formed in citrate buffer by using PVL with a molecular weight of 2000, indicating that the PVLs had more advantageous properties as guest polymers for the formation of the inclusion complexes in the citrate buffer compared with the PCLs.

The IR spectrum of the PVL in the product was compared with that of the original PVL, and showed that no crystalline PVL existed in the product, due to the inclusion of the PVL chain into the cavity of amylose. This was in good agreement with the XRD data, in which the XRD profile of PVL showed strong crystalline peaks, whereas no such peaks were observed at all in the XRD profile of the product. The same results were obtained in the IR and XRD analyses of the product from PCL.

As the other guest polyester, a hydrophobic poly(ester-ether) (PEE) was employed in the formation of the inclusion complex [11]. The preparation of the inclusion complex using PEE was attempted in an experimental manner similar to that using the above polyesters (Scheme 2e). The structure of the product was determined by means of the ^{1}H NMR and XRD measurements to be the inclusion complex. When the hydrophilic poly(ester-ether) (-CH$_2$CH$_2$C(=O)OCH$_2$CH$_2$O-) was used as the guest polymer, no inclusion complex was formed. This result indicated that the hydrophobicity of the guest polymers strongly affected the formation of the inclusion complexes.

2.5. Preparation of Amylose-Polycarbonate Inclusion Complexes

The aforementioned guest polymers contain the relative polar linkages in the main-chains and do not have the side groups. Taking into this consideration, aliphatic polycarbonates as guest polymers for the vine-twining polymerization were employed, giving rise to the corresponding inclusion complexes with amylose [15]. The four hydrophobic polycarbonates with different methylene chain lengths were used in this study, which were poly(tetramethylene carbonate) (PC-4), poly(octamethylene carbonate) (PC-8), poly(decamethylene carbonate) (PC-10), and poly(dodecamethylene carbonate) (PC-12).

First, PC-4 was selected as a guest polycarbonate for the vine-twining polymerization. The preparation of an amylose-PC-4 inclusion complex was carried out under the conditions almost same as those using polyethers and polyesters as guest polymers described above (Scheme 2f). The precipitated product was characterized by means of the ^{1}H NMR, XRD, and IR measurements, which supported the structures of amylose-PC-4 inclusion complex. In addition, the structure of the product was investigated by means of the differential scanning calorimetry (DSC) measurement. In the DSC thermogram of PC-4, an endothermic peak corresponding to the melting (T_m) was observed, whereas any thermal transition is not observed in the DSC trace of the product. These DSC traces supported that no crystalline PC-4 existed in the product.

The effect of the methylene chain lengths in the polycarbonates was investigated on the formation of the inclusion complexes in this polymerization system using PC-4, PC-8, PC-10, and PC-12, as the guest molecules. One helical turn of amylose is composed of ca. 6 repeating glucose units and the repeat distance of the helix of amylose has been reported as ca. 0.795 nm as described above [34], whereas the lengths of one unit of PC-4, PC-8, PC-10, and PC-12 were calculated to be ca. 0.84, 1.38, 1.65, and 1.92 nm, respectively (Figure 4). On the basis of the above calculations, the integrated ratios of the signal due to H-1 proton of amylose to the signal due to β or γ of PC-4, PC-8, PC-10, and PC-12 in the ^{1}H NMR spectra of the inclusion complexes were found to be 0.63, 0.77, 0.97, and 1.11, respectively. The integrated ratio of these two signals in the ^{1}H NMR spectrum of the product obtained using PC-4 was 0.65, which was in good agreement with the calculated value. However, that of the product obtained using PC-8 (0.55) was slightly smaller than the calculated value. On the other hand, the values of the products obtained using PC-10 and PC-12 were 0.23 and 0.20, respectively, which were obviously lower than the calculated values. The guest polycarbonates having longer methylene chain lengths were probably aggregated in the aqueous buffer/acetone mixed solvent, and accordingly separated from the cavity of amylose, causing difficulty of complex formation.

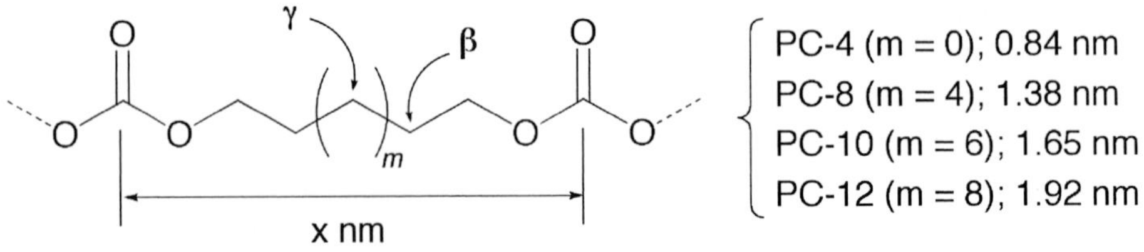

Figure 4. Lengths of one unit of polycarbonates.

2.6. Selective Inclusion of Amylose toward Resemblant Guest Polymers

As described above, hydrophobic polyethers, polyesters, a poly(ester-ether), and polycarbonates having appropriate methylene chain lengths were employed as the guest polymers for the vine-twining polymerization to form the corresponding inclusion complexes with amylose. On the basis of the results in the studies using these guest polymers, the suitable hydrophobicity of the guest polymers had been considered to be a very important factor in whether amylose was include them or not. The aforementioned property of amylose that included limited guest polymers with appropriate hydrophobicity in the vine-twining polymerization was applied to perform the selective inclusion toward two resemblant guest polymers. For example, the vine-twining polymerization was performed in the presence of a mixture of POXT (M_n = ~1800) and PTHF (M_n = ~1600) in sodium acetate buffer for 6 h at 40-45 °C (Scheme 3) [14].

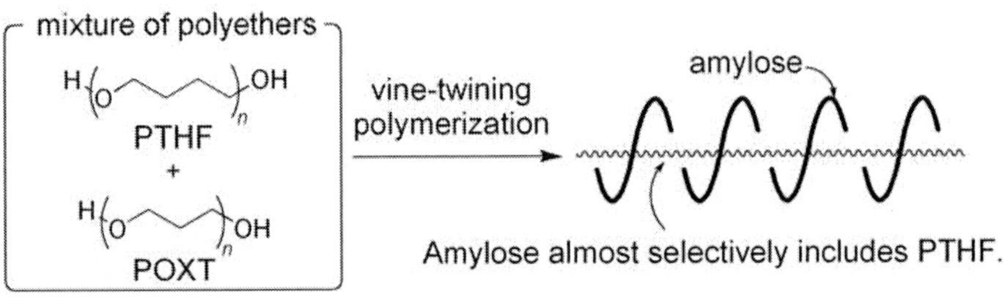

Scheme 3. Selective inclusion of amylose in vine-twining polymerization.

The ^{1}H NMR spectrum of the employed mixture of the guest polyethers (CDCl$_3$) showed that the unit ratio of POXT/PTHF in feed was assessed to be 0.90:1.00. On the other hand, in the ^{1}H NMR spectrum of the product obtained by the vine-twining polymerization (DMSO-d_6), the signals due to PTHF and amylose were prominently observed and the signal due to POXT slightly appeared (POXT/PTHF = 0.02:1.00). When POXT or PTHF was independently used as the guest for the vine-twining polymerization, the inclusion complex was formed well as described above [10]. These results indicated that amylose almost selectively included PTHF from a mixture of the two resemblant polyethers in the vine-twining polymerization. The slight difference in the hydrophobicities of two polyethers probably caused the difference in the inclusion by amylose toward them.

The concentrations of G-1-P and G$_7$ in feed strongly affected the selectivity on inclusion of amylose in this polymerization manner. When the vine-twining polymerization was carried out in higher concentrations of G-1-P and G$_7$ than those described above, for example 15 times, the unit ratio of POXT/PTHF in the products was increased (0.21:1.00) upon increasing the yields of the products (ca. 20 times). Because the concentration of PTHF decreased with the progress of the polymerization due to the predominant inclusion, amylose probably started to include POXT at a later stage of the polymerization.

3. PREPARATION OF INCLUSION COMPLEXES COMPOSED OF AMYLOSE AND STRONGLY HYDROPHOBIC POLYESTERS IN PARALLEL ENZYMATIC POLYMERIZATION SYSTEM

In the vine-twining polymerization described above, the hydrophobicity of the guest polymers is a very important factor because the driving force for the formation of the inclusion complexes is probably hydrophobic interaction. However, in addition to no formation of the inclusion complex from hydrophilic polymer, e.g., PEG, the preparation of the inclusion complexes had not been achieved from the polymers with strong hydrophobicity, e.g., poly(oxepane), attributed to their aggregation in the aqueous buffer of the solvent for the enzymatic polymerization.

To obtain the inclusion complex from a strongly hydrophobic guest polymer, a parallel enzymatic polymerization system was investigated, i.e., two enzymatic polymerizations, which were the phosphorylase-catalyzed polymerization of G-1-P from G$_7$, giving rise to amylose, and the lipase-catalyzed polycondensation of dicarboxylic acids and diols, leading to the aliphatic polyesters, were simultaneously performed (Scheme 4) [16]. As the monomers for the guest polyesters, the dicarboxylic acid and the diol having methylene units of 8, hereafter denoted as Diacid-8 and Diol-8, were firstly employed. The isolated product was characterized by means of the ^{1}H NMR and XRD measurements, which supported the structures of amylose-polyester inclusion complex.

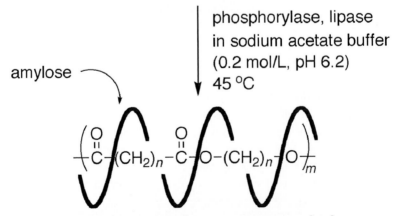

$$HOOC-(CH_2)_n-COOH \quad + \quad HO-(CH_2)_n-OH$$

$\begin{cases} n = 8; \quad \text{Diacid-8} \\ n = 10; \text{ Diacid-10} \\ n = 12; \text{ Diacid-12} \end{cases}$ $\begin{cases} n = 8; \quad \text{Diol-8} \\ n = 10; \text{ Diol-10} \\ n = 12; \text{ Diol-12} \end{cases}$

+

G_7

+

G-1-P

phosphorylase, lipase
in sodium acetate buffer
(0.2 mol/L, pH 6.2)
45 °C

amylose

$$\left(\overset{O}{\underset{\parallel}{C}} - (CH_2)_n - \overset{O}{\underset{\parallel}{C}} - O - (CH_2)_n - O \right)_m$$

**inclusion complexes composed of
amylose and strongly hydrophobic polyesters**

Scheme 4. Preparation of inclusion complexes composed of amylose and strongly hydrophobic polyesters in the parallel enzymatic polymerization system.

To demonstrate that the inclusion complex composed of amylose and the strongly hydrophobic polyester could be prepared by only the present parallel enzymatic polymerization system, following two experiments were performed. First, amylose-forming polymerization was performed in the presence of the polyester having methylene units of 8. In ^{1}H NMR spectrum of the product, the signals due to only amylose were observed, indicating that this polymerization method did not afford corresponding inclusion complex. On the other hand, polyester-forming polycondensation was carried out in the presence of amylose. Consequently, although the monomers for the polyester were included in the cavity of amylose, the polyester was not included in it, confirmed by means of ^{1}H NMR measurement. These results indicated that the inclusion complex composed of amylose and the strongly hydrophobic polyester was prepared by only the present parallel enzymatic polymerization system.

To investigate the effect of the methylene chain length of the dicarboxylic acids and the diols on the formation of inclusion complexes, the dicarboxylic acids and the diols having methylene units of 10 and 12, hereafter denoted as Diacid-10, 12 and Diol-10, 12, were employed for this polymerization system. The characterizations of the obtained products were performed by the same manners as those for the product from Diacid-8 and Diol-8 as described above. Consequently, it was indicated that the polyesters having methylene units of 10 and 12 were hardly included in the cavity of amylose. Furthermore, the ^{1}H NMR results of the products suggested that relatively larger amounts of the Diacids-10, -12 were included in the cavity of amylose compared with the case of the inclusion complex obtained using Diacid-8 and Diol-8. Because the hydrophobicities of Diacids-10 and -12 are stronger than those of Diacid-8, they would be readily included in the cavity of amylose. Actually, when amylose-forming polymerization were investigated in the presence of Diacid-8, Diacid-10, or Diacid-12, individually, the dicarboxylic acids having larger numbers of methylene units were easily included in the cavity of amylose. These results indicated that Diacids-10 and -12 were predominantly included in the cavity of amylose in the parallel enzymatic polymerization system to disturb the inclusion of the polyesters. In addition, the polyesters obtained using Diacids-10, -12 and Diols-10, -12 would be aggregated in the aqueous buffer more than those obtained using Diacid-8 and Diol-8 due to stronger hydrophobicity, and accordingly separated from the cavity of amylose.

4. CONTROL OF CRYSTALLINE STRUCTURE OF SYNTHETIC CELLULOSE IN CELLULASE-CATALYZED POLYMERIZATION

4.1. Chemical Synthesis of Cellulose

Cellulose, a natural linear polysaccharide linked through $(1\rightarrow4)$-β-glycosidic linkages of a dehydrated D-glucose repeating unit, is the most abundant organic substance on the earth. Despite a number of efforts at chemically synthesis of cellulose, it had not yet been achieved to control the structure of cellulose because of difficulty in stereocontrol of the anomeric C-1 carbon and difficulty in regioselectively control of many hydroxy groups with similar reactivity.

Chemical synthesis of cellulose was first achieved by the enzymatic polymerization of β-CF monomer catalyzed by cellulase in a mixed solvent of acetonitrile/acetate buffer (0.05 mol/L, pH 5) (5:1) at 30 °C for 12 h (Scheme 5) [17,18,35]. The resulting product was characterized by means of the solid-state CP/MAS ^{13}C NMR and IR measurements. Consequently, these spectra of synthetic cellulose were similar to those of natural cellulose. The resulting synthetic cellulose was readily hydrolyzed by cellulase to produce glucose and cellobiose. In addition, the thermal behavior of synthetic cellulose was completely like that of the natural sample; it decomposed ca at 260 °C without showing a melting point. These data indicated the structure of an exclusive (1→4)-β-glycosidic linkage, i.e., the stereochemistry at the C-1 carbon and the regioselectivity of the hydroxy groups were perfectly controlled without the protection of specific hydroxy groups in β-CF during the polymerization.

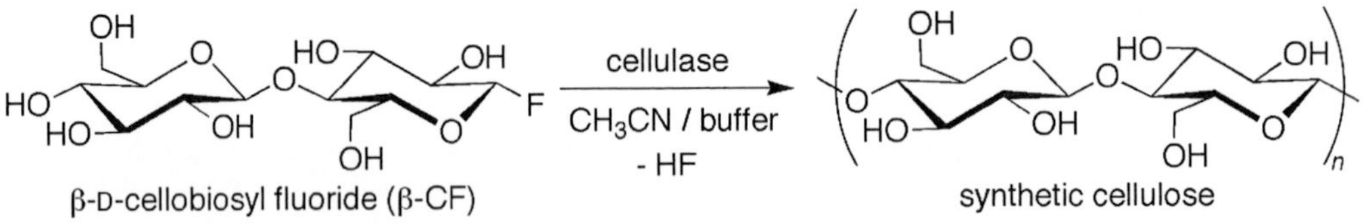

Scheme 5. Enzymatic polymerization of β-CF monomer catalyzed by cellulase.

4.2. Control of Crystalline Structures of Synthetic Cellulose

In the biosynthetic pathway for synthesis of cellulose, uridine 5'-diphospho-α-glucose is used as the monomer. Its polycondensation is catalyzed by the cellulose synthase enzyme. Cellulose forms two-types of different crystalline structures. One is cellulose I showing a parallel structure of glucan chains, and the other one is cellulose II showing an antiparallel structure. Generally, living cells produce only cellulose I, which is a thermodynamically metastable form, and it had long been believed to be produce only in living systems. In contrast, cellulose II is thermodynamically more stable, and so once cellulose I is converted into cellulose II, it never reverts back to cellulose I.

On the other hand, in the preparation of the artificial synthetic cellulose, the polymerization of β-CF by a crude cellulase in an acetonitrile/buffer (5:1) mixed solvent yielded cellulose II, whereas use of a partially purified cellulase as a catalyst in the same polymerization in acetonitrile/buffer (2:1) mixed solvent led to cellulose I (Figure 5) [17,18]. These results indicated that the formation of the synthetic cellulose I was achieved because of the purification and enrichment of the enzyme protein as catalyst for the polymerization and micelle formation due to the optimized acetonitrile/buffer ratio. There was an indication of microscopic micellar aggregates for organizing catalytic subunits to assemble synthetic cellulose I, although they were not as perfectly organized as the natural system (Figure 5). This was the first successful formation of cellulose I in a nonbiosynthetic way. Thus, the control of a nanostructure of synthetic cellulose was achieved by the polymerization of β-CF using cellulase with the different purification grades.

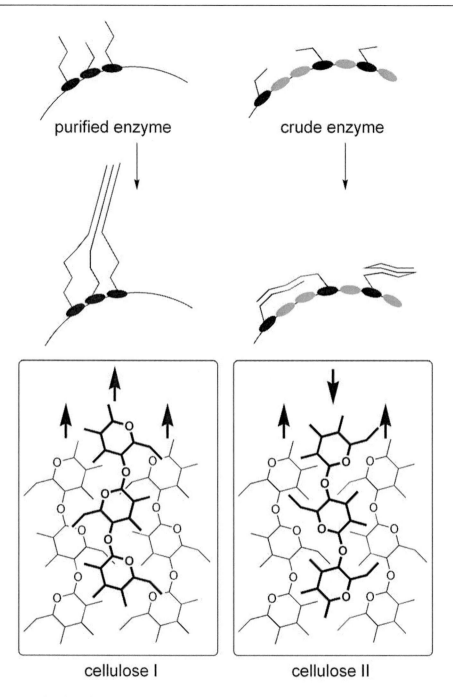

Figure 5. Postulated models for the formation of cellulose I by the purified enzyme (left) and the formation of cellulose II by the crude enzyme (right).

5. CONTROL OF CRYSTALLINE STRUCTURE OF POLYETHYLENE IN THE POLYMERIZATION WITHIN MESOPOROUS SILICA

In the preparation of biopolymer such as cellulose as described above, control of crystalline structure has been achieved by using an enzymatic catalyst, i.e., cellulase. On the other hand, recent developments in synthetic polymer chemistry have also enabled the control of not only primary structures such as molecular weight, co-monomer sequence, and stereo structure, but also nanostructures such as crystalline structure and morphology. In this part, we describe the preparation of crystalline polyethylene fibers by polymerization of ethylene with titanocene (Cp_2Ti, where Cp is the cyclopentadienyl ligand) supported by a mesoporous silica fiber (MSF) in the presence of methylalumoxane (MAO) as a cocatalyst (Figure 6) [19].

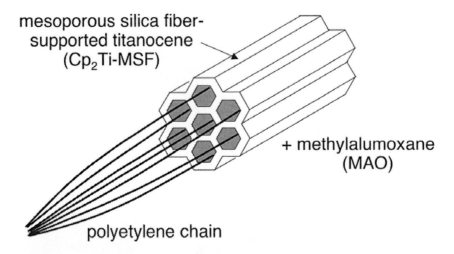

Figure 6. Concept for preparation of crystalline fibers of polyethylene by mesoporous silica-assisted extrusion polymerization

The MFS-supported titanocene (Cp_2Ti-MSF) was prepared by reaction of MSF with Cp_2TiCl_2 in the presence of triethylamine in a dichloromethane at 20 °C. The polymerization of ethylene with Cp_2Ti-MSF in the presence of MAO gave a cocoon-like solid mass consisting of fibrous polyethylene. The obtained polyethylene had an ultrahigh viscometric molecular weight (M_v = 6,200,000) and a higher density (1.01 g·cm^{-3}) than that of a polyethylene obtained with a homogeneous system, Cp_2TiCl_2-MAO, under similar conditions (0.97 g·cm^{-3}). The ^{13}C NMR spectrum of the polyethylene in 1,2,4-trichlorobenzene-C_6D_6 at 130 °C showed a CH_2 signal at δ 30.1 ppm, which indicates a linear sequence of the repeating ethylene units without any branch structures.

The scanning electron microscope (SEM) images of the freeze-dried polyethylene revealed bundles of polyethylene fibers. In addition, further magnified SEM image showed ultra-thin discrete fibers with a smooth surface and 30-50 nm in diameter. The XRD pattern showed typical crystalline structure of polyethylene, whereas an amorphous halo was negligibly small. The small-angle X-ray scattering (SAXS) of the crystalline polyethylene fibers showed only a diffuse scattering without any diffraction peaks due to folded-chain crystals, indicating that the polyethylene fibers consisted predominantly of extended-chain crystals. The DSC analysis of the obtained polyethylene showed a main endotherm at 140 °C, where the heat of fusion (ΔH) was evaluated to be 350 J•g^{-1}, whereas that of the polyethylene prepared with the homogeneous Cp$_2$TiCl$_2$-MAO system showed a endotherm at a lower temperature (130 $^{\circ}$C) with a much smaller ΔH of 207 J•g^{-1}. From these observations, it was indicated that each polyethylene fiber consisted of extended-chain crystals. These results probably indicated that polyethylene chain, formed at the activated titanocene sites within the individual mesopores, were extruded into the solvent phase and assembled to form extended-chain crystalline fibers.

6. CONCLUSION

Based on the viewpoint from the combination of polymerization and supramolecular chemistries, in this chapter, we reviewed the developments for control of the nanostructures of polymeric assemblies using the polymerization field. This type of research has often been compared with the phenomena taken place in biopolymeric assemblies. For example, because the high performances and the specific functions of the biological macromolecules are generally appeared by the control of their nanostructures, the researches described in this review have the important potentials to give the nanomaterials having the new functions. We conclude that this review hopefully suggests one of the future ways to the developing new research areas in nanomaterials chemistry.

REFERENCES

[1]   Stryer, L. (1995). *Biochemistry*. New York: W. H. Freeman & Company.
[2]   Kimura, T. (2003). Study on the effect of magnetic fields on polymeric materials and its application. *Polym. J.*, *35*, 823-843.
[3]   Bazbouz, M. B. & Stylios, G. K. (2008). Alignment and optimization of nylon 6 nanofibers by electrospinning. *J. Appl. Polym. Sci.*, *107*, 3023-3032.
[4]   Seguela, R. (1999). Processing - structure - property relationships in polymeric materials. *Rev. Metall.-Cah. Inf. Techn.*, *96*, 1477-1487.
[5]   Vögtle, F. (1989). *Supramolecular Chemistry: An Introduction*. Chichester: Wiley.
[6]   Lehn, J. M. (1995). *Supramolecular Chemistry*. Weinheim: VCH.
[7]   Lehn, J. M. (1990). Perspectives in supramolecular chemistry - from molecular recognition towards molecular information-processing and self-organization. *Angew. Chem.-Int. Edit.*, *29*, 1304-1319.
[8]   Kadokawa, J., Kaneko, Y., Tagaya, H., & Chiba, K. (2001). Synthesis of an amylose-polymer inclusion complex by enzymatic polymerization of glucose 1-phosphate catalyzed by phosphorylase enzyme in the presence of polyTHF: a new method for synthesis of polymer-polymer inclusion complexes. *Chem. Commun.*, 449-450.
[9]   Kadokawa, J., Kaneko, Y., Nakaya, A., & Tagaya, H. (2001). Formation of an amylose-polyester inclusion complex by means of phosphorylase-catalyzed enzymatic polymerization of α-D-glucose 1-phosphate monomer in the presence of poly(ε-caprolactone). *Macromolecules*, *34*, 6536-6528.

[10] Kadokawa, J., Kaneko, Y., Nagase, S., Takahashi, T., & Tagaya, H. (2002). Vine-twining polymerization: Amylose twines around polyethers to form amylose - Polyether inclusion complexes. *Chem. Eur. J.*, *8*, 3321-3326.

[11] Kadokawa, J., Nakaya, A., Kaneko, Y., & Tagaya, H. (2003). Preparation of inclusion complexes between amylose and ester-containing polymers by means of vine-twining polymerization. *Macromol. Chem. Phys.*, *204*, 1451-1457.

[12] Kaneko, Y. & Kadokawa, J. (2005). Vine-twining polymerization: A new preparation method for well-defined supramolecules composed of amylose and synthetic polymers. *Chem. Rec.*, *5*, 36-46.

[13] Kaneko, Y. & Kadokawa, J. (2006). Synthesis of nanostructured bio-related materials by hybridization of synthetic polymers with polysaccharides or saccharide residues. *J. Biomater. Sci., Polymer Edn.*, *17*, 1269-1284.

[14] Kaneko, Y., Beppu, K., & Kadokawa, J. (2007). Amylose selectively includes one from a mixture of two resemblant polyethers in vine-twining polymerization. *Biomacromolecules*, *8*, 2983-2985.

[15] Kaneko, Y., Beppu, K., & Kadokawa, J. (2008). Preparation of amylose/polycarbonate inclusion complexes by means of vine-twining polymerization. *Macromol. Chem. Phys.*, *209*, 1037-1042.

[16] Kaneko, Y., Saito, Y., Nakaya, A., Kadokawa, J., & Tagaya, H. (2008). Preparation of inclusion complexes composed of amylose and strongly hydrophobic polyesters in parallel enzymatic polymerization system. *Macromolecules*, *41*, 5665-5670.

[17] Kobayashi, S., Shoda, S., Lee, J., Okuda, K., & Brown, R. M. (1994). Direct visualization of synthetic cellulose formation via enzymatic polymerization using transmission electron-microscopy. *Macromol. Chem. Phys.*, *195*, 1319-1326.

[18] Kobayashi, S. (2005). Challenge of synthetic cellulose. *J. Polym. Sci. Pol. Chem.*, *43*, 693-710.

[19] Kageyama, K., Tamazawa, J., & Aida, T. (1999). Extrusion polymerization: Catalyzed synthesis of crystalline linear polyethylene nanofibers within a mesoporous silica. *Science*, *285*, 2113-2115.

[20] Kim, O. K., Choi, L. S., Zhang, H. Y., He, X. H., & Shih, Y. H. (1996). Second-harmonic generation by spontaneous self-poling of supramolecular thin films of an amylose-dye inclusion complex. *J. Am. Chem. Soc.*, *118*, 12220-12221.

[21] Choi, L. S. & Kim, O. K. (1998). Unusual thermochromic behavior of photoreactive dyes confined in helical amylose as inclusion complex. *Macromolecules*, *31*, 9406-9408.

[22] Lalush, I., Bar, H., Zakaria, I., Eichler, S., & Shimoni, E. (2005). Utilization of amylose-lipid complexes as molecular nanocapsules for conjugated linoleic acid. *Biomacromolecules*, *6*, 121-130.

[23] Sanji, T., Kato, N., Kato, M., & Tanaka, M. (2005). Helical folding in a helical channel: Chiroptical transcription of helical information through chiral wrapping. *Angew. Chem.-Int. Edit.*, *44*, 7301-7304.

[24] Sanji, T., Kato, N., & Tanaka, M. (2006). Chirality control in oligothiophene through chiral wrapping. *Org. Lett.*, *8*, 235-238.

[25] Kim, O. K., Je, J., & Melinger, J. S. (2006). One-dimensional energy/electron transfer through a helical channel. *J. Am. Chem. Soc.*, *128*, 4532-4533.

[26] Sanji, T., Kato, N., & Tanaka, M. (2006). Switching of optical activity in oligosilane through pH-responsive chiral wrapping with amylose. *Macromolecules*, *39*, 7508-7512.

[27] Shogren, R. L., Greene, R. V., & Wu, Y. V. (1991). Complexes of starch polysaccharides and poly(ethylene-co-acrylic acid) - structure and stability in solution. *J. Appl. Polym. Sci.*, *42*, 1701-1709.

[28] Shogren, R. L. (1993). Complexes of starch with telechelic poly(ε-caprolactone) phosphate. *Carbohydr. Polym.*, *22*, 93-98.

[29] Star, A., Steuerman, D. W., Heath, J. R., & Stoddart, J. F. (2002). Starched carbon nanotubes. *Angew. Chem.-Int. Edit.*, *41*, 2508-2512.

[30] Ikeda, M., Furusho, Y., Okoshi, K., Tanahara, S., Maeda, K., Nishino, S., Mori, T., & Yashima, E. (2006). A luminescent poly(phenylenevinylene)-amylose composite with supramolecular liquid crystallinity. *Angew. Chem.-Int. Edit.*, *45*, 6491-6495.

[31] Kida, T., Minabe, T., Okabe, S., & Akashi, M. (2007). Partially-methylated amyloses as effective hosts for inclusion complex formation with polymeric guests. *Chem. Commun.*, 1559-1561.

[32] Ziegast, G. & Pfannemüller, B. (1987). Phosphorolytic syntheses with di-, oligo- and multi-functional primers. *Carbohydr. Res.*, *160*, 185-204.

[33] Seneviratne, H. D. & Biliaderis, C. G. (1991). Action of α-amylases on amylose-lipid complex superstructures. *J. Cereal Sci.*, *13*, 129-143.

[34] Zobel, H. F. (1988). Starch crystal transformations and their industrial importance. *Starch*, *40*, 1-7.

[35] Kobayashi, S., Kashiwa, K., Kawasaki, T., & Shoda, S. (1991). Novel method for polysaccharide synthesis using an enzyme - the 1st in vitro synthesis of cellulose via a nonbiosynthetic path utilizing cellulase as catalyst. *J. Am. Chem. Soc.*, *113*, 3079-3084.

In: Modern Trends in Macromolecular Chemistry
Editor: Jon N. Lee

ISBN: 978-1-60741-252-6
© 2009 Nova Science Publishers, Inc.

Chapter 9

APPLICATION OF THE DIFFERENTIAL SCANNING CALORIMETRY IN MEASURING OF POLYMERIZATION KINETICS. ADVANTAGES AND LIMITATIONS

Dimitris S. Achilias[*]

Laboratory of Organic Chemical Technology, Department of Chemistry,
Aristotle University of Thessaloniki, Thessaloniki, Greece

ABSTRACT

In this study, a detailed investigation on the application of the Differential Scanning Calorimetry (DSC) in measuring polymerization kinetics is presented. The advantages of this technique are highlighted and the limitations are explained. Use of this method in a wide diversity of polymerization reactions resulting in linear (i.e. poly(methyl methacrylate), polystyrene, etc.), branched (poly(vinyl acetate), poly(butyl acrylate)) and crosslinked (formed from bis-phenol A glycidyl dimethacrylate, urethane dimethacrylate and triethylene glycol dimethacrylate) macromolecules is explored. Polymerization techniques such as in bulk or solution are also included.

The main advantages of the DSC include the use of a small sample mass, which assures that the reaction will be carried out isothermally during the whole conversion range and especially in the autoacceleration region. In advance, the measurements are continuous and not in discrete time intervals and reactions difficult-to-study with other techniques, such as those leading to crosslinked macromolecules can be equally well investigated. Finally, one of the greatest advantages of DSC is that it provides a direct measure of instantaneous rate of polymerization rather than conversion. Conversion histories are readily obtained from integration of the raw data. This process is inherently more accurate than evaluating rates from the slope of the conversion curve. Among the limitations of the method could be mentioned the non-adequate removal of oxygen in some cases and the possible loss of monomer during reaction due to evaporation at relatively high temperatures.

---

[*] Correspondence to: e-mail: axilias@chem.auth.gr

INTRODUCTION

Free radical polymerization is a process of major financial and scientific interest. It is for this reason that the free radical polymerization of vinyl monomers is one of the most extensively and thoroughly investigated organic reactions [1-2]. Early kinetic studies date back for about six decades. They revealed the radical chain nature of the reaction mechanism with its three essential elementary steps: chain initiation, propagation and termination.

The central equation of free-radical polymerization kinetics related the rate of polymerization, R_p, with the concentrations of the reactants, i.e. monomer [M] and initiator concentration, [I], as

$$R_p = -\frac{d[M]}{dt} = K[M][I]^{1/2} \tag{1}$$

From equation (1) it is noted that the rate of polymerization is a differential, instantaneous quantity. With only a few exceptions the primary data extracted from kinetic experiments on the other hand are cumulative, e.g. the monomer conversion, X.

The estimation of monomer conversion can in principle be performed in two distinct ways: the direct method of measuring the concentration of unreacted monomer by chemical analysis or by collecting and weighing the porduced polymer and on the other hand the indirect mesurement of monomer conversion by monitoring the change with time of a physical property of the reaction mixture.

Doubtless the application of the standard and tested methods of analytical chemistry to withdrawn and quenched samples is the most reliable way of following the course of a polymerization. In advance using this technique besides monomer conversion other properties of the polymer formed (i.e. molecular weight distribution, etc.) could be simultaneously recorded. However, these discontinuously working analytical procedures are often relatively laborious and time consuming. So, to avoid this expenditure of time and effort, indirect methods of conversion measurement without sampling are frequently used. The methods used to follow conversion in free radical polymerization reactions include: [3]

1. Gravimetry

The most direct way of obtaining conversion data in free-radical polymerization is the gravimetric determination of the produced polymer. Generally, sealed tubes are used for this purpose and the polymer is isolated by precipitation. This technique has sometimes been criticized as being to laborious extremely clumsy and susceptible to several errors, e.g. inclusion of monomer and/or solvent with the precipitated polymer, loss of low-molecular weight material, uncontrolled polymerization conditions for the different sealing tubes of a series (purity, induction time), etc. On the other hand gravimetry needs no calibration which might introduce a systematic error.

2. Dilatometry

Polymerization in the liquid phase is generally accompanied by a considerable decrease in volume. So, dilatometry is a highly sensitive and convenient indirect way of recording the progress of monomer conversion. The application of convntional dilatometers in polymerization is limited by two sources of error. If, as is often the case, the monomer itself is used to record the volume decrease, the increasing viscosity of the mixture soon causes a distortion of the meniscus in the capillary, which makes accurate reading impossible. In addition, the dissipation of heat by either convection or conduction is hindered by the viscous solution. It is for these reasons that special dilatometer devises have been built [3].

3. Measurement of Physical Properties Changing during Reaction such as Density, Viscosity, Refractive Index

Density

It requires a digital density measuring device and is based on a correlation of the monomer conversion with the change of the density of the reaction mixture. During bulk polymerization at high conversions the verry sticky reaction mixture plugs the measuring tube.

Viscosity

Viscosity measurements have been used in following the coarse of polymerization since 80 years, but as the viscosity of a polymer solution at a given temperature depends on both the concentration and the molecular weight of the produced polymer the evaluation of kinetic data from such experiments is rather complicated.

Refractive Index

Concentration changes during polymerization can be followed by measuring the corresponding changes in refractive index of the reaction mixture. This method is in many respects quite similar to the dilatometric technique and both methods are sensitive to temperature variations. But, while volume contraction for complete conversion are usually about 20%, the change in refractive index is considerably less (about 10%).

4. Chromatographic Methods

Chromatographic techniques can be very efficient tools for studying radical polymerization kinetics. Determination of residual monomer is by far the most important application of gas chromatography (GC). Other research areas are concerned with by-products or with the mechanism of chain initiation or termination. Under appropriate experimental conditions both conversion and molecular weight distribution of the polymer could be obtained in a single run using Size Exclusion Chromatography (SEC).

5. Spectroscopic Methods

Among the spectroscopic methods available, those operating in UV, or IR region are the most convenientand have a long tradition. Sophisticated mathematical methods of data analysis have been developed even for very complex reaction systems and Fourier Transformed-IR has been widely used. Raman spectroscopy has been also used alone or as a complementary analytical tool to the FTIR method.

6. Light Scattering

Rayleigh scattering of a monomer/polymer mixture has been also proposed to follow polymerization and estimate kinetic parameters. There is some similarity between this method and the viscosity method in as much as with these two techniques the signal obtained is a function of both polymer concentration and molecular weight. Simultaneous application of both static and dynamic light scattering has been proved a very promising technique.

7. Thermochemical Methods

Monomers used in free radical polymerization reactions include a double bond in their molecule. Polymerization of these vinyl monomers is accompanied by a significant heat release (polymerization enthalpy) due to the addition reaction to the double bond. Among thermochemical methods the most commonly employed one is the differential scanning calorimetry (DSC), where the output signal is proportional to the rate of heat production (dH/dt). DSC is a very sensitive and precise technique for measurement of the polymerization rate as a function of time, by monitoring the rate at which energy (heat) is released from the polymerizing sample. For homo-polymerizations the reaction rate is directly proportional to this. One of the greatest advantages of this method is that it provides a direct measure of the instantaneous reaction rate rather than conversion. The degree of monomer conversion (in terms of double bond conversion) is calculated by integrating the area between the DSC thermograms and the baseline established by extrapolation from the trace produced after complete polymerization (no change in the heat produced during the reaction). This process is inherently more accurate than evaluating rates from the slope of the conversion curve. In addition, the final conversion can be calculated, together with the maximum polymerization rate and the time to achieve it. Measurements can be easily carried out in a variety of experimental conditions including reaction temperature, initial initiator concentration, type of initiator used and monomer(s) chemical structure. Thus, overall kinetic rate constants and the effective activation energy can be easily estimated. Polymerizations with one, two or multi monomers initially present in the reaction mixture can be easily performed. The use of a DSC for the kinetic study of the isothermal bulk polymerization of methyl methacrylate (MMA) has been initially presented by Horie et al. [4]. Since then, several studies have dealt with the use of DSC for the study of different polymerization reactions [5-11]. The DSC has been an especially useful tool when applied in studies of the gel-effect (autoacceleration) as well as in crosslinking reactions [12].

In the following sections, results are presented on the application of the DSC in measuring the free radical polymerization kinetics. A number of monomers were examined including methyl methacrylate, styrene, vinyl acetate, butyl acrylate, bis-phenol A glycidyl dimethacrylate, bis-phenol A ethoxylated dimethacrylate and triethylene glycol dimethacrylate resulting in linear, branched or crosslinked macromolecules. Polymerization techniques such as in bulk or solution were also investigated. Moreover, results on the variation of conversion with time obtained from DSC were compared to measurements from other traditional techniques, such as gravimetry.

2. FREE-RADICAL POLYMERIZATION KINETICS

The chemistry and kinetics of free-radical polymerization has been described in detail in several textbooks [1,2]. A simple mechanism of free-radical polymerization can be derived in terms of the following four elementary reactions:

Initiation:

$$I \xrightarrow{k_d} 2I^\bullet \tag{2}$$

$$I^\bullet + M \xrightarrow{k_I} R_1^\bullet \tag{3}$$

Propagation:

$$R_n^\bullet + M \xrightarrow{k_p} R_{n+1}^\bullet \tag{4}$$

Chain transfer to monomer:

$$R_n^\bullet + M \xrightarrow{k_{tr,m}} R_1^\bullet + D_n \tag{5}$$

Termination by combination / disproportionation:

$$R_n^\bullet + R_m^\bullet \begin{cases} \xrightarrow{k_{tc}} D_{n+m} \\ \\ \xrightarrow{k_{td}} D_n + D_m \end{cases} \tag{6}$$

In the above kinetic scheme, the symbols I, $I^\bullet$ and M denote the initiator, radicals formed by the fragmentation of the initiator and monomer molecules, respectively. The symbols

$R_n^\bullet$ $and\, D_n$ are used to identify the respective "live" macroradicals and the "dead" polymer chains, containing n monomer structural units. Finally, k_d, k_p, $k_{tr,m}$, k_{tc} and k_{td}, denote the respective rate constants of the initiator decomposition, propagation, chain transfer to monomer, termination by combination and termination by disproportionation reactions.

The reaction starts with the thermal decomposition of an initiator, such as azo-bis-isobutyronitrile (AIBN), or benzoyl peroxide (BPO) and the formation of a primary radical. These radicals find monomer molecules and react to form macroradicals, which increase in length through the propagation reaction. They eventually find one another to terminate and create 'dead' macromolecules. As an example the free radical polymerization of methyl methacrylate using azo-bis-isobutyronitrile (AIBN) as an initiator is presented next.

Initiation reaction

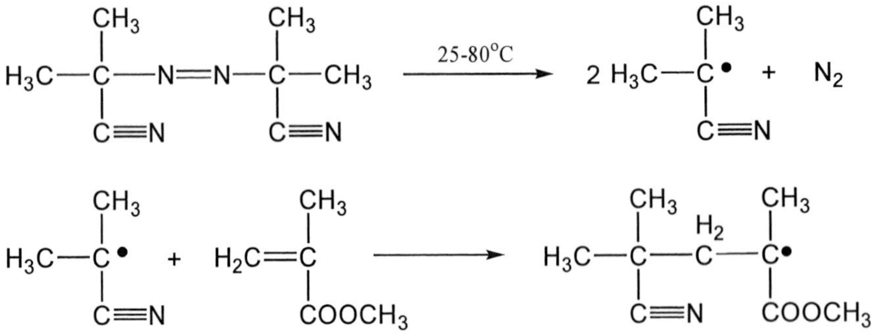

Propagation

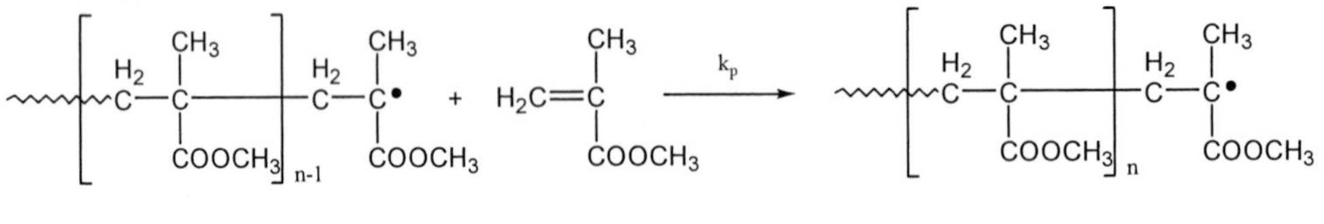

Termination by combination

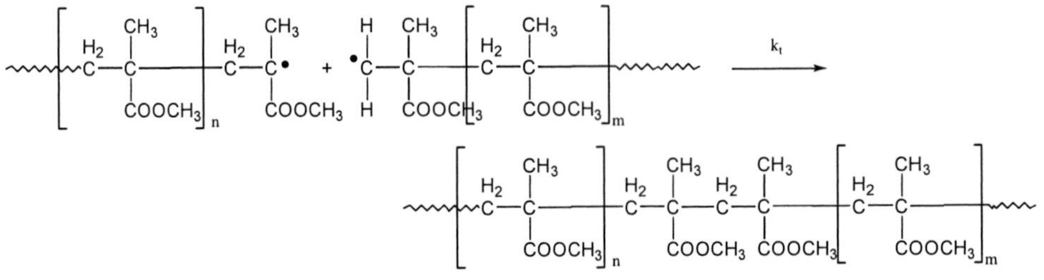

Termination by disproportionation

To describe the progress of the reaction and molecular weight or chain structural developments during polymerization, population mass balance equations are derived for all chemical species present in the reactor. These constitute a set of simultaneous differential equations which are usually solved numerically provided that the appropriate rate constants of every elementary reaction are known.[13,14] However, the term 'rate constants' in equations (2)-(6) is somewhat of a misnomer,[15] as these so-called rate constants vary during the course of any polymerization. This variation was assumed in order to quantitatively describe the effect of diffusion-controlled phenomena on the polymerization kinetics. That means that k_i s' appearing in equations (2) to (6), which are influenced by diffusional phenomena are to be regarded as "apparent rate constants" or "rate coefficients" rather than "rate constants".[16-18] In free-radical polymerization, the presence of diffusion limitations is so well documented that different manifestations have been given particular names. The impact of diffusion on the termination step is labelled as the Trommsdorff or gel-effect, while the effects on the propagation and initiation reactions are known as the glass and the cage-effect, respectively. Different theories for the modelling of such effects have been proposed [19-22]. A short description of the theoretical process follows.

As conversion increases, the polymer chains and the macroradicals begin to form entanglements and translational as well as segmental diffusion are significantly retarded. The situation becomes more complex since the radical besides diffusion it also continuously increases with time via the propagation step. As a consequence, during the lifetime of a radical, no single molecular weight can realistically represent the radical, making it difficult to apply the scaling concepts. Also, at high degrees of conversion, besides center-of-mass diffusion and segmental reorientation, the description of the diffusion behavior of propagating polymer coils can be further complicated by other modes of diffusion, such as reaction diffusion. The later is associated with the diffusive motion of the macroradical chain end as a result of propagation, which lengthens the radical chain and eventually moves the radical end in spatial position [20]. All these different modes of diffusion however are not equally important over the entire range of conversion and therefore the polymerization is divided into three[19] or sometimes four conversion regimes.[20] These could be classified from macroscopic measurements of the rate of polymerization or monomer conversion versus time (Figure 1). As it has been reported,[20] in the first stage of polymerization (low conversions), the conversion – time curve as well as polymerization rate Rp, versus time t, follows the 'classical' free-radical kinetics and all kinetic rate coefficients remain constant. A plot of – ln(1-X) versus t is almost linear. The crossover between regime I and II denotes the onset of the gel-effect and it corresponds to 10-40% conversion. Regime II is characterized by a sharp increase in the polymerization rate followed by an increase in conversion. The maximum in the Rp versus time curve marks the crossover between regimes II and III. In regime III, the reaction rate falls significantly and the curvature of the conversion versus time changes. Finally, at very high conversions beyond 80 to 90% the reaction rate tends asymptotically to zero and the reaction almost stops before the full consumption of the monomer (regime IV). It is a situation happening when the polymerization temperature is below the glass transition temperature and at this point the T_g of the monomer-polymer mixture approaches reaction temperature, thus a glassy state appears and it corresponds to the well known glass-effect.

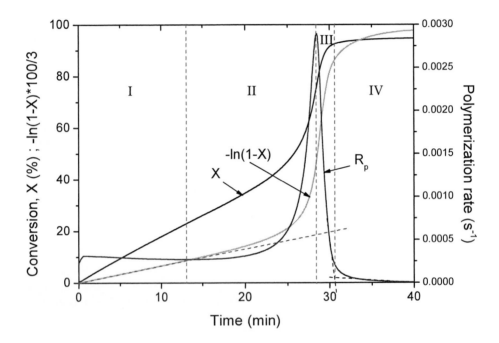

Figure 1. Indicative results on polymerization rate, R_p, conversion, X and -ln(1-X) versus time from polymerization of MMA at 80°C with AIBN 0.03 mol/L, presenting the classification of reaction into four regimes [22].

Simple mass balance equations describing the time variation of the initiator, monomer and total radical concentration follow. Initiator is consumed only in equation (1), then, according to the kinetic scheme, the initiator concentration, [I], can be estimated by integrating the corresponding differential mass balance equation:

$$-\frac{d[I]}{dt} = k_d[I] \quad \Rightarrow \quad [I] = [I]_0 e^{-k_d t} \tag{7}$$

The rate of monomer consumption, which is equal to the polymerization rate, is calculated from the following differential equation, according to the kinetic scheme (eqs 3-5). Equation (8) is further simplified using the long-chain assumption which states that the rate of monomer consumption in the initiation reaction is much smaller compared to that in the propagation reaction:

$$-\frac{d[M]}{dt} = k_p[M][R^\bullet] + k_{tr,m}[M][R^\bullet] + k_i[M][I^\bullet] \cong k_p[M][R^\bullet] \tag{8}$$

where $[R^\bullet]$ denotes the total radical concentration and its time variation is given by.

$$\frac{d[R^\bullet]}{dt} = 2fk_d[I] - 2k_t[R^\bullet]^2 \tag{9}$$

Assuming that all kinetic rate-constant are chain-length and conversion-independent and applying the steady state approximation for the sum of the free radicals, equation (9) is transformed to

$$\frac{d[R^\bullet]}{dt} = 0 \quad \Rightarrow \quad 2fk_d[I] = 2k_t[R^\bullet]^2 \quad \Rightarrow \quad [R^\bullet] = \left(\frac{fk_d[I]}{k_t}\right)^{1/2} \tag{10}$$

Substituting equation (10) into equation (8) results

$$R_p = -\frac{d[M]}{dt} = k_p[M]\left(\frac{fk_d[I]}{k_t}\right)^{1/2} \quad \Rightarrow \quad \frac{dX}{dt} = k_p(1-X)\left(\frac{fk_d[I]}{k_t}\right)^{1/2} \tag{11}$$

where X denotes fractional monomer conversion

Equation (11) indicates a first order dependence of the rate of polymerization on the monomer concentration and a square-root dependence on the concentration of the initiator.

EXPERIMENTAL

Materials

The initiators used were, either 2,2'-azo-bis-isobutyronitrile (AIBN) (Akzo Chemie Ltd) or a combination of benzoyl peroxide (BPO) with an amine and they were recrystallized twice from methanol. The monomers used were methyl methacrylate (MMA), styrene, vinyl acetate (VAC), butyl acrylate (BuA), vinyl neodecanoate (VnD) and triethylene glycol dimethacrylate, from Aldrich and most of them contained hydroquinone. In order to remove the inhibitor, the monomers were passed at least twice through a disposable packing bed column obtained from Aldrich and stored in the refrigerator until use. The monomers were also degassed immediately prior to polymerization.

Procedure

Polymerization was investigated using the DSC, Pyris 1 (from Perkin-Elmer) equipped with the Pyris software for windows. Indium was used for the enthalpy and temperature calibration of the instrument. Isothermal polymerizations were carried out at different temperatures, circulating oxygen-free nitrogen in the DSC cell outside the pans in order to avoid atmospheric oxygen supply into the sample. The reaction temperature was recorded and maintained constant (within ± 0.01 °C) during the whole conversion range. Reaction mixtures were prepared by weighing the appropriate amount of initiator and dissolving it into the monomer. Samples of these solutions were placed in aluminium Perkin-Elmer sample pans, accurately weighted (10 to 20 mg) sealed and then placed into the appropriate position of the instrument.

The reaction exotherm (in normalized values, W/g) at a constant temperature was recorded as a function of time. The rate of heat release ($d(\Delta H)/dt$, W/g) measured by the DSC was directly converted into the overall reaction rate (dX/dt, s^{-1}) using the following formula:

$$R_p = \frac{dX}{dt} = \frac{1}{\Delta H_T}\frac{d(\Delta H)}{dt} \tag{12}$$

In which ΔH_T [J/g] denotes the total reaction enthalpy released from the reaction of all double bonds in the monomer molecule and is calculated from the product of the number of double bonds per monomer molecule ($n = 1$ or 2) times the standard heat of polymerization of a methacrylate double bond ($\Delta H_0 = 54.9$ kJ/mol) or in general of the monomer studied double bond (i.e. $\Delta H_{VAC} = 88$ kJ/mol) over the monomer mixture molecular weight, i.e. $\Delta H_T = n\, \Delta H_0$ / MW_m.

The polymerization enthalpy and conversion were calculated by integrating the area between the DSC thermograms and the baseline established by extrapolation from the trace produced after complete polymerization (no change in the heat produced during the reaction). The residual monomer content and the total reaction enthalpy can be determined by heating the sample from the polymerization temperature to 180°C at a rate of 10 K/min. The sum of enthalpies of the isothermal plus the dynamic experiment was the total reaction enthalpy. After the end of the polymerization the pans were weighted again and a negligible loss of monomer (less than 0.2 mg) was observed only in a few experiments.

In order to compare the conversion-time measurements obtained from the DSC, additional polymerizations of styrene with AIBN was also carried out in test tubes. A known amount of a monomer-initiator solution was placed into several test tubes. The tubes were sealed, purged with nitrogen and placed in a constant temperature water bath. The tubes were removed from the bath at different times and frozen to stop the reaction. The contents were dissolved in methylene chloride, precipitated in methanol or hexane, dried in a vacuum oven and weighed for conversion.

All the experimental results reported in the following section were taken from an average of at least two experiments.

RESULTS

1. Comparison of DSC Data with Other Experimental Techniques

In order to test the reliability of the experimental measurements obtained from DSC, results on monomer conversion were initially compared to corresponding from the standard and absolute method of gravimetry. In order to fuldil this task exactly the same experiment was repeated twice one in the DSC and the second using test-tubes using exactly the same experimental conditions and materials. The bulk polymerization of styrene (which is perhaps the most thoroughly documentated vinyl polymerization) was investigated as a model polymer and polymerization technique. AIBN was used as a free radical initiator at a concentration of 0.1 M and the polymerization temperature was kept constant at 90°C. Polymerization rate data taken from the DSC were converted to conversion versus time and appear in Figure 2, together with the gravimetric measurements.

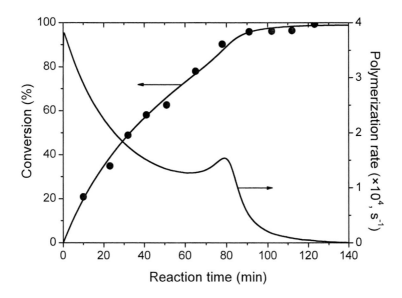

Figure 2. Polymerization rate and monomer conversion versus time obtained from DSC measurements (continuous line) or gravimetry discrete data points, during bulk polymerization of styrene at 90°C using AIBN initiator at 0.1 M.

Despite the fact that DSC provides double bond conversion, while gravimetric measurements represent direct polymer weight, results are in very good agreement and almost identical. Therefore, it was concluded that results obtained could be considered accurate and reliable. An interesting point coming from the DSC measurements on polymerization rate is related to the peak observed at approximately 80 min. This is attributed to the effect of diffusion-controlled phenomena on the reaction rate, according to what it has been mentioned in the previous section. However, using only the conversion measurements from garvimetry, the gel-effect is not so clear. This is a first indication on the amount of information that can be gained from the DSC measurements.

2. Bulk Polymerization – Formation of Linear Polymers

Test of the Isothermal Hypothesis

Before proceeding with the study of the effect of several parameters on the polymerization rate data, the isothermal assumption was examined. As it was mentioned previously, one serious problem associated with the free radical polymerization of vinyl monomers is the production of a large amount of energy (heat) during the course of the reaction. This is especially pronounced with bulk polymerizations in the region of the so-called auto-acceleration, or gel-effect, or Trommsdorff-Norrish effect. Therefore, a significant temperature rise was measured during these polymerizations, questioning the results carried out under isothermal conditions. MMA is such a monomer presenting the auto-acceleration effect in a great extent and generating a large amount of heat during polymerization. A number of papers have been published questioning the isothermal conditions during the reaction [8,22]. Results on polymerization rate and reaction temperature during synthesis of PMMA at two temperatures 80 and 90°C appear in Figure 3. As it can be seen from this Figure, although the reaction rate increases during autoacceleration by an order of magnitude, the temperature in the reaction medium does not increase more than 0.05°C. Therefore, it can be postulated that DSC is able to keep strictly isothermal conditions during the whole course of polymerization.

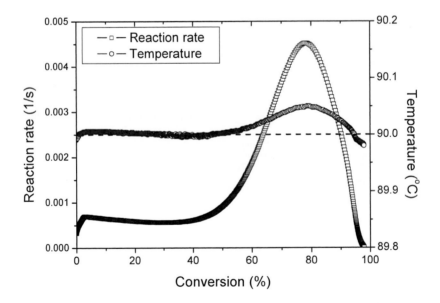

Figure 3. (Continued)

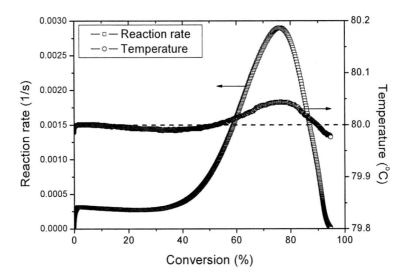

Figure 3. Polymerization rate and reaction temperature versus monomer conversion during bulk polymerization of MMA at 90°C (a) and 80°C (b) using AIBN as a free radical initiator at an initial concentration of 0.03 M.

Determination of the Onset of Diffusion-Controlled Phenomena

Furthermore, the application of DSC in determining the onset of the effect of diffusion-controlled phenomena on polymerization rate is demonstrated. From a rearrangement of equation (11) the following eq. (13) results:

$$R_p = \frac{dX}{dt} = k_p \left(\frac{fk_d}{k_t} \right)^{1/2} [I]^{1/2}(1-X) = k'[I]^{1/2}(1-X) \cong k(1-X) \Rightarrow \frac{R_p}{(1-X)} = k \quad (13)$$

where

$$k' = k_p \left(\frac{fk_d}{k_t} \right)^{1/2} \qquad and \qquad k = k'[I]^{1/2}$$

Then, in the absence of diffusion-controlled phenomena on the propagation and termination rate constants, as well as negligible initiator consumption by plotting $R_p/(1-X)$ versus conversion or time a straight line parralel to the x-axis should be obtained. Such a plot appears in Figure 4. As it can be seen in the embedded figure, initially a constant value holds for k with increased values with reaction temperature. The point where this line deviats from the initial constant value denotes the onset of diffusion-controlled phenomena. This value ranges between 10 to 12% depending on polymerization temperature and is lower to the corresponding values estimated by other techniques (for example by taking the deviation from linearity of the $-\ln(1-X)$ versus time curve).

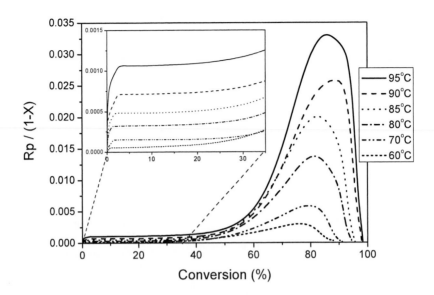

Figure 4. Plot of Rp/(1-X) versus conversion for the free radical polymerization of MMA at different temperatures.

Effect of Temperature

Furhermore, results are presented on the bulk polymerization of MMA, a model polymer extensively studied in literature. Kinetic data for this system are readily available in literature from numerous independent investigators [4-6, 8-10]. Polymerization rate data were collected at different reaction temperatures and initial initiator concentrations. In particular the following set of experiments was carried out:

1. Isothermal bulk polymerization at 0.03 M initial AIBN concentration and temperatures 60, 70, 80 ,85, 90 and 95°C and
2. Isothermal bulk polymerization at 80°C and initial AIBN concentrations of 0.01, 0.03, 0.05, 0.07 and 0.09 mol/L, presented in the following section.

The effect of temperature on the reaction rate and conversion variation with time appears in Figure 5a & b.

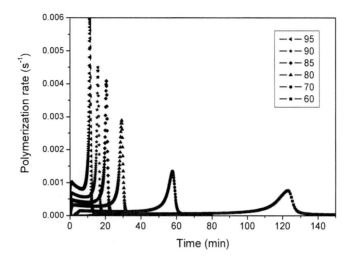

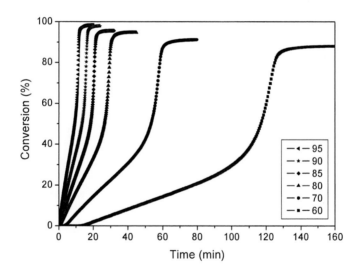

Figure 5. Polymerization rate (a) and conversion (b) versus time during bulk polymerization of MMA at different temperatures. Initial initiator concentration 0.03 M.

As it is expected an increase in reaction temperature results in increased reaction rates and higher final conversion values. Polymerization lasts for 160 min at 60°C, while only for 15 min at 95°C. Higher reaction temperatures also result in higher final degrees of conversion as well as higher maximum reaction rates.

All polymerizations showed a short induction period before the onset of the reaction. The induction period was found to decrease with increasing polymerization temperature and becomes significant at the lowest temperature (i.e. 60°C). The induction period appears to be caused by oxygen inhibition of the polymerization caused by small amounts of oxygen encapsulated inside the sealed sample pans. The presence of an inhibition time is sometimes beneficial, since it enables stabilization of the sample temperature in DSC and achievement of a good baseline before the beginning of polymerization.

In order to quantify the effect of temperature on the reaction kinetics, equation (11) can be integrated assuming that all kinetic rate constants and initiator efficiency are constant to yield an expression which directly correlates the monomer conversion with an observed overall kinetic rate coefficient, k. It should be noted that equation (13) is valid only for low degrees of monomer conversion:

$$-\ln(1-X) = kt \tag{13}$$

The overall kinetic rate constant, k can be obtained from the slope of the initial linear part of the plot of -ln(1-X) versus t. Such plots at conversion values in between 1% and 10% (i.e 1% < X < 10%) appear in Figure 6. The lowest limit was taken such as to eliminate the inhibition period, while the highest value was well below the onset of diffusion-controlled phenomena. The experimental data fit very well to straight lines at all different temperatures, indicating the validity of equation (13) in the specific conversion interval.

Considering the temperature dependence of the polymerization rate, it is given by the temperature dependence of the individual rate coefficients. Each rate coefficient follows its own Arrhenious law, $k_i = A_i \exp(-E_i/RT)$, where A_i is the pre-exponential factor and E_i denotes the activation energy. According to the definition of the observed overall kinetic rate coefficient, k:

$$k = k_p \left(\frac{fk_d}{k_t} \right)^{1/2} [I]^{1/2} \tag{14}$$

the overall activation energy of the polymerization rate, E_R should be given by the activation energies of the elementary reactions, propagation (E_p), initiation (E_i) and termination (E_t), according to:

$$E_R = E_p + \frac{1}{2}(E_i - E_t) \tag{15}$$

Accordingly, from the slope of ln(k) versus 1/T the polymerization reaction overall activation energy could be obtained. As it can be seen in Figure 7, all data follow a good straight line with a slope providing an activation energy equal to 84 kJ/mol and an overall kinetic rate coefficient, assuming constant initiator concentration, equal to:

$$k' \ (L^{1/2} mol^{-1/2} s^{-1}) = 5.2 \ 10^9 \ \exp(-84100/RT)$$

The values of E_R and A_R are very close to the literature values reported in Table 1. Some differences appear especially with earlier publications, since these authors had not repoved the inhibitor from the monomer or the accuracy of the instrument used was not so high.

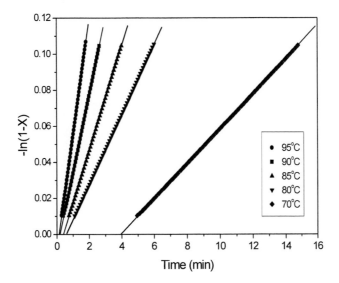

Figure 6. Plot of –ln(1-X) versus time for the bulk polymerization of MMA at different temperatures

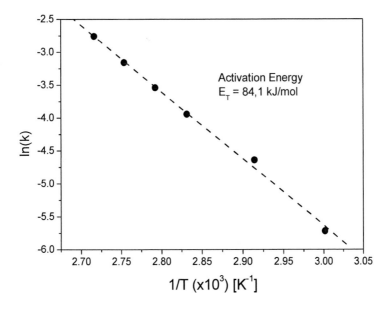

Figure 7. Arrhenius-type plot for the estimation of the overall activation energy during bulk free radical polymerization of MMA

Table 1. Experimental and literature values of the overall pre-exponential factor and activation energy together with the overall kinetic rate constant at 70°C for the free radical polymerization of MMA.

Method used	Pre-exponential factor, A ($*10^{-9}L^{1/2}mol^{-1/2}min^{-1}$)	Activation energy, E (kJ/mol)	k' (at 70°C) ($*10^2 L^{1/2}mol^{-1/2}min^{-1}$)	Reference
DSC	0.0894	62.0	3.24	[4]
DSC	6.18	74.0	3.36	[5]
DSC	318	84.9	3.78	[8]
DSC	804	85.5	6.90	[9]
DSC	245	83.1	5.34	[10]
DSC	29.4	76.8	5.82	[6]
Dilatometer	186	82.6	4.92	[23]
DSC	314	84.1	5.58	This study

Effect of Initial Initiator Concentration

In another series of experiments, the effect of the initial initiator concentration on reaction rate and conversion as a function of time was investigated. Results appear in Figure 8.

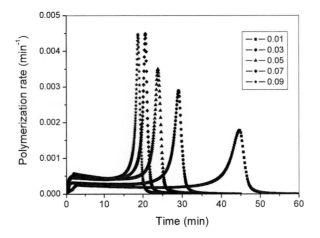

Figure 8 (Continued).

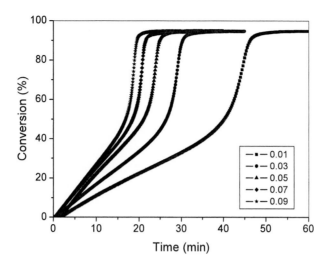

Figure 8. Polymerization rate (a) and conversion (b) versus time during bulk polymerization of MMA at different initial initiator concentrations, T=80°C.

An increase in the initial initiator concentration results in the production of a larger amount of free radical increasing thus the polymerization rate and resulting in completion of the reaction in a shorter time. The final double bond conversion does not seem to depend on the initial initiator concentration. The gel-effect and the glass-effect are obvious at all experimental conditions. A short induction period was measured only at the lower initial initiator concentrations. Furthermore, a plot of the overall kinetic rate coefficient, k versus $[I_0]^{1/2}$ appears in Figure 9. As it can be seen a satisfactory straight line was obtained verifying again equation (13).

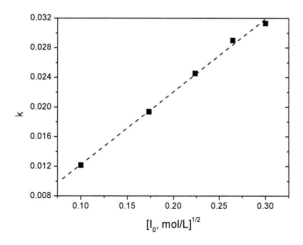

Figure 9. Dependence of k on $[I_0]^{1/2}$ during free radical polymerization of MMA at 80°C

Effect of Monomer Chemical Structure

The next step was to investigate the effect of the monomer chemical structure on the shape of the polymerization rate curve. Comparative data for the polymerization of MMA, BuA, VAC and VnD appear in Figure 10a&b.

Depending on monomer type significantly different reaction rate curves were obtained. BuA presented the longest induction time, while gel-effect started immediately after the beginning of polymerization. VAC exhibited the higher initial reaction rate which was further increased due to the effect of diffusion-controlled phenomena on the termination rate constant. The strong autoacceleration effect appeared in MMA polymerization does not appear in VnD, while in VAC appeared with a lower intensity. A limiting conversion was obtained after 20 min of polymerization in BuA and VAC, whereas after 30 min for MMA and more than 50 min in VnD reaction.

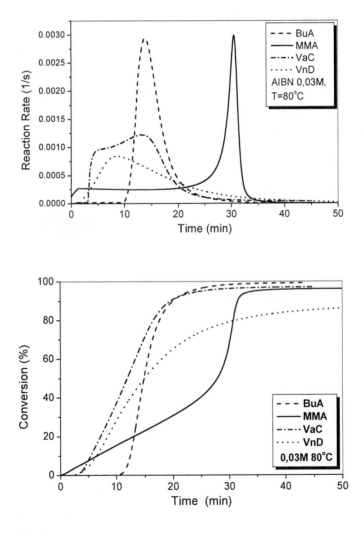

Figure 10. Polymerization rate (a) and conversion (b) versus time during bulk polymerization of MMA, BuA, VAc and VnD at 0.03M AIBN and T=80°C.

3. Bulk Polymerization – Formation of Branched Polymers

Furthermore, polymerization resulting in branched polymers was investigated. As a model polymer in this category the polymerization of VAC with AIBN initiator was studied. Results on the effect of temperature on polymerization rate and double bond conversion appear in Figure 11a&b. Again an increase in temperature promotes the reaction, which is completed at 15 min at 90°C compared to more than 100 min at 60°C.

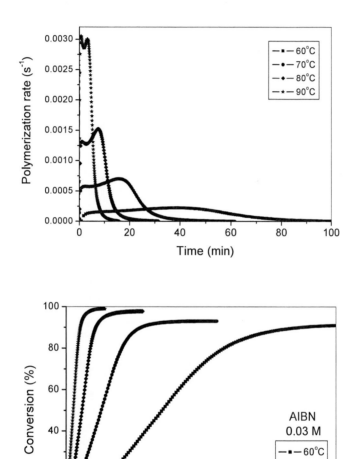

Figure 11. Effect of temperature on polymerization rate (a) and conversion (b) versus time during bulk polymerization of VAC with AIBN initiator at an initial concentration of 0.03 M.

It is obvious that the DSC technique could be equally well applied to polymerizations leading to branched macromolecules.

4. Bulk Polymerization – Formation of Crosslinked Networks

In this section some results are presented on the application of the DSC technique in polymerizations leading in crosslinked network structures. As such macromolecules those used as biomaterials in dental applications were investigated. The most widely used resin in dental composites is that based on the copolymer prepared from a combination of 2,2-bis[p-(2'-hydroxy-3'methacryloxypropoxy)phenylene]propane (Bis-GMA) and triethylene glycol dimethacrylate (TEGDMA) [24]. TEGDMA is usually added to Bis-GMA in order to achieve workable viscosity limits, since the latter monomer possesses very high viscosity due to the intermolecular hydrogen bonding. Urethane dimethacrylate (UDMA) is another monomer used in combination or in replacement of Bis-GMA. In order to carry out the reaction in a short time-period at body temperature (37°C), a redox initiation system is employed consisted of a tertiary aromatic amine together with benzoyl peroxide (BPO). Many amines have been suggested as accelerators, but esthetic and biocompatibility requirements have greatly limited the number of compounds, which can be used for dental or medical applications. The most commonly used amine is N,N dimethyl-p-toluidine (DMT) which is a suspected, but yet not proven, carcinogen. For this purpose several other amines have been proposed with better biocompatibility than DMT. Among them the 4-N,N dimethylaminophenethyl alcohol (DMPOH), has been proposed as a more biocompatible and highly reactive accelerator for the polymerization of dental composites [25]. With low amine concentration, nearly colorless restorations having good color stability were obtained. The kinetics of the BPO/DMPOH redox system compared to BPO/DMT for the polymerization of MMA have been recently studied [25]. A point of concern in BPO/amine initiated polymerizations is whether the maximum rate occurs when equimolar, or not equimolar initial initiator concentrations are used. In a previous publication it was found that when the product of [BPO][Amine] was equal to 0.001 M then the maximum rate occurs when the ratio of [BPO]/[Amine] is approximately equal to 1.5 [25]. It is worthy to note that generally it is undesirable to have an excess of amine remaining in the polymer, because of possible adverse effects on its properties and biocompatibility.

In this section the kinetics of the BPO/DMPOH initiated free radical polymerization of TEGDMA are investigated using DSC for different initial initiator concentrations. The initiator combinations were selected such as the reaction to be completed in approximately 10 min since the specifications for dental direct filling resins requires a minimum working time of 1.5 min and a maximum hardening time of 8 min. The amine concentration was not exceeded 50 mM (due to material yellowing and biocompatibility reasons).

Figure 12a shows the measured polymerization rate versus time for the polymerization of TEGDMA at 37°C obtained under different initial initiator concentrations. As it can be seen, an increase in the initial initiator concentrations leads to higher reaction rates, shortening of the inhibition time and completion of polymerization in only 8 min. At such fast reaction rates it is very difficult for other experimental techniques to follow the reaction. By integrating the reaction rate the double bond conversion with respect to time can be obtained which is plotted in Figure 12b.

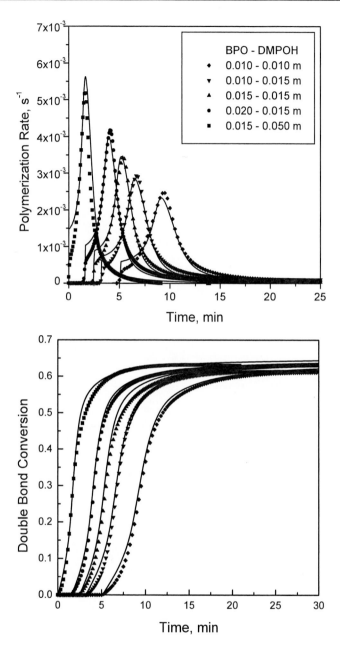

Figure 12. Effect of the initial ratio of the two initiators BPO/DMPOH on the polymerization rate (a) and double bond conversion (b) versus time during bulk polymerization of TEGDMA at T=37°C [24].

Moreover, the effect of the type of the initiator system used was investigated. Thus, different amines were employed and the profiles of polymerization rate versus time or conversion for Bis-GMA appear in Figure 13. It was observed that the type of amine does not influence much the reaction characteristics. Polymerization showed a sharp maximum rate soon after the beginning (< 2min) and a very low conversion (< 5%). It seems that the structure of amine has a slight influence on kinetic parameters, with the 4-N,N-dimethylaminophenylacetic acid (DMAPAA) giving better results followed by DMPOH. The polymerization rate is a function of the product of initiator efficiency and decomposition constant (fk_d). Not all the free radicals produced are capable of initiating polymerization, because alternate pathways exist which do not lead to initiation of polymerization. Any such wastage reactions lower the efficiency. It was found that efficiency of the BPO/amine system is enhanced by the presence of electron-withdrawing substituents on the aniline. Also it was found that electron-donating substituents which increase the nucleophilicity of the amine increase the rate constant of decomposition. The substituents p-CH_2CH_2OH and p-CH_2COOH of DMPOH and DMAPAA correspondingly have electron donating character as the p-CH_3 group of DMT. Therefore it is difficult to explain the observed slightly higher reactivity of BPO/DMPOH and BPO/DMAPAA initiating system compared to that of BPO/DMT for the polymerization of Bis-GMA.

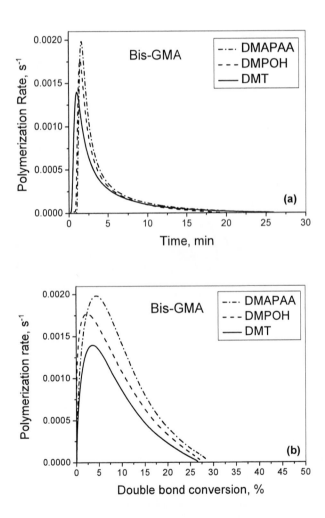

Figure 13. Effect of the amine chemical structure on the profile of the polymerization rate versus time (a) or double bond conversion (b) for the Bis- phenol-A-bis(glycidyl-methacrylate) (Bis-GMA) polymerization at 37°C, initiated by the benzoyl peroxide (BPO)/amine system.[BPO]=[amine]=0.5mol% [26]

Finally, Figure 14 shows the effect of the monomer chemical structure on the rate profiles of polymerizations initiated by the same BPO/DMPOH couple. UDMA compared with Bis-GMA showed a higher value for R_p^{max} ($\sim 4 \times 10^{-3}$ s^{-1}), while the R_p^{max} of Bis-GMA was about 2×10^{-3} s^{-1}. The much higher reactivity of UDMA than Bis-GMA monomer was attributed to both the greater flexibility of the UDMA molecular structure and the possible hydrogen abstraction and a chain transfer reaction mechanism, which enhances radical initiation or causes an alternate polymerization pathway. Moreover, TEGDMA showed the highest values of degree of conversion followed by UDMA and Bis-GMA. It seems that the degree of conversion is mainly affected by the chemical structure of monomer, which defines the mobility of the polymer network being formed and the system viscosity and it is not affected significantly by the structure of the amine used.

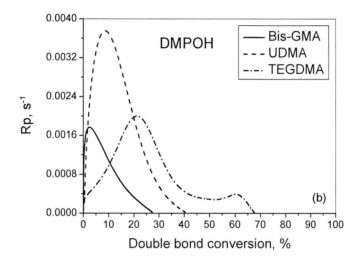

Figure 14. Effect of the monomer chemical structure on the profile of the polymerization rate versus conversion for the BPO/DMPOH initiated polymerization at 37°C. [DMPOH]=[BPO]= 0.5mol% [26]

5. Solution Polymerization

Differential scanning calorimetry could be also applied in monitoring polymerization reactions taking place in solution. As such the solution polymerization of vinyl acetate was investigated in two different solvents, namely toluene and t-butanol. The effect of the amount of solvent on the polymerization rate and double bond conversion during solution polymerization of VAC in t-butanol and toluene appear in Figures 15a&b and 16a&b, respectively.

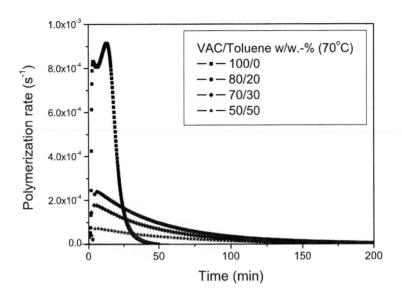

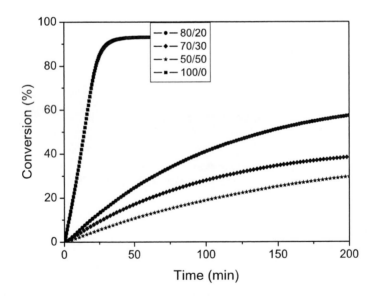

Figure 15. Effect of the amount of solvent on polymerization rate (a) and conversion (b) versus time during solution polymerization of VAC in Toluene at 70°C using AIBN initiator 0.03 M.

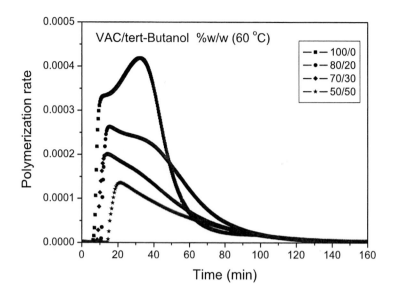

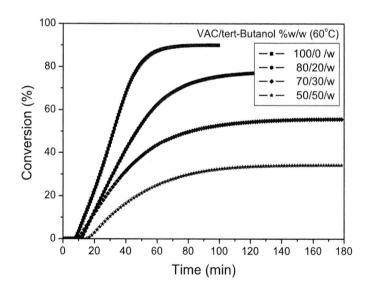

Figure 16. Effect of the amount of solvent on polymerization rate (a) and conversion (b) versus time during solution polymerization of VAC in t-butanol at 60°C using AIBN initiator 0.03 M.

It is seen that as the amount of solvent is increased the gel-effect is suppresed the reaction rate is much lowered and the reaction lasts longer with lower degrees of conversion. An important point here is that since most solvents used are volatile, during solution polymerization a great care shoud be put on to carry out the reaction at temperatures enough lower than the boiling point of the solvent. Otherwise possible solvent evapoation could take place altering the mass of the sample and the precision of the results.

CONCLUSION

During application of the DSC in monitoring the free radical polymerization reaction kinetics, as conclusion, the following advantages and limitations can be postulated.

Advantages

- Achievement of strictly isothermal conditions during polymerization even in the region of strong effect of diffusion-controlled phenomena on the reaction kinetics.
- Use of very small amounts (a few mg) of reaction mixtures (monomer/initiator).
- Ability of continuous monitoring the reaction and not taking only discrete experimental data at specific time intervals.
- Easy application in polymerizations leading in crosslinked networks where the use of other techniques is very difficult.
- Intrinsic phenomena like gel effect are more easily distinguishable and DSC has been proved a very useful tool when applied in studies of the gel-effect.
- Very fast polymerization reactions completed in a few minutes can be easily followed.
- Possibility of performing non-isothermal experiments which could lead in process optimization.
- Finally, one of the greatest advantages of DSC is that it provides a direct measure of instantaneous rate of polymerization rather than conversion. Conversion histories are readily obtained from integration of the raw data. This process is inherently more accurate than evaluating rates from the slope of the conversion curve.

Limitations

- Not adequate removal of the amount of oxygen in monomer or trapped in the reaction pan may lead to the appearance of an induction time due to the inhibition effect.
- Use of volatile solvents limits the reaction temperature values since possible evaporation could lead in a mass loss and alteration of the results. Use of specially designed pans hermetically sealed and reaction temperatures well belowe the boiling point could avoid these limitations.
- From the method limitations it could be considered that since samples are not collected at discrete time intervals, the method is not adequate when the variation of other properties changing during reaction (i.e. molecular weight distribution, or copolymer composition, etc.) is needed.

ACKNOWLEDGEMENTS

I would like to thank my coleague Associate Professor I. D. Sideridou for fruitful discussions.

REFERENCES

[1] Moad, G.; Solomon, D.H. The Chemistry of radical polymerization. Second fully revised edition; Elsevier: The Netherlands, 2006.

[2] Matyjaszewski, K.; Davis, T. P. Handbook of Radical Polymerization; Wiley-Interscience: Hoboken, 2002.

[3] Stickler, M. Makromol. *Chem. Macromol. Symp.* 1987, 10/11, 17-69.

[4] Horie, K.; Mita, I.; Kambe, H. *J. Polym. Sci. Part A-1 1968*, 6, 2663.

[5] Malavasic, T.; Vizovisek, I.; Lapanje, S.; Moze, A. *Makromol. Chem.* 1974, 175, 873-880.

[6] Malavasic, T.; Osredkar, U.; Anzur, I.; Vizovisek, I. *J. Macromol. Sci. Chem.* 1986, A23, 853-860.

[7] Sebastian, D. H.; Biesenberger, J. A. *J. Macromol. Sci. Chem.* 1981, A15, 553-584.

[8] Armitage, P. D.; Hill, S.; Johnson, A.F.; Mykytiuk, J.; Turner, *J.M.C. Polymer* 1988, 29, 2221.

[9] Bauduin, G.; Pietrasanta, Y.; Rousseau, A.; Granier-Azema, D. *Eur. Polym. J.* 1992, 28, 923.

[10] Feliu, J.A.; Sottile, C.; Bassani, C.; Ligthart, J.; Maschio, G. *Chem. Eng. Sci.* 1996, 51, 2793-2798.

[11] Feliu, J.A.; Sottile, C.; Bassani, C.; Maschio, G. In 5th International Workshop on Polymer Reaction Engineering; Reichert, K.-H.; Moritz, H.-U.; Eds; *DECHEMA Monographs Vol. 131;* Dechema: Frankfurt am Main, GE, 1995, 131, 531-546.

[12] Horie, K.; Otagawa, A.; Muraoka, M.; Mita, I. *J. Polym. Sci. Polym. Chem. Ed.* 1975, 13, 445-454.

[13] Achilias, D. S.; Kiparissides, C. *Polymer 1994*, 35(8), 1714.

[14] Verros, G.D.; Latsos, T.; Achilias, D.S. *Polymer 2005*, 46, 539-552.

[15] Dube´, M.; Soares, J.B.P.; Penlidis, A.; Hamielec, A. E. Ind. *Eng. Chem. Res.* 1997, 36, 966.

[16] Andrzejewska, E. Prog. *Polym. Sci.* 2001, 26, 605.

[17] Russell, G. T. Aust. J. *Chem. 2002,* 55, 399.

[18] Buback, M.; Egorov, M.; Gilbert, R.G.; Kaminsky, V.; Olaj, O.F.; Russell, G.T.; Vana, P.; Zifferer, G. *Macromol. Chem. Phys.* 2002, 203, 2570.

[19] de Kock, J. B. L.; Van Herk, A. M.; German, A. L. J. *Macromol. Sci., Polymer Rev.* 2001, C41, 199.

[20] Achilias, D. S.; Kiparissides, C. *Macromolecules 1992,* 25, 3739.

[21] Achilias, D. S.; Kiparissides, C. *J. Appl. Polym. Sci.* 1988, 35, 1303-1323.

[22] Achilias, D. S. Macromol. *Theory Simul.* 2007, 16 (4), 319-347.

[23] Maschio, G.; Bello, T.; Scali, C. *Chem. Eng. Sci.* 1994, 49, 5071.

[24] Achilias, D. S.; Sideridou, I. *Macromolecules 2004*, 37, 4254.

[25] Achilias, D. S.; Sideridou, I. J. *Macromol. Sci. Pure Appl. Chem.* 2002, A39, 1435.

[26]   Sideridou, I.D.; Achilias, D. S.; Karava, O. *Macromolecules* 2006, 39, 2072-80.

In: Modern Trends in Macromolecular Chemistry
Editor: Jon N. Lee

ISBN: 978-1-60741-252-6
© 2009 Nova Science Publishers, Inc.

Chapter 10

TRIMERIZATION AND TETRAMERIZATION OF ETHYLENE

Evgenii P. Talsi and Konstantin P. Bryliakov

G. K. Boreskov Institute of Catalysis, Siberean Branch of the Russian Academy of
Sciences, Novosibirsk, Russian Federation

ABSTRACT

In recent years the interest to the selective trimerization and tetramerization of ethylene has grown significantly. In this review paper we discuss the current state-of-the-art of the experimental mechanistic studies of the catalytic systems for trimerization and tetramerization of ethylene based on chromium and titanium compounds.

INTRODUCTION

Linear α-olefins such as 1-hexene and 1-octene are important comonomers for the copolymerization with ethylene to generate linear low density polyethylene. They are generated industrially predominantly via nonselective oligomerization of ethylene [1]. This oligomerization generally produces a broad range of olefins characterized by a Schulz-Flory distribution. Such distribution does not closely match the market demand. Therefore, catalytic systems that are selective for specific desirable alkenes would be of great industrial and academic interest. The first process for the selective production of 1-hexene has been commercialized in 2003 by Chevron–Phillips in Mesaieed, Qatar [2]. Consisting of substituted pyrrole ligand, chromium source and aluminium activator, Phillips catalytic systems achieve selectivities to 1-hexene in excess of 90 wt%. Due to the ill-defined structure of paramagnetic chromium species present in Phillips catalytic systems, their mechanistic studies are complicated.

In recent years the interest to the selective trimerization and tetramerization of ethylene has grown significantly after the discovery of new, highly active catalysts based on better characterized chromium complexes [3, 4]. Considerable number of new chromium catalysts was found and important studies of the mechanism of their performance were published [5-34]. Besides chromium catalysts, titanium based trimerization systems have attracted substantial interest [35-37]. Despite the fact that chromium catalysts are more promising for the practical application, titanium based systems seem to be more suitable for detailed mechanistic studies, since in contrast to paramagnetic chromium species, diamagnetic titanium(IV) species display well resolved ^{1}H NMR spectra in solution. In this review paper we discuss the current state-of-the-art of the experimental mechanistic studies of the catalytic systems for trimerization and tetramerization of ethylene based on chromium and titanium compounds.

PHILLIPS ETHYLENE TRIMERIZATION CATALYST

The key discoveries relating to selective trimerization of ethylene were made by Manyik, Walker and Wilson [38]. They observed that during the polymerization of ethylene using Cr(III)-2-ethylhexanoate (Cr(III)-2-EH) activated by partially hydrolysed tri-isobutylaluminium (PIBAO), some of the ethylene trimerizes to produce 1-hexene. The rate of 1-hexene formation was dependent on the square of the ethylene pressure. The authors concluded that 1-hexene was produced via mechanism other than linear chain growth, and they proposed a metallacycle mechanism to explain the second order dependence. This mechanism was modified by Briggs and his scheme is still accepted today [39]. We will discuss the metallacycle mechanism in the following section. The Phillips trimerization catalyst systems were discovered in the late 1980's when the catalytic properties of chromium pyrrolyl compounds were investigated [8, 40]. Later it was found that it is possible to prepare an active catalyst by simply combining a chromium(III) alkanoanate, such as Cr(III)-2-EH, with pyrrole and triethylaluminium (TEA) in cyclohexane [41]. 2,5-Dimethylpyrrole is Phillips preferred ligand due to its high air, light and temperature stability. It was also found that the presence of a halogen compound, such as diethyl aluminium chloride (DEAC) during the catalyst preparation led to marked improvements in both the catalyst activity and selectivity towards 1-hexene formation [41].

Phillips catalyst prepared by combining Cr(III)-2-EH, 2,5-dimethylpyrrole, diethylaluminium chloride and TEA in molar ratios 1:3.3:7.8:10.8 at room temperature showed activities exceeding $1.57 \cdot 10^5$ g/g Cr per hour at a reaction temperature of 115 °C and 100 bar reaction pressure. The trimerization reaction yielded 93% of 1-hexene. The main side products were a mixture of linear and branched decenes (5.38%) [2,8]. The first commercialized trimerization process is based on this type catalyst system.

Mitsubishi developed there own protocol of Phillips catalyst using Cr(III)-2-EH, 2,5-dimethylpyrrole, hexachloroethane and triethylaluminium in molar ratios of 1:6:4:40, and by carefully controlling the molar ratio of 1-hexene to ethylene inside the reactor (at less than 0.5:1) they managed to obtain unprecedented catalyst activities $3.78 \cdot 10^6$ g/g Cr per hour at 105 °C and 50 bar (95.4% overall 1-hexene) [42].

Despite the practical importance of Phillips trimerization catalyst system, the fundamental aspects of its performance are unknown. The precise nature of the active species in these catalysts is unclear, and it is difficult to control the catalyst performance by well-directed catalyst modifications. Based on results of X-ray photoelectron spectroscopy, it is proposed that Cr^{3+} oxidation state is responsible for trimerization of ethylene by Phillips catalyst system [43].

CROMIUM CATALYSTS WITH AR₂PN(ME)PAR₂ DIPHOSPHINE LIGANDS

Trimerization of Ethylene

British Petroleum published the first chromium based ethylene trimerization systems with diphosphine ligands of the type $Ar_2PN(Me)PAr_2$, where Ar is an o-methoxy-substituted aryl group (Scheme 1) [3, 44]. When activated with MAO, these systems provide extremely active and selective trimerization catalysts. At 20 bar ethylene pressure, the productivity is over 1 million g/g Cr per hour. This catalytic activity exceeds reported values for the Phillips catalysts at least by order of magnitude at a given pressure. The kinetic studies reveal a second order dependence of productivity on ethylene pressure, as is observed for Phillips catalyst [3]. The catalyst proved extremely stable and no deactivation was observed over the run time. Good selectivity to 1-hexene (typically > 85%) were achieved, the main byproducts being C_{10} olefins. In contrast to Phillips catalyst systems, the starting chromium(III) complexes with $Ar_2PN(Me)PAr_2$ ligands can be well characterized. Figure 1 shows X-ray structure of the catalyst precursor formed upon interaction of $CrCl_3(THF)_3$ with $Ar_2PN(Me)PAr_2$ in CH_2Cl_2, where Ar is an o-OCD_3-substituted aryl group [15].

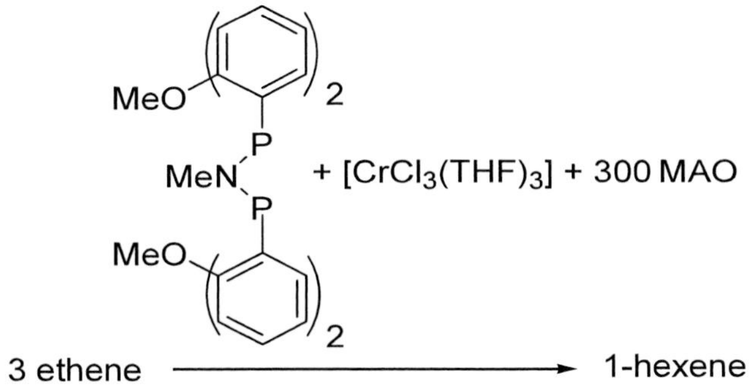

Scheme 1. Chromium bis(diarylphosphino)amine catalyst.

It is assumed that trimerization follows the mechanism previously proposed by Briggs [39] (Scheme 2). Two ethylene molecules coordinate to the chromium center and oxidatively couple to form a metallacyclopentane. A further ethylene is coordinated and inserts to form a metallacycloheptane. This metallacycle can undergo β-elimination to form an alkyl-hydride species which after reductive elimination, produces 1-hexene and regenerates the initial chromium species. The above mechanism is based on shuttling of chromium species between Cr^n and Cr^{n+2} valent states. In support of the mechanism, involving metallacyclic intermediates, Jolly and co-workers have reported well-characterized η^5-cyclopentadienyl-stabilized chromacyclopentane and chromacycloheptane complexes; the latter decomposes more readily and yields 1-hexene [45].

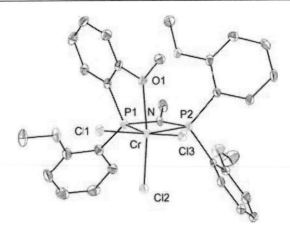

Figure 1. Structural drawing of mononuclear chromium(III) complex formed upon interaction of $CrCl_3(THF)_3$ with $Ar_2PN(Me)PAr_2$ in CH_2Cl_2 (Ar is an o-OCD_3-substituted aryl group) (reproduced with permission from [15]).

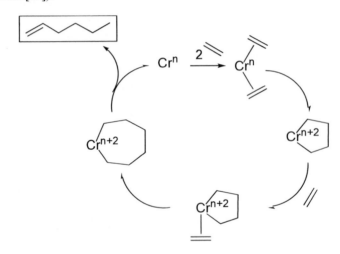

Scheme 2. Metallacyclic trimerization mechanism [39].

The compelling evidence for the metallacycle mechanism of trimerization was provided by Bercaw and co-workers [6, 21]. It was shown that trimerization of 1:1 mixture of C_2D_4 and C_2H_4 allows to distinguish between Cosee-type (insertion/β-H elimination) and metallacycle mechanisms. The metallacyclic route would give no H/D scrambling – only C_6D_{12}, $C_6D_8H_4$, $C_6D_4H_8$ and C_6H_{12} should be observed – while the Cosee – type mechanism would lead to H/D scrambling and formation of isotopomers containing an odd number of deuterons. Addition of equimolar mixture of C_2D_4/C_2H_4 to the catalyst obtained by activating $CrPh_3(Ar_2PN(Me)PAr_2)$ with $[H(Et_2O)_2]^+\{B[3,5-C_6H_3(CF_3)_4]\}^-$ produces only the isotopomers C_6D_{12}, $C_6D_8H_4$, $C_6D_4H_8$ and C_6H_{12} in a 1:3:3:1 ratio in agreement with metallacycle mechanism [6, 21].

According to the results of Bercaw and co-workers, treatment of the complex $[CrPh_3(Ar_2PN(Me)PAr_2)]$ with $[H(Et_2O)_2]^+\{B[3,5-C_6H_3(CF_3)_4]\}^-$ generates cationic complex with proposed structure $[CrPh_2(Ar_2PN(Me)PAr_2)]^+\{B[3,5-C_6H_3(CF_3)_4]\}^-$. By-products include benzene and biphenyl, in agreement with protonation of one phenyl ligand followed by reductive elimination of the remaining aryl fragments leading to Cr(I) active site. On the basis of these data, it was proposed that the active catalyst is cationic $(Ar_2PN(Me)PAr_2)$ chromium species that shuttles between Cr(III) and Cr(I) oxidation state (Scheme 2, n = 1) [15-17]. The study of Wass and coworkers of $[(diphosphine)Cr^0(CO)_4]$ complexes activated with $AlR_3/[NAr_3][B(C_6F_5)_4]$ systems is also in line with Cr^I-Cr^{III} mechanism of trimerization [28].

An alternative hypothesis was proposed by Gambarotta and co-workers. It was found that when complex $[CrCl_3(Ph_2PN(Cy)PPh_2]$ was exposed to $AlMe_3$, reduction to Cr(II) and formation of a cationic complex ensued. These results led the authors to the assumption that Cr(III) is reduced by the alkyl aluminum activator at the early stages of catalyst formation, and cationic Cr(II) species is the active catalyst [18]. At this stage, there are no compelling evidences which allow to distinguish between Cr^I-Cr^{III} and Cr^{II}-Cr^{IV} mechanisms of selective trimerization of ethylene by catalyst systems based on chromium complexes with $Ar_2PN(Me)PAr_2$ ligands.

Tetramerization of Ethylene

In 2004, a group from Sasol reported a selective ethylene tetramerization reaction for the first time [7]. This was a significant breakthrough, since for a long time it was suggested that selective ethylene tetramerization is hardly possible, since, a metallacycle-based tetramerization of ethylene to yield 1-octene requires discrimination between 7- and 9-membered intermediates. It seemed improbable that such discrimination can be achieved. In contrast to trimerization systems reported by Waas and co-workers (Scheme 1, [3]), the ligands $Ar_2PN(R)PAr_2$ used for tetramerization systems had no o-methoxy-substituents in aryl group (e. g. Ar = Ph). Operating the catalysts at 40 bar ethylene pressure, up to 70 wt % 1-octene was obtained, the main by-product being 1-hexene. While catalyst preparations were typically conducted in situ (i.e., the ligand, Cr(III) precursor, and activator (MAO) were added separately to the reactor), it was possible to conduct catalytic reactions with the same activity and selectivity using preformed $(R^2)_2PN(R^1)P(R^2)_2$-Cr(III) complexes. For example, a $(Ph_2P)N(Ph)(PPh_2)$-Cr(III) chloride complex was synthesized. An X-ray structure determination of this complex revealed the chloride-bridged dimer $\{Cr[(Ph_2P)N(Ph)(PPh_2)]Cl_2(\mu\text{-}Cl)\}_2$ (Figure 2) [7]. This dimer, activated with MAO in toluene, gave similar C_8 selectivity and productivity as the catalyst prepared by mixing $Cr(THF)_3Cl_3$, ligand and MAO in toluene.

Through the use of deuterium labeling, the Sasol group have confirmed the metallacycle mechanism for ethylene tetramerization [11]. The high selectivity to 1-octene is a consequence of both the relative stability of the metallacycloheptane intermediate and the instability of the metallacyclononane species [11]. The synthesis of cationic Cr^I carbonyl complex containing $Ph_2PN(^iPr)PPh_2$ ligand which can be easily converted to the active tetramerization catalyst by triethylaluminum evidenced in favor of Cr^I-Cr^{III} mechanism of tetramerization (Scheme 3, n = 1) [27]. However, the alternative Cr^{II}-Cr^{IV} mechanism (Scheme 3, n = 2) can not be unambiguously excluded.

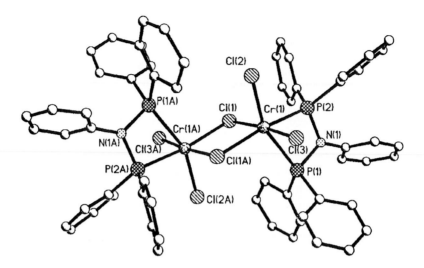

Figure 2. Molecular structure of $\{Cr[(Ph_2P)N(Ph)(PPh_2)]Cl_2](\mu\text{-}Cl)\}_2$ (reproduced with permission from [7]).

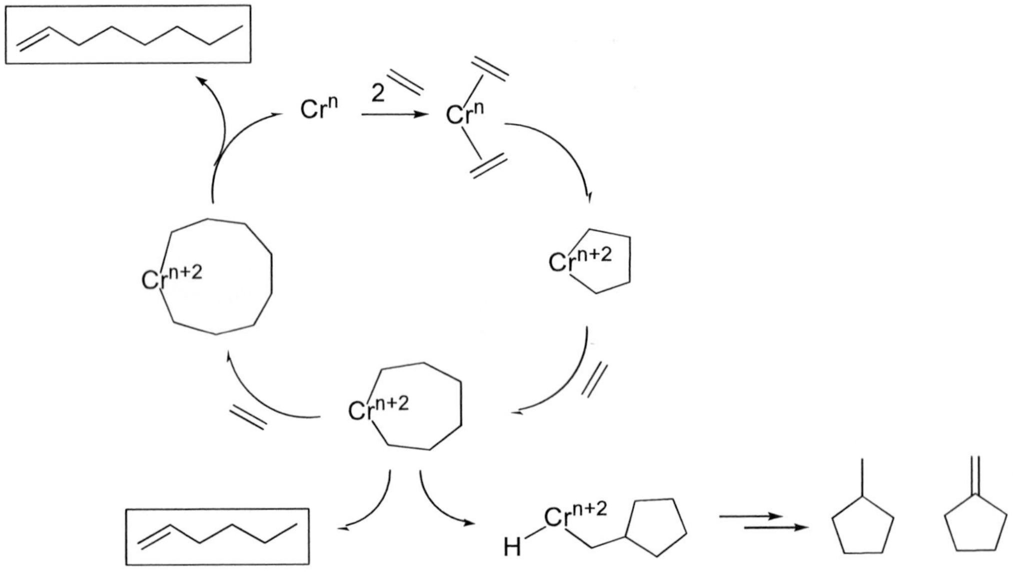

Scheme 3. Tetramerization mechanism [5, 22].

Very recently, a $Cr(acac)_3/Ph_2PN(^iPr)PPh_2/MMAO$ mixture was monitored by *in situ* EPR in cyclohexane and toluene, during oligomerization of ethylene. Reduction of Cr^{3+} into low-spin Cr^+ exhibiting characteristic EPR signal was observed. However, the rate of Cr^+ formation was much slower than the rate of Cr^{3+} reduction, suggesting that the major active species can be EPR-silent, possibly an antiferromagnetic Cr^+ dimer and/or Cr^{2+} [31]. Therefore, further studies are needed to discriminate between Cr^+/Cr^{3+} and Cr^{2+}/Cr^{4+} redox couples in the metallacycle mechanism of selective trimerization and tetramerization with chromium $Ar_2PN(R)PAr_2$ complexes.

One intriguing side reaction in the selective tetramerization of ethylene is the formation of equal amounts of methylcyclopentane and methylenecyclopentane as the major byproducts in the C_6 fraction (Scheme 3). A mechanistic study of this side reaction has lead to a proposal of binuclear alkyl-hydride bridged chromium(II) species (Scheme 4) [11]. Key to this mechanism is the high stability of this binuclear species toward reductive elimination. Theopold and co-workers have recently reported the crystal structure of very stable bimetallic chromium complex that contains a bridging hydride and phenyl ligands between two Cr^{II} centers [29]. The high stability of this complex can be caused by various spin states of the starting complex and the product of the reductive elimination of its phenyl and hydride ligands [33]. The existence of stable binuclear alkyl-hydride chromium complexes supports the proposed mechanism of methylcyclopentane and methylenecyclopentane formation (Scheme 4).

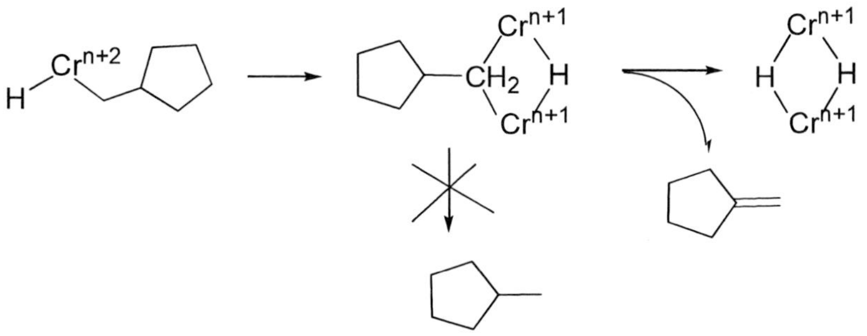

Scheme 4. Postulated binuclear mechanism for the formation of the C_6 cyclics [11].

In order to elucidate factors determining the activity and selectivity of bis(diarylphosphino)amine chromium catalysts towards 1-hexene and 1-octene, a wide range of ligands $(R^2)_2PN(R^1)P(R^2)_2$ has been probed [9,10,17,24,25,34]. It was shown that ortho-substitution at the diphosphinoamine aryl rings is crucial for switching between tetramerization and trimerization selectivity. In general, more sterically encumbered ligands favor trimerization to tetramerization. In other words, the relative ratio of 1-hexene to 1-octene was found to be dependent on the total amount of steric bulkiness, as measured by the number of ortho substituents and the nature of the N substituent [9, 10, 17, 22].

Methoxy substituents also have a special effect. The first chromium based ethylene trimerization systems $Ar_2PN(Me)PAr_2$-Cr(III)/MAO contained Ar groups with o-methoxy-substituents (Scheme 1) [3]. It was proposed that the ortho-methoxy groups acted as pendant donors to the chromium metal center and were a prerequisite for catalytic activity. This assumption was supported by the fact that the catalyst containing o-Et substituents instead of o-methoxy ones was completely inactive [3]. Moreover, X-ray study of a series of Cr(III)-complexes of the ligand $Ar_2PN(Me)PAr_2$ (Ar = 2-methoxyphenyl) showed that these complexes have octahedral geometry, with one bound methoxy group (Figure 1, [15]). NMR studies confirmed hemilabile coordination of the o-methoxy group in these catalysts in solution [15, 16]. However, further studies of Blann and co-workers demonstrated that, under the appropriate conditions (high ethylene pressure), pendant coordination was not mandatory for selective ethylene trimerization [9]. Nevertheless, o-methoxy substitution can dramatically affect the catalyst activity and selectivity. Only one o-methoxy substituent is required to strongly favour trimerization [10]. Besides, methoxy groups are required to observe noticeable activity at low ethylene pressure. Therefore, it was suggested that tetramerization and trimerization selectivity may be mediated by either steric crowding around the catalytic centre (non-polar substitution) or by pendant coordination of a donor substituent (ortho-methoxy substitution) [9, 10, 22].

To elucidate the effect of cocatalyst on the relative selectivity of ethylene conversion to 1-hexene and 1-octene, the products formed in the system $CrCl_3(thf)_3/[Ph_2PN(iPr)PPh_2]/AlEt_3/cocatalyst/ethylene$ were studied. The following cocatalyts were evaluated: $B(C_6F_5)_3$, $Al(OC_6F_5)_3$, $[(Et_2O)H][Al(OC_6F_5)_4]$, $[Ph_3C][Ta(OC_6F_5)_6]$, $(Et_2O)Al\{OCH(C_6F_5)_2\}_3$, $(Et_2O)Al\{OC(CF_3)_3\}_3$, $[Ph_3C][Al\{OC(CF_3)_3\}_4]$, $[Ph_3C][AlF\{OC(CF_3)_3\}_3]$, $[Ph_3C][\{(F_3C)_3CO\}_3Al-F-Al\{OC(CF_3)_3\}]$ and $[Ph_3C][CB_{11}H_6Br_6]$. It was found that cocatalysts incorporating less coordinating anions shift the selectivity toward tetramerization. For example, $Al(OC_6F_5)_3$ promotes formation of 1-hexene (selectivity 90%), whereas $[Ph_3C][Al\{OC(CF_3)_3\}_4]$ favors formation of 1-octene (selectivity 72%). The overall productivity and selectivity of the catalyst is dependent upon both cocatalyst stability and the nature of the anion present [26].

SASOL MIXED HETEROATOMIC CHROMIUM CATALYSTS

Another large family of ethylene trimerization catalysts based on chromium complexes with linear mixed-donor tridentate ligands $HN(CH_2CH_2PR_2)_2$ and $HN(CH_2CH_2SR)_2$ was discovered by researches from Sasol (Scheme 5) [4,5]. Complex with $HN(CH_2CH_2PPh_2)_2$ ligand was the first prepared and tested for ethylene trimerization. A single crystal of complex $CrCl_3[HN(CH_2CH_2PPh_2)_2]$ was grown. It displays a slightly distored octahedral geometry with tridentate ligand coordinated in meridonal fashion. When activated with MAO, it gives reasonable activity with excellent selectivity toward 1-hexene (99.2%). The replacement of Ph group by less sterically demanding Et group led to two-fold increase in activity, whereas bulky cyclohexyl substituent greatly attenuated activity, and the main product formed was polyethylene [4]. The presence of NH functionality is essential for catalytic activity: substitution of an alkyl group for the proton results in a drastic decrease in activity and selectivity [12].

Later the same research group have prepared complexes $CrCl_3[HN(CH_2CH_2SR)_2]$ (R = Me, Et, ^tBu, ^nBu, ^ndecyl) [5,12,14]. Along with MAO as cocatalyst, all complexes gave rise to highly active and selective catalysts for ethylene trimerization to 1-hexene. For instance, for the catalyst $CrCl_3[HN(CH_2CH_2SEt)_2]$ with 280 equiv of MAO, an activity of $1.61 \cdot 10^5$ g/g of Cr/h was obtained, with selectivity 98.4% toward C_6 (of which 99.7% was 1-hexene). Single crystal of $CrCl_3[HN(CH_2CH_2SEt)_2]$ was obtained. As in the case of PNP complexes, Cr displays the expected octahedral coordination geometry, with the tridentate ligand coordinated in a meridonal fashion (Figure 3) [5]. The introduction of bulky substituent R (^tBu instead of Et) attenuated the activity of the SNS catalyst system, as expected [5,14].

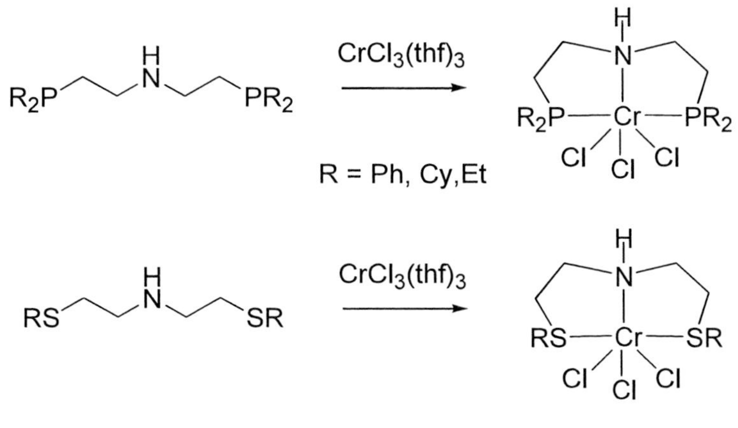

R = Ph, Cy, Et

R = Me, Et, ^nBu, ^tBu, n-decyl

Scheme 5. Sasol Cr(III)-PNP and Cr(III)-SNS ethylene trimerization catalysts.

The coordination chemistry of selected $HN(CH_2CH_2PR_2)_2$ and $HN(CH_2CH_2SR)_2$ ligands with Cr^{III} and Cr^{II} has been studied, and the resulting complexes were evaluated for ethylene trimerization [14]. The performance of Cr^{II} precatalysts is comparable with the Cr^{III} counterparts when MAO activation is used. Complexes $CrCl_3[HN(CH_2CH_2PR_2)_2]$ and $CrCl_3[HN(CH_2CH_2SR)_2]$ can be activated with $AlR_3/B(C_6F_5)_3$ system, suggesting that a formally cationic active species are responsible for trimerization [14,23]. On the basis of these data, McGuinness and coworkers proposed a Cr^{II}-Cr^{IV} mechanistic cycle for 1-hexene formation [14].

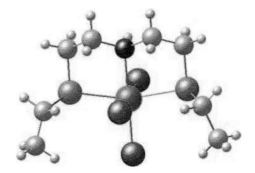

Figure 3. Molecular structure of $CrCl_3[HN(CH_2CH_2SEt)_2]$ (reproduced with permission from [5]).

Gambarotta and coworkers have attempted to determine the oxidation state of the active species of the catalytic system based on chromium complex with $(CySCH_2CH_2)_2NH$ ligand. It was concluded that the initial trivalent Cr(III) species are precursors to the active Cr(II) counterparts, and Cr^{II}-Cr^{IV} mechanistic cycle can be suggested. The main supporting evidence for this was the isolation of a divalent cationic complex resulting from the reaction of $[CrCl_3(CySCH_2CH_2)_2NH]$ with isobutylaluminoxane [19]. In the following publication, the same authors have discovered that the alkyl aluminum activator may indirectly induce re-oxidation of Cr(II) towards the more active and selective Cr(III) catalyst precursor [20]. These data evidence in favor of Cr^{III}-Cr^I mechanistic cycle for 1-hexene formation. Thus, despite extensive studies, the oxidation state of the active species of chromium based trimerization catalysts with $HN(CH_2CH_2PR_2)_2$ and $HN(CH_2CH_2SR)_2$ ligands is still unclear. The last studies on chromium based ethylene oligomerization are directed to the search and testing of new catalysts [30, 32]. However, catalysts which can surpass the discussed ones are still not found.

TITANIUM ETHYLENE TRIMERIZATION CATALYSTS

Hessen and coworkers have shown, that a mono(cyclopentadienyl)titanium catalyst system activated by MAO, $(\eta^5\text{-}C_5H_4CMe_2Ph)TiCl_3/MAO$ (Ar = Ph, $3,5\text{-}Me_2C_6H_3$), is able to trimerize ethylene with high activity [35]. This catalyst was first prepared by Bochmann and coworkers (Figure 4), but was examined in propylene polymerization [46]. Catalytic ethylene conversion experiments with the catalyst system $(\eta^5\text{-}C_5H_4CMe_2Ar)TiCl_3/MAO$ (toluene solvent, Ti:Al ratio of 1:1000, T = 30 °C) show that under these conditions, olefin trimerization products are produced with high selectivity (>95 wt% overall). These trimerization products consist of two fractions: C_6 (trimers of ethylene, 86-87%) and C_{10} (cotrimers of ethylene and 1-hexene, 9-10%). In addition to the trimerization products, smaller amounts of C_8 (1 wt%) and polyethylene (PE, 1-3 wt%) are produced [35, 36].

The rate of production of 1-hexene increases with increasing ethylene pressure. The roughly linear dependence between the amount of C_6 product and the ethylene pressure was observed (the run time of 30 min). The C_6 productivity is around 550-600 kg/((mol of Ti)bar h) over a range of 2-10 bar of ethylene pressure. The C_6 product fraction consists predominantly of 1-hexene (99.5%), with the remaining 0.5% being a mixture of 2- and 3-hexenes. GC/MS indicates that the C_{10} fraction mainly consists of 5-methylnon-1-ene (83%). The only detectable product in the C_8 fraction (by GC) is 1-octene [35-36].

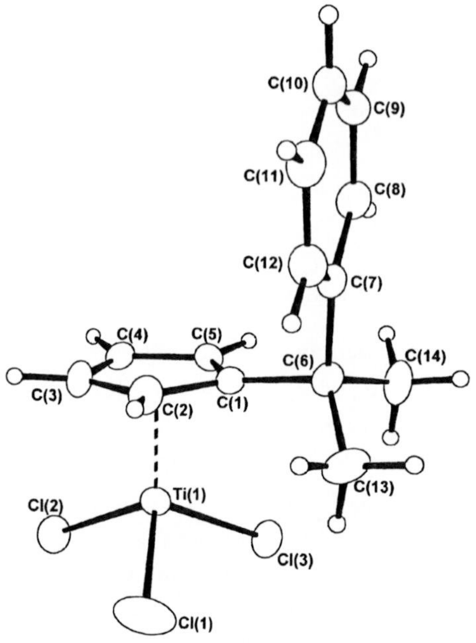

Figure 4. Molecular structure of $(C_5H_4CMe_2Ph)TiCl_3$ (reproduced with permission from [46]).

The thermal stability of the catalyst system $(\eta^5\text{-}C_5H_4CMe_2Ph)TiCl_3/MAO$ is modest: increasing the reaction temperature from 30 to 80 °C decreases the overall catalyst productivity (by factor of 5) and increases the relative amount of PE produced (by factor of 10). This is likely to be associated with catalyst degradation [36].

The study of ethylene conversion by the $(\eta^5\text{-}C_5H_4CMe_2Ar)TiCl_3/MAO$ catalysts (Ar = Ph, 4-MeC$_6$H$_4$, 3,5–Me$_2$C$_6$H$_3$) shows that introduction of methyl substituents into phenyl group significantly diminishes the productivity of the catalyst (by factor of 2), whereas the selectivity for trimerization is retained. The absence of the pendant aryl group leads to the predominant formation of polyethylene. Thus, the pendant arene group is necessary for obtaining selective trimerization [35-36].

It is generally accepted that chromium based trimerization catalysts operate through a mechanism involving metallacyclic intermediates (Scheme 2). This Scheme explains the lack of 1-butene formation, as the chromacyclopentane is expected to be much more stable toward β-H abstraction than the more flexible chromacycloheptane [45]. It is likely that a similar mechanism is operative in the selective trimerization performed by the (cyclopentadienyl-arene) titanium catalysts. This mechanism involves cationic Ti(IV) metallocyclic intermediates and the Ti(II)/Ti(IV) couple for oxidative coupling/reductive elimination (Scheme 6) [36]. It is still not entirely clear how the cationic low-valent Ti(II) species are generated in the system $(\eta^5\text{-}C_5H_4CMe_2Ar)TiCl_3/MAO$. The possible pathway for this transformation is presented in Scheme 6. Initially, (Cp-arene)TiCl$_3$ is alkylated by MAO. This alkylation is followed by alkyl anion abstraction and sequential ethylene insertions into the Ti-Me bonds. The obtained cationic $[(Cp\text{-}arene)Ti(CH_2CH_2R)_2]^+$ complex is in equilibrium with its hydride-olefin isomer $[(Cp\text{-}arene)Ti(H)(\eta^2\text{-}CH_2=CHR)(CH_2CH_2R)]^+$. This may be followed either by dissociation of the alkene (to yield a cationic hydrido alkyl Ti(IV) intermediate that is likely to undergo subsequent reductive elimination of the alkane to give a cationic Ti(II) species) or by reductive elimination of alkane (to yield a cationic Ti(II) olefin complex) [36]. The first order dependence of the rate of 1-hexene formation in ethylene pressure implies that the insertion of the third ethylene molecule into one of the metal-carbon bonds of the proposed cationic titanacyclopentane intermediate is the rate-determining step. Kinetic studies of chromium based trimerization catalysts discussed above have shown a second-order rate dependence on ethylene concentrations, indicating that in these systems the formation of the chromacyclopentane is likely to be rate-determining. However, the selective trimerization of 1-hexene with triazacyclohexane chromium catalysts was recently reported to have a rate-determining the insertion of the third monomer molecule [47].

To evaluate the effect of the cocatalyst on the trimerization catalysis, the catalytic conversion of ethylene using $(\eta^5\text{-}C_5H_4CMe_2Ph)TiMe_3$ catalyst with the cocatalysts $[PhNMe_2H][B(C_6F_5)_4]$, $B(C_6F_5)_3$ and MAO was investigated. $[PhNMe_2H][B(C_6F_5)_4]$ proved to be efficient cocatalyst, affording active and highly selective (> 97 wt%) trimerization catalyst. The selectivity of the catalyst formed upon activation of $(\eta^5\text{-}C_5H_4CMe_2Ph)TiMe_3$ with $B(C_6F_5)_3$ was slightly worse (about 93 wt% of trimerization products).

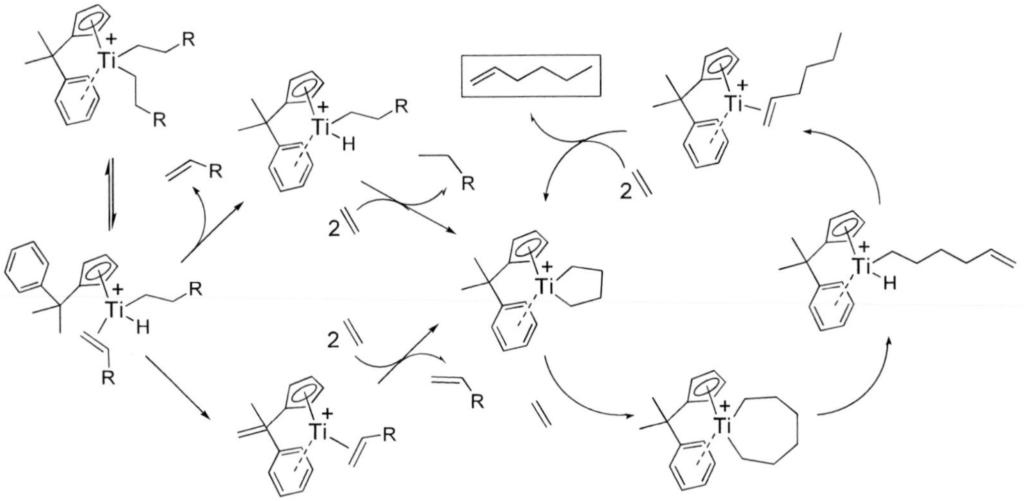

Scheme 6. Metallacyclic trimerization mechanism [36].

The effect of the bridge between the cyclopentadienyl ligand and the aryl group on the catalytic ethylene trimerization was also investigated. The precatalysts $(\eta^5\text{-}C_5H_4CH_2Ph)TiCl_3$, $(\eta^5\text{-}C_5H_4SiMe_2Ph)TiCl_3$ and $(\eta^5\text{-}C_5H_4CMe_2CH_2Ph)TiCl_3$ activated with MAO were tested and compared with the reference catalyst $(\eta^5\text{-}C_5H_4CMe_2Ph)TiCl_3$. The catalysts with CH_2 and $SiMe_2$ bridges have a very poor selectivity for trimerization. The catalyst with CMe_2CH_2 bridge produces mainly 1-hexene, but at a very slow rate. Thus, a disubstituted C_1 bridge between the cyclopentadienyl and aryl moieties gives the best selectivity and activity in catalytic ethylene trimerization [36]. The formation of trimerization catalyst involves the coordination of aryl moiety to the electron-deficient titanium center to generate the ansa-Cp-arene Ti(II) species (Scheme 6).

The presented results show that the catalytic ethylene trimerization with the titanium catalyst/MAO system is a highly efficient and very selective process, resulting in more than 90 wt. % of C_6 product from the converted ethylene, with an excellent 1-hexene selectivity of over 99%. CR_2-bridge between the Cp and aryl moieties is essential for the trimerization selectivity. The selectivity of the titanium-based system is higher than for most of the other trimerization processes. However, like all other trimerization catalysts, the titanium based catalyst systems produce 2-5% of high molecular weight polyethylene. The presence of high molecular weight PE in the reaction mixture can cause reactor fouling. This is especially true for the titanium based process, which currently runs at lower temperatures than for the chromium based systems (30-80 °C versus 110-125 °C for the Philips catalyst) [37]. It is, therefore, important to know which factors determine the formation of polymeric products and how the formation of these products can be reduced or prevented. The PE formation turns out to be catalyzed by at least two different species. A significant amount of PE is formed at the early stages of the reaction, caused by the presence of partly alkylated cationic titanium species namely, $[(\eta^5\text{-}C_5H_4CH_2Ph)TiClMe^+][Me\text{-}MAO]^-$. The PE formation during later stages of the reaction is due to degraded catalyst species. The best results with respect to both productivity and selectivity were obtained by starting from trimethyltitanium compounds, e.g. $(\eta^5\text{-}C_5H_4CMe_2Ph)TiMe_3$. For a low yield of PE it is essential to use MAO that does not contain additives of triisobutylaluminum, containing aluminum hydride species. The polymer output can be reduced by premixing $(\eta^5\text{-}C_5H_4CMe_2Ph)TiCl_3$ and MAO prior to injection into the reactor [37].

CONCLUDING REMARKS

During the last five years, considerable progress in development of new catalysts for selective trimerization and tetramerization of ethylene was achieved. The most practically promising catalysts are based on complexes of chromium. Through the use of deuterium labeling, the compelling evidence for the metallacycle mechanism of trimerization and tetramerization of ethylene with chromium based catalysts was obtained. This mechanism is based on shuttling of chromium species between Cr^n and Cr^{n+2} valent states. The Cr^I/Cr^{III} redox couple is suggested in the majority of the last publications. However, further studies are needed to exclude the Cr^{II}/Cr^{IV} alternative. More extensive EPR studies of various chromium based ethylene trimerization and tetramerization systems can provide additional evidences in favor of Cr^I/Cr^{III} redox couple. It is difficult to expect that more detailed data on the structure of the active species of trimerization or tetramerization will be obtained for chromium systems, due to ill-resolved NMR spectra of paramagnetic chromium species in the reaction solution. One can expect that titanium system can be more fruitful for the NMR identification of the active species of trimerization, since the proposed titanium(IV) metallacycles are diamagnetic. Further low temperature ^1H NMR studies of titanium based trimerization systems are needed to verify this possibility.

REFERENCES

[1]   Greiner, E.; Gubler, R.; Inoguchi, Y. Linear Alpha Olefins, *CEH Marketing Research Report*, Chemical Economics Handbook, SRI International, Menlo Park, SF, May 2004.

[2]   Freeman, J. W.; Buster, J. L.; Knudsen, R. D. U. S. Patent 5, 856, 257 (Phillips Petroleum Company), January 5, 1999.

[3]   Carter, A.; Cohen, S. A.; Cooley, N. A.; Murphy, A.; Scutt, J.; Wass, D. F. *Chem. Commun.* 2002, 858-859.

[4]   McGuinness, D. S.; Wasserscheid, P.; Keim, W.; Hu, C. H.; Englert, U.; Dixon, J. T.; Grove,C. *Chem. Commun.* 2003, 334-335.

[5]   McGuinness, D. S.; Wasserscheid, P.; Keim, W.; Morgan, D.; Dixon, J. T.; Bollmann, A.; Maumela, H.; Hess, F.; Englert, U. *J. Am. Chem. Soc.* 2003, *125*, 5272-5273.

[6]   Agapie, T.; Day, M. W.; Schofer, S. J.; Labinger, J. A.; Bercaw, J. E. *J. Amer. Chem. Soc.* 2004, *126*, 1304-1305.

[7]   Bollmann, A.; Blann, K; Dixon, J. T.; Hess, F.; Killian, E.; Maumela, H.; McGuinness, D. S.; Morgan, D. H.; Neveling, A.; Otto, S.; Overett, M. J.; Slawin, A. M. Z.; Wasserscheid, P.; Kuhlmann, S. *J. Am. Chem. Soc.* 2004, *126*, 14712-14713.

[8]   Dixon, J. T.; Green, M. J.; Hess, F. M.; Morgan, D. H. *J. Organomet. Chem.* 2004, *689*, 3641-3668.

[9]   Blann, K; Bollmann, A.; Dixon, J. T.; Hess, F.; Killian, E.; Maumela, H.; Morgan, D. H.; Neveling, A.; Otto, S.; Overett, M. J. *Chem. Commun.* 2005, 620-621.

[10] Overett, M. J.; Blann, K; Bollmann, A.; Dixon, J. T.; Hess, F. ; Killian, E.; Maumela, H.; Morgan, D. H.; Neveling, A.; Otto, S. *Chem. Commun.* 2005, 622-624.

[11] Overett, M. J.; Blann, K; Bollmann, A.; Dixon, J. T.; Haasbroek, D.; Killian, E.; Maumela, H.; McGuinness, D. S.; Morgan, D. H. *J. Am. Chem. Soc.* 2005, *127*, 10723-10730.

[12] McGuinness, D. S.; Wasserscheid, P.; Morgan, D. H.; Dixon, J. T. *Organometallics*, 2005, *24*, 552-556.

[13] Crewdson, P.; Gambarotta, S.; Djoman, M. C; Korobkov, I.; Duchateau, R. *Organometallics* 2005, *24*, 5214-5216.

[14] McGuinness, D. S.; Brown, D. B.; Tooze, R. P.; Hess, F. M.; Dixon, J. T.; Slawin, A. M. Z. *Organometallics* 2006, *25*, 3605-3610.

[15] Agapie, T.; Schofer, S. J.; Labinger, J. A.; Bercaw J. E. *Organometallics* 2006, *25*, 2733-2742.

[16] Schofer, S. J.; Day, M. W.; Henling, L. M.; Labinger, J. A.; Bercaw J. E. *Organometallics* 2006, *25*, 2743-2749.

[17] Elove, P. R.; McCann, C.; Pringle, P. J.; Spitzmesser, S. K.; Bercaw J. E. *Organometallics* 2006, *25*, 5255-5260.

[18] Jabri, A.; Crewdson, P.; Gambarotta, S.; Korobkov, I.; Duchateau, R. *Organometallics* 2006, *25*, 715-718.

[19] Jabri, A. ; Temple, C. ; Crewdson, P. ; Gambarotta, S. ; Korobkov, I. ; Duchateau, R. *J. Am. Chem. Soc.* 2006, *128*, 9238-9247.

[20] Temple, C. ; Jabri, A. ; Crewdson, P. ; Gambarotta, S. ; Korobkov, I. ; Duchateau, R. *Angew. Chem. Int. Ed.* 2006, *45*. 7050-7053.

[21] Agapie, T. ; Labinger, J. A. ; Bercaw, J. E. *J. Am. Chem. Soc.* 2007, *129*, 14281-14295.

[22] Wass, D. F. *Dalton Trans.* 2007, 816-819.

[23] McGuinness, D. S.; Overett, M.; Tooze, R. P.; Blann, K.; Dixon, J. T.; Slawin, A. M. Z. *Organometallics*, 2007, *26*, 1108-1111.

[24] Kuhlmann, S.; Blann, K.; Bollmann, A.; Dixon, J. T.; Killian, E.; Maumela, M. C.; Maumela, H.; Morgan, D. H. ; Prétorius, M. ; Taccardi, N. ; Wasserscheid, P. *J. Catal.* 2007, *245*, 279-284.

[25] Blann, K.; Bollmann, A.; de Bod, H.; Dixon, J. T. ; Killian, E. ; Nongodlwana, P. ; Maumela, M. C.; Maumela, H.; McConnell, A. E. ; Morgan, D. H.; Overett, M. J,; Prétorius, M.; Kuhlmann, S.; Wasserscheid, P. *J. Catal.* 2007, *249*, 244-249.

[26] McGuinness, D. S.; Rucklidge, A. J.; Tooze, R. P.; Slawin, A. M. Z. *Organometallics*, 2007, 26, 2561-2569.

[27] Rucklidge, A. J.; McGuinness, D. S.; Tooze, R. P.; Slawin, A. M. Z.; Pelletier, J. D. A.; Hanton, M. J.; Webb, P. B. *Organometallics* 2007, *26*, 2782-2787.

[28] Bowen, L. E.; Haddow, M. F.; Orpen, A. G.; Wass, D. F. *Dalton Trans.* 2007, 1160-1168.

[29] Monillas, W. H.; Yap, G. P. A.; Theopold, K. H. *Angew. Chem. Int. Ed.* 2007, *46*, 6692-6694.

[30] Moulin, J. O.; Evans, J.; McGuinness, D. S.; Reid, G.; Rucklidge, A. J.; Tooze, R. P.; Tromp, M. *Dalton Trans.* 2008, 1177-1185.

[31] Bruckner, A.; Jabor, J. K.; McConnell, A. E. C.; Webb, P. B. *Organometallics*, 2008, *27*, 3849-3856.

[32] Albahily, K.; Koç, E.; Al-Baldawi, D.; Savard, D.; Gambarotta, S.; Burchell, T. J.; Duchateau, R. *Angew. Chem. Int. Ed.* 2008, 47, 1-5.

[33] Köhn, R. D. *Angew. Chem. Int. Ed.* 2008, *47*, 245-247.

[34] Jiang, T.; Zhang, S.; Jiang, X.; Yang, C. ; Niu, B. ; Ning, Y. *J. Mol. Catal., A: Chem.* 2008, *279*, 90-93.

[35] Deckers, P. J. W.; Hessen, B.; Teuben, J. H. *Angew. Chem. Int. Ed.* 2001, *40*, 2516-2519.

[36] Deckers, P. J. W.; Hessen, B.; Teuben, J. H. Organometallics 2002, *21*, 5122-5135.

[37] Hessen, B. *J. Mol. Catal. A : Chem.* 2004, *213*, 129-135.

[38] Manyik, R. M.; Walker, W. E.; Wilson, T. P. *J. Catal.* 1977, *47*, 197-180.

[39] Briggs, J. R. *Chem. Commun.* 1989, 674-675.

[40] Reagan, W. K. EP Patent, 0, 417, 477 (Phillips Petroleum Company), March 20, 1991.

[41] Reagan, W. K.; Freeman, J. W.; Conroy, B. K.; Pettijohn, T. M.; Benham, E. A. EP Patent, 0, 608, 447 (Phillips Petroleum Company) August 3, 1994.

[42] Araki, Y.; Nakamura, H.; Nanda, Y.; Okanu, T. U. S. Patent, 5, 856, 612 (Mitsubishi Chemical Corporation), January 5, 1999.

[43] Fang, Y. ; Liu, Y. ; Ke, Y. ; Guo, C.; Zhu, N.; Mi, X.; Ma, Z.; Hu, Y. *Appl. Catal. A,* 2002, *235*, 33-38.

[44] Wass, D. F. WO 02/04119 (BP Chemical Ltd), January 17, 2002.

[45] Emrich, R.; Heinemann, O.; Jolly, P. W.; Krüger, C.; Verhovnik, G. P. J. *Organometallics*, 1997, *16*, 1511-1513.

[46] Saßmannshausen, J.; Powell, A. K.; Anson, C. E.; Wocadlo, S.; Bochmann, M. J. Organomet. Chem. 1999, *592*, 84-94.

[47] Köhn, R. D.; Haufe, M.; Kociok-Köhn, G.; Grimm, S.; Wasserscheid, P.; Keim, W. *Angew. Chem. Int. Ed.* 2000, *39*, 4337-4339.

In: Modern Trends in Macromolecular Chemistry
Editor: Jon N. Lee

ISBN: 978-1-60741-252-6
© 2009 Nova Science Publishers, Inc.

Chapter 11

HYBRID MACROMOLECULES BUILDUP FROM ORGANOIMIDO SUBSTITUTED POLYOXOMETALATES[*]

Pingfan Wu[1, 2], Yongge Wei[†, 1, 3], Yuan Wang[3] and Meicheng Shao[3]

[1] Department of Chemistry, Tsinghua University, Beijing 100080, China
[2] School of Chemical and Environmental Engineering, Hubei University of Technology, Wuhan 430068, China
[3] State Key Laboratory for Structural Chemistry of Unstable and Stable Species, College of Chemistry and Molecular Engineering, Peking University, Beijing 100871, China

ABSTRACT

Polyoxometalates (POMs) are a class of unique inorganic oxide macromolecules with nanosized geometry. Recently, organoimido derivatives of POMs have received increasing interest in the supramolecular chemistry and chemistry of materials, since they are valuable building blocks for constructing novel nanostructured organic-inorganic hybrid molecular materials with the so-called ''value-adding properties'' and possible synergistic effects, including unique catalytic, photo and electronic properties dramatically different from the corresponding parent materials, which results from the π electrons in the organic component of such molecules may extend their conjugation to the inorganic framework and dramatically modify the electronic structure and redox properties of the corresponding parent POMs. The surface modification of POMs, that is, the substitution of one or more terminal or bridged oxo groups with organoimido ligand, is the fundamental method to synthesize organoimido derivatives of POMs via common organic chemistry reactions. In this chapter of the book, the synthetic chemistry of organoimido derivatives of POMs using different organic reagents will be reviewed, of which the novel DCC-dehydrating protocol to prepare organoimido derivatives of POMs recently developed by us and co-authors will be concentrated. As it can be expected, the reaction chemistry of organoimido derivatives of POMs stands for the fascinating future of the chemistry of organoimido derivatives of POMs since it opens not only a new road

---

[*] A version of this chapter was also published in Research in Hybrid Materials, edited by Simon J. Brunner and Julian W. Egger, Nova Science Publishers. It was submitted for appropriate modifications in an effort to encourage wider dissemination of research.

[†] Correspondence to: Tel: +86-10-62797852, E-mail: yonggewei@mail.tsinghua.edu.cn; ygwei@pku.edu.cn

to the chemical modification of POMs, but also an exciting research arena where a variety of hybrid materials containing covalently bonded POM clusters and organic conjugated segments can be prepared in a more controllable and rational manner. Hence, in the rest of this chapter, Pd-catalyzed coupling reactions of the functionalized organoimido derivatives of POMs will be introduced in detail for this purpose. Especially, different rationally assembled hybrid macromolecules and polymers where the organic molecules and inorganic clusters are bonded together by an imido nitrogen atom will then be highlighted.

1. INTRODUCTION

Polyoxometalates (POMs), formed by the early transition metal ions such as vanadium, molybdenum and tungsten in their high oxidization states with bridged oxide anions, are a unique class of well-defined inorganic nanocluster compounds with much diversity in size, composition, structure and function, which can be, in fact, considered as the "oligomeric" or molecular states and model compounds of the corresponding "polymeric" metal oxides with extended solid state structure [1]. Since the first salt of POM, ammonium 12-molybdophosphate, $(NH_4)_3[PMo_{12}O_{40}]$ was reported by Berzelius in 1826, as a yellow precipitate from the aqueous solution of phosphoric acid and ammonium molybdate [2], the chemistry of POMs has gained dramatic development and is remarkable for its significance in quite diverse disciplines including catalysis, medicine and materials science [3- 5]. Especially, during the last decade, a lot of nanoscale huge clusters of polyoxometalates and their assemblies have been prepared and structurally characterized by our research team at Peking University [6- 9], A. Müller's group at University of Bielefeld [10- 24] and others [25- 28]. As examples, in figures 1-3, there are shown three typical nanoscale polyoxomolybdates, $i.$ $e.$, tyre-like $\{Mo_{176}\}$ (1) [7, 12], ball-like $\{Mo_{132}\}$ (2) [9, 15], and chain-like $\{Mo_{38}\}$ (3) self-assembled from $\{Mo_{36}\}$ [6], synthesized by A. Müller et $al.$ and us, respectively. These high nuclearity huge clusters are not only ideal theoretical models for mesocopic physics but also provide us an exciting research arena for nanochemistry and materials science— The nanotyres 1 are potential building blocks for fabricating novel nanotubes if they are arranged to stack along their symmetrical axes. The nanoballs 2 have a spherical cavity of ca 8 nm^3 with accessible openings, indicating that they are a good candidate for nanocontainers and may be used in controllable drug-releasing. The nanochains 3 show significantly red-shif in its UV-vis spectra, which implies that they are potential applications in making unique nanoscale molecular quantum wells or one-dimensional conducting materials. Furthermore, T. B. Liu et $al.$ have recently observed and confirmed, according to laser light scattering studies, that in their solution, such nanoclusters of polyoxometalates spontaneously assemble into almost monodisperse, well-defined vesicle-like nanoaggregates with spherical structures [29, 30], via an unusually slow self-association process [31]; and more interestingly, in the NaCl-containing aqueous solution, dissolution and precipitation, two opposite processes, of these hydrophilic nanoanions automatically and subsequently occur without changing external conditions [32].

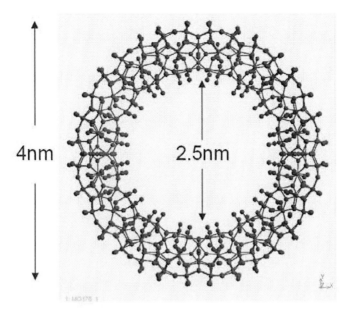

Figure 1. Structure of the nanotyre 1 with 176 Mo atoms in a cluster anion (view parallel to the C_8 axis of the cluster). Also see Ref. 7 and 12.

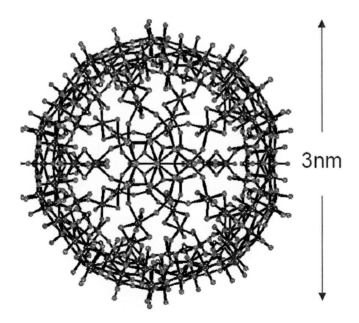

Figure 2. A view of the {Mo_{132}} nanoball 2 along its C_5 axis. Also see Ref. 9 and 15.

Figure 3. Drawing of the nanochain 3 extended along the 21 screw axis. Also see Ref. 6.

Parallel to the rapid and great progress made on these nanosized inorganic clusters of metal-oxide anions, studies on conjugated organic and polymeric materials have also exploded and flourished [33- 36], since such molecular materials have exciting technological applications in various photic and electronic devices, for instance, organic light emitting diodes (OLEDs) [37- 38], organic field-effect transistors (OFETs) or organic thin film transistors (OTFTs) [39- 42], and solid-state lasers [43, 44]. The fact that the 2000's Nobel Prize in chemistry was bestowed upon three preeminent scientists working on conjugated polymers is the ultimate testimony to the important developments in organic conjugated systems [45- 48].

Although they are vastly different in molecular structures, POMs and conjugated organic molecules/ polymers are both electrically active materials with similar electrical and optical properties such as photochromism, electrochromism, and conductivity [34, 49, 50]. The underlying mechanisms of these properties are, however, different for the two types of materials, with $d\pi$ electrons responsible for the inorganic POM clusters, and delocalized $p\pi$ electrons responsible for the organic counterpart. While both areas have been enjoying considerable success, there has been little success in bringing these two types of materials together through covalent bonds to make novel POM-based organic-inorganic hybrids [51] owing to the lack of reliable synthetical chemistry tools to functionalize POMs and prepare their organic derivatives in high throughput. As could be expected, however, such organic-inorganic hybrid materials will not only combine the advantages of organic materials, such as good processability and fine-regulable structure and electronic properties [34], with those of inorganic POM clusters, such as good chemical stability and strong electron acceptability [1], to produce so-called "value-adding properties" [52], but also may bring exciting synergistic effects due to the close interaction of delocalized organic $p\pi$ orbits with the inorganic POM cluster's $d\pi$ orbits. With their unique structures and potential novel properties, originating from the functional nanobuilding blocks of POMs [51], these sorts of hybrid materials are extremely interesting not only to synthetic chemists and materials scientists, but also to theoretical and experimental physicists.

Organically functionalized POMs, $i.e.$, species where one or some oxo or $\{Mo_x\}^{n+}$ groups have been replaced with organic functional groups, have been extensively studied and now form the largest subclass of POM derivatives [53], since the pioneering work at du pont, of which alkylated derivatives or esters of Keggin-structure phosphomolybdic and phosphotungstic acid, such as $(n\text{-}Bu_4N)_3[PMo_{12}O_{39}(OMe)]$ (4) and $(n\text{-}Bu_4N)_3[PW_{12}O_{39}(OMe)]$ (5), were prepared by Knoth et al in the early 1980s [54, 55]. There are a lot of reasons for the current interest in these derivatized POMs [53]: 1) Due to the perceived structural analogies of POMs to the surfaces of metal oxides, they can be viewed as soluble analogues of metal oxides and therefore are of special interest as model compounds for investigating the reactions and properties of metal oxides. Given the versatility of metal oxides in catalyzing organic conversions, and the difficulty in determining the intimate mechanism of these reactions, the study of the stoichiometric reactivity of well-defined surface POMs might provide an understanding of the elementary steps of heterogeneous reactions, particularly with respect to surface-bound intermediates. Therefore, the characterization of organic derivatives of POMs is relevant to the molecule-level modeling of catalytic reactions occurring at metal oxide surfaces. Furthermore, POM-supported transition-metal catalysts represent a new class of oxide-supported catalyst materials that can be fully studied at the atomic level, both structurally and mechanistically. 2) Derivatization may result in the activation of surface oxo groups of POMs, thus giving more chances to be modified. 3) Functionalization can conduce to the stabilization of otherwise labile POM frameworks in the derivatives, providing novel building blocks for the assembly of larger systems. 4) The construction of interconnected POM networks could be built up through the incorporation of organic groups with functionality and polysubstituted derivatives could be utilized in the context of the development of novel dendrimers. 5) Such derivatives can probably provide multifunctional oxidation or acidification catalysts that display selective recognition of substrates, thus higher selectivities, and might develop targeting of POMs in antiviral and anticancer chemotherapy.

Among many organic derivatives of POMs, organoimido derivatives have attracted increasing interest and recently, a lot of arylimido derivatives including polymers of the Lindqvist hexamolybdate cluster, $[Mo_6O_{19}]^{2-}$, have been synthesized continuously [56], since the π electrons in the organic component of such derivatives may extend their conjugation to the inorganic POM framework via the N atoms of the imido groups, thus result in strong d–pπ interactions and dramatically modify the electronic structure and redox properties of the corresponding parent POMs. In addition, organoimido derivatives of POMs with a remote active functional group can be exploited as building blocks to conveniently and controllably fabricate the complicated covalently-linked POM-based organic-inorganic hybrids, including the nano-dumbbells, polymeric chains and even networks of POMs [56]. This modular building block approach brings rational design and structure regulation into the synthesis of organic-inorganic hybrid molecular materials and has just been reviewed by one of our co-workers at UMKC [56]. In this chapter, concentrating on the novel DCC-dehydrating protocol to prepare organoimido derivatives of POMs recently developed in this laboratory and others, various synthetic approaches to organoimido derivatives of POMs using different organic reagents will be reviewed. As it can be expected, the reaction chemistry of organoimido derivatives of POMs stands for the fascinating future of the chemistry of organoimido derivatives of POMs since it opens not only a new road to the chemical modification of POMs, but also an exciting research arena where a variety of hybrid materials containing covalently bonded POM clusters and organic conjugated segments can be prepared in more controllable and rational manner. Hence, in the rest of this chapter, Pd-catalyzed coupling reactions of the functionalized organoimido derivatives of POMs will be introduce in details. Furthermore, various rationally assembled hybrid molecules and polymers where the organic molecules and inorganic clusters are bonded together by an imido nitrogen atom will then be highlighted.

2. Synthetic Chemistry of Organoimido
Derivatives of POMs

As an important subclass of oxo-imido complexes of early transition metals, organoimido derivatives of POMs have only been synthesized and characterized very recently. They were mentioned first by J. Zubieta *et al* in an outstanding communication in 1988 [57], however, there was lack of a reliable and detailed structural characterization in this paper, probably due to the purification problem of their products. Three years later, using a different synthetic methodology, E. A. Maatta's group finished the synthesis and single crystal X-ray structural determination of the first organoimido derivative of POMs [58]. Since then, the organoimido derivatives of POMs have been extensively investigated and a number of organoimido derivatives of the Lindqvist hexamolybdate ion, $[Mo_6O_{19}]^{2-}$, have been reported [53]. However, little work has been conducted on other POMs including the other Lindqvist POMs, the Keggin POMs, and especially, the recently discovered nanoscale huge clusters of POMs such as compounds 1-3 [6-28], the synthesis of their organoimido derivatives is still a formidable challenge. Furthermore, although the bridging oxo groups of POMs can be replaced, in principle, with isoelectronic organoimido ligands, the substitution of terminal oxo groups features largely in the reported organoimido derivatives of POMs, in which the imido groups attached to only one metal center in a linear fashion and the N-atom in the imido groups can be considered sp hybridized as σ-π donor ligand to form one σ and two π bonds with the bound metal atom.

In general, organoimido derivatives of POMs can be obtained in the following ways [53]:

- Imido metathesis via Wittig-like (net) [2 + 2] exchange reactions of M=O bonds with imido-releasing agents, including isocyanates, phosphinimines, or sufinylamines.
- The α-hydrogen transfer reactions with amines.
- The displacement of labile ligands from an imido-containing metal complex by active oxometalates, followed by aggregation, or by coordination-unsaturated lacunary POMs.
- The addition or displacement reactions of active functional groups at the imido ligands from existing organoimido precursors of POMs.

Apparently, the first two strategies create the new metal-nitrogen multiple bonds and the latter ones deal with other chemical bonds from known metal imido complexes. In this section of the chapter, the synthetic methodologies in respect to the former three strategies will be surveyed. The fourth strategy will be introduced in Section 3 because it is involved in the reaction chemistry of organoimido derivatives of POMs.

2.1. Organoimido Derivatives of the Lindqvist Hexamolybadte

The hexamolybdate ion, $[Mo_6O_{19}]^{2-}$, is amongst the most well-known POM clusters for its thermal and chemical robust and ease to be prepared [59], which has the so-called Lindqvist structure [60]. As it is depicted in figure 4, the Lindqvist structure consists of a central oxyanion surrounded in an octahedral cage formed by six metal atoms. All the six metal atoms also have an octahedral environment. Besides the central oxygen atom, they each coordinate triply to one terminal oxygen atom, forming a terminal metal-oxo group (M≡O), and share an additional four doubly-bridging oxygen atoms (μ_2-O atoms) with neighboring metal atoms. Generally speaking, a Lindqvist ion has a superoctahedral structure approach to O_h point group and features its six terminal metal-oxo groups aligned along the Cartesian axes. For the hexamolybdate, these molybdyl groups (Mo≡O) are reactive enough for the terminal oxygen atoms to be directly replaced by various nitrogenous species, for example, diazenido [61], diazoalkyl [62], and imido groups [58]. In the case of imido derivatives of the hexmolybdate, so far, three types of reactions have been developed. These include reactions with phosphinimines [Eq. (1)] [58, 63], isocyanates [Eq. (2)] [64, 65], and primary amines [Eq. (3)] [66- 68], which are given in the following equations, respectively.

$$(n\text{-}Bu_4N)_2[Mo_6O_{19}] + x\ Ph_3P{=}NAr \longrightarrow (n\text{-}Bu_4N)_2[Mo_6O_{19-x}(NAr)_x] + x\ Ph_3P{=}O \quad (1)$$

$$(n\text{-}Bu_4N)_2[Mo_6O_{19}] + x\ RNCO \longrightarrow (n\text{-}Bu_4N)_2[Mo_6O_{19-x}(NAr)_x] + x\ CO_2 \quad (2)$$

$$(n\text{-}Bu_4N)_2[Mo_6O_{19}] + x\ ArNH_2 \longrightarrow (n\text{-}Bu_4N)_2[Mo_6O_{19-x}(NAr)_x] + x\ H_2O \quad (3)$$

With these approaches, the six terminal oxo (in certain case, some bridging-oxo groups) in the hexamolybdate cluster can be partially or completely substituted with organoimido ligands. Nowadays, a large number of monosubstituted [58, 63- 78], disubstituted [63, 65, 66, 71, 76, 79- 83] and polysubstituted [65, 71, 84, 85] organoimido derivatives of the hexamolybdate have been synthesized and structurally characterized.

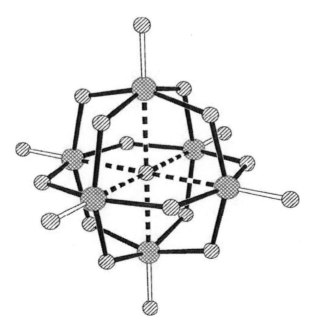

Figure 4. Structure of the Lindqvist-type hexamolybdate ion, [Mo6O19]2-.

2.1.1. Monosubstituted Derivatives

Monofunctionalized organoimido derivatives of the hexamolybdate were firstly obtained by E. A. Maatta *et al* with phosphinimines as imido-releasing reagents [Eq. (1)] in 1991 [58]. They discovered that the reaction of equimolar amounts of $(n\text{-Bu}_4\text{N})_2[\text{Mo}_6\text{O}_{19}]$ (6) and $\text{Ph}_3\text{P}=\text{Ntol}$ (prepared from the reaction of p-tolyl azide and Ph_3P in ether) in strictly anhydrous pyridine at 85 °C yields $(n\text{-Bu}_4\text{N})_2[\text{Mo}_5\text{O}_{18}(\text{MoNC}_6\text{H}_4\text{CH}_3)]$ (7) within 2 days. Two years later, A. Proust and co-workers founded that this type of reaction also works well in refluxed anhydrous acetonitrile under dried nitrogen gas for 60 hours and synthesized the monophenylimido derivative, $(n\text{-Bu}_4\text{N})_2[\text{Mo}_6\text{O}_{18}(\text{NPh})]$ (8) [63]. Although these reactions involve the not-easily accessible phosphinimine reagents, proceed slow, and the triphenylphosphine oxide byproduct is very difficult to separate from the desired reaction products, they nonetheless demonstrate that direct functionalization can be carried out on the hexamolybdate salt itself in a straightforward manner. Moreover, phosphinimines are powerful reagents to introduce C=N bond into organic compounds from carbonyl substrates via the famous aza-Wittig reactions in classic organic syntheses [86], the reactions thus imply that the molybdyl in the hexamolybdate behaves like a carbonyl of aldehydes, ketones or esters in some extent and may have the chemistry similar to that of a carbonyl.

The phosphinimine approach is very useful for the synthesis of arylimido monosubstituents of the hexamolybdate cluster. Recently, an interesting styrylimido-hexamolybdate complex, $(n\text{-Bu}_4\text{N})_2[\text{Mo}_6\text{O}_{18}(\text{NC}_6\text{H}_4\text{CH}=\text{CH}_2)]$ (9), shown in figure 5, which is the important building block in fabricating polystyrene compositions with the POM as a pendant group, has also been prepared in *ca* 90% high yield by this approach [72]. However, attempts to make the alkylimido analogues using this approach are unsuccessful. For instance, the reaction of $(n\text{-Bu}_4\text{N})_2[\text{Mo}_6\text{O}_{19}]$ (6) with $\text{Bu}^n_3\text{P}=\text{NBu}^t$ failed to give the tertbutylimdo derivative [64] because phosphinalkylimines are more reactive than the corresponding phosphinarylimines and result in the reduction and rearrangement of the hexamolybdate cluster.

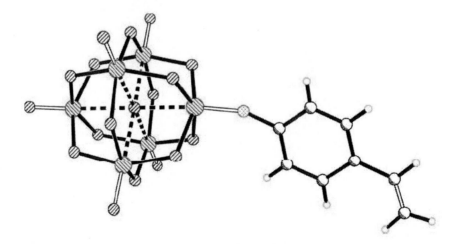

Figure 5. Ball-and-stick model of the styrylimido-hexamolybdate complex 9.

The second approach [Eq. (2)] to produce organoimido derivatives of the hexamolybdate was first set up by R. J. Errington and co-workers a little over a decade ago [64], who have shown that both of the corresponding monoimido-substituted derivatives could be afforded by reactions of alkyl and aryl isocyanates with $[Mo_6O_{19}]^{2-}$ at elevated temperatures in acetonitrile or benzonitrile, and reported the first structurally characterized tertbutylimdo derivative of the hexamolybdate, $(n\text{-}Bu_4N)_2[Mo_6O_{18}(NBu^t)]$ (10) [64]. Maatta and co-workers have subsequently developed this approach successfully using dry pyridine as the solvent [65]. They discovered that alkyl isocyanates are more reactive with the hexamolybdate clusters in dry pyridine and the reaction between them can occur at room temperature and finish in 2 to 3 days [69, 71]. Nevertheless, the reaction of aryl isocyanates with $[Mo_6O_{19}]^{2-}$ in dry pyridine still needs to be carried out at high temperature (110 °C) and must be refluxed under strictly anhydrous conditions for a prolonged time [69, 71]. It is worthwhile to point out here that this reaction is also analogous to that of isocyanates with aldehydes or ketones to form imines [87], which again suggests the comparability between the molybdyl and the carbonyl.

Using this improved route, E. A. Maatta *et al* have further synthesized a lot of monosubstituted organoimido derivatives [65, 69- 71], including the charge-transfer ferrocenylimido complex, $(n\text{-}Bu_4N)_2[(FcN)Mo_6O_{18}]$ (11), $FcN = (\eta^5\text{-}C_5H_5)Fe(\mu\text{-}\eta^5:\sigma\text{-}C_5H_4N)$, (figure 6) [70] and the organodi-imido-bridged bis(hexamolybdate) derivative, $(n\text{-}Bu_4N)_2$ $[(Mo_5O_{18})Mo\equiv N\text{-}Z\text{-}N\equiv Mo\,(O_{18}Mo_5)]$ (12), $Z = 1,4\text{-}cyclo\text{-}C_6H_{10}$, (figure 7) [69]. Compound 11 features the covalent attachment of an oxidizable ferrocenyl donor to a reducible POM acceptor system. Compound 12 stands for the fist example of the hexamolybdate complexes of diimido ligands, which are coordinated to two hexamolybdates and links them as a bridge.

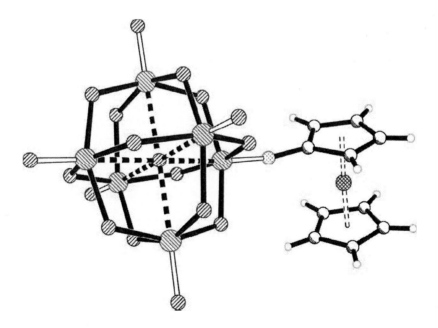

Figure 6. Ball-and-stick model of the charge-transfer ferrocenylimido-hexamolybdate complex 11.

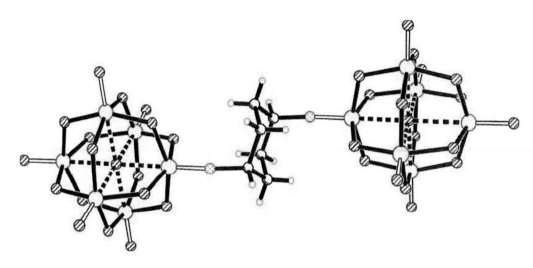

Figure 7. Ball-and-stick model of the organodi-imido-bridged bis(hexamolybdate) derivative 12.

The last approach given in Eq(3) was also put forward by R. J. Errington and co-workers firstly. They observed that in the presence of triethylamine, an aromatic primary amine, for example 1,4-diaminobenzene, reacted directly with the hexamolybdate ions in benzonitrile solvent at *ca* 150 °C [66]. Albeit this reaction occurs under rather harsh conditions and only a mixture that includes mono- and di-functionalized derivatives, and various oligomers is obtained, from which pure imido derivatives are difficult to isolate from each other, it indeed provides us a chance to prepare organoimido derivatives with cheaper and more accessible arylamines as the imido-releasing reagents. Additionally, this reaction confirms further the similarity between the molybdyl and carbonyl groups since it extremely resembles the Schiff-base reaction taken place between primary amines with aldehydes or ketones, which involves of nucleophilic addition of the amine to the carbonyl followed by elimination of water [88].

Several years later, R. A. Roesner et al improved this approach with dry pyridine as the solvent instead of benzonitrile. However, only aromatic amines with the strongly electron-donating substituents, such as the alkoxy groups, promise this reaction, under heating and dry N_2-gas protecting conditions, to create the corresponding mono-imido derivatives of the hexamolybdate in moderate yield [67].

We have now a long time interest in the chemistry of derivatives of POMs [6, 8, 9, 68, 73- 84], including new synthetic methodologies for the organoimido derivatives [68, 73- 83]. In the light of the pioneering investigations of E. A. Maatta and R. J. Errington's groups on this field, we strongly feel that the molybdyl is, under certain cases, one of the inorganic analogues, in some extent, of the carbonyl in organic chemistry [88], and some organic synthetic techniques can be extended into the chemistry of derivatives of POMs. For the primary amine approach, for example, if the resulting water is able to remove from the reactive mixture successively, just like common dehydrating reactions in organic chemistry, the reaction of Eq(3) will proceed rapidly and irreversibly, and complete thoroughly at the end.

Five years ago, together with our co-workers at UMKC, we discovered that using carbodiimides, which are among the most widely-used reagents in common organic synthesis [89-91], such as DPC (N,N'-diisopropylcarbodiimide) and DCC (N,N'-dicyclohexylcarbodiimide), as the dehydrating reagent, the reaction of $[Mo_6O_{19}]^{2-}$ with aromatic primary amines can be dramatically promoted, and have developed a novel DCC-dehydrating synthetic protocol to afford arylimido derivatives of POMs in the context of the primary amine approach [68]. As shown in Eq(4), in the presence of one equivalent of DCC, one equivalent of aromatic primary amines reacts smoothly with an excess of $(n\text{-}Bu_4N)_2[Mo_6O_{19}]$ (6) under dry N_2 gas and refluxing acetonitrile, pyridine or hot N,N'-dimethylformamide, to give the corresponding mono-substituted imido derivatives usually in excellent yield of more than 90%. Not only a variety of aromatic primary amines occur this reaction, but also it is usually completed in less than 12 hours, and pure *crystalline* products can be readily obtained with convenient bench manipulations. Moreover, this reaction can be carried out in the open air without nitrogen protection if slightly over one equivalent of DCC (1.2 equivalent) is added. Best of all, the reaction also tolerates additional remote functional groups on the aromatic ring, including iodo, ethynyl and terpyridyl groups. Figures 8-10 show the X-ray structures of three imido derivatives containing such three functional groups, $(n\text{-}Bu_4N)_2[Mo_6O_{18}(NAr)]$, Ar = 4-iodo-2,6-dimethyl-C_6H_2 (13) [68], 4-ethynyl-2,6-dimethyl-C_6H_2 (14) [68], 4-(4'-terpyridylethynyl)-2,6-dimethyl-C_6H_2 (15) [73], respectively, of which compounds 13 and 14 both have two stable conformational isomers differing by a rotation of the aromatic ring relative to the hexamolybdate skeleton. All the three complexes are valuable building blocks to construct POM-based organic-inorganic hybrid molecular materials in our controllable molecule-module-assembly strategy [56].

$$[Mo_6O_{19}]^{2-} \text{ (excess)} + ArNH_2 + DCC \longrightarrow [Mo_6O_{18}(NAr)]^{2-} + (cyclo\text{-}C_6H_{11}NH)_2CO \quad (4)$$

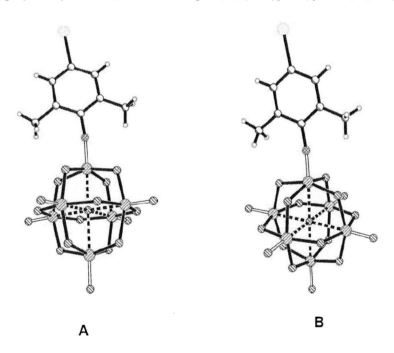

A   **B**

Figure 8. Structure of iodo-substituted derivative 13. A and B stand for its two conformational isomers differing by a rotation of the aromatic ring relative to the hexamolybdate skeleton.

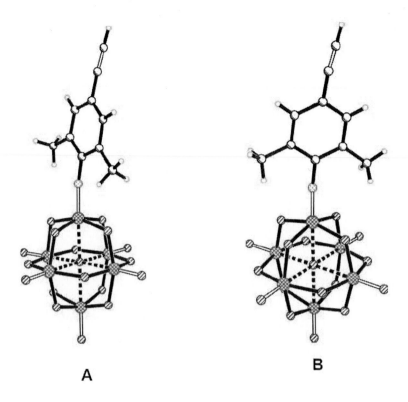

Figure 9. Structure of ethynyl-functionalized derivative 14. A and B stand for its two conformational isomers differing by a rotation of the aromatic ring relative to the hexamolybdate skeleton.

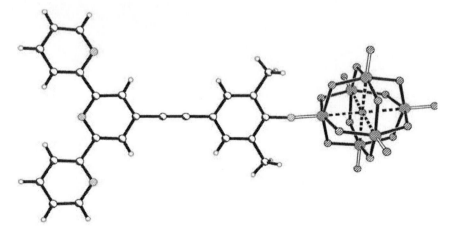

Figure 10. Ball-and-stick drawing of the organoimido hexamolybdate compound 15 containing a terpyridyl group.

In our investigations, other drying or dehydrating agents have also been tested instead of carbodiimides, including magnesium sulfate, calcium chloride, 4Å zeolite, phosphorus pentoxide, N'N-carbonyldiimidazole or even calcium hydride. However, they do not show similar activation effect like that of DCC [68]. This suggests that DCC does not simply serve as a dehydration agent to remove the water generated in the reaction. The likely role of DCC is presumedly to activate the terminal Mo-O bond and hence to increase the electrophilicity of the molybdenum atom, similar to its activating effect on the carboxyl group in amide or peptide synthesis [90]. In addition, an extremely excess of DCC is harmful to this reaction because DCC can undergo the side reactions directly with the molybdate ions and the initially formed products; for example, addition of eight equivalent of DCC, reaction of 2,6-dimethylaniline with $(n\text{-}Bu_4N)_2[Mo_6O_{19}]$ (6) under 12 hours of refluxing in dry acetonitrile affords, in place of the expected imido products, only an octamolybdtate compound, (*cyclo*-$C_6H_{11}NH_3)_4$ β-$[Mo_8O_{26}]$ (16), which is usually the major byproduct found in this reaction [68].

Recently, we have succeeded in extending this synthetic route from aniline homologs and their derivatives to other aromatic primary amines containing fused rings, including naphthylamine, anthramine, phenanthramine and aminopyrene. For instance, 1-naphthylamine undergoes this reaction effectively, in *ca* 70% yield, to afford the corresponding imido derivative (figure 11), $(n\text{-}Bu_4N)_2[Mo_6O_{18}(NC_{10}H_7\text{-}1)]$ (17) [74].

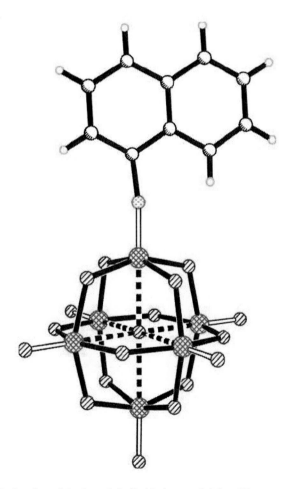

Figure 11. Ball-and-stick showing of the 1-naphthylimido hexamolybdate 17.

It should be pointed out here that, aromatic amines only with electron-withdrawing substituents such as nitro, fluoro and chloro groups do not undertake this reaction, neither do the aminopyridines, due to the weak nucleophilicity or basicity and easy oxidization of these kinds of amines [75]. We have also failed to prepare the alkylimido derivatives using this route with the corresponding aliphatic primary amines, such as benzylamine, since they violently reduce the hexamolybdate ion, usually resulting the single electron reduction produce of the parent cluster, $[Mo_6O_{19}]^{3-}$ [68, 75]. However, much effort to broaden the applicability of the DCC-dehydrating protocol to a broad spectrum of amines is still being taken in our laboratory.

Moreover, in our recent attempts to functionalize the octamolybdtates with organoimido groups, we discovered that in the presence of DCC, a proton can dramatically speed up the reaction of α-$[Mo_8O_{26}]^{4-}$ with aromatic primary amines under much milder conditions and, in the meanwhile, monofunctionalized organoimido derivatives of $[Mo_6O_{19}]^{2-}$ are selectively synthesized in high purity and moderate yield with easy workup [75]. This acid-assisting route has allowed the facile synthesis of a large number of mono-imido hexamolybdates, particularly, including those even bearing electron-withdrawing substituents such as bromo, chloro and fluoro groups [75- 78]. As shown in Eq(5), on addition of DCC (2 mmol), a variety of halogenated aromatic amines and their hydrochloride salts, in equivalent amounts (1.34 mmol), react with $(n\text{-}Bu_4N)_2$·α-$[Mo_8O_{26}]$ (1.0 mmol) smoothly, under nitrogen in hot acetonitrile (10 mL), obtaining the corresponding monosubstituted imido derivative in ca 50% yield after about 6 hours refluxing. Two structurally characterized halogenated imido derivatives, $(n\text{-}Bu_4N)_2[Mo_6O_{18}(NAr)]$, Ar = p-chlorophenyl (18) [75], p-bromophenyl (19) [77], respectively, are exemplified in figures 12 and 13. Nitroanilines also undergo this reaction, however, we have not successfully separated the corresponding derivatives from the reaction mixtures yet due to their decomposition in solution.

$$[Mo_8O_{26}]^{4-} + ArNH_2 + ArNH_3Cl \xrightarrow[\text{Refluxing, 6h}]{\text{2DCC, CH}_3\text{CN}} [Mo_6O_{18}(NAr)]^{2-} \tag{5}$$

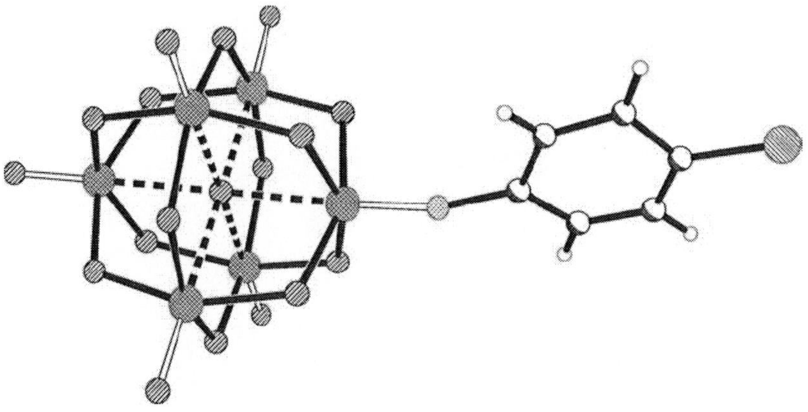

Figure 12. Ball-and-stick showing of the p-chlorophenylimido hexamolybdate 18.

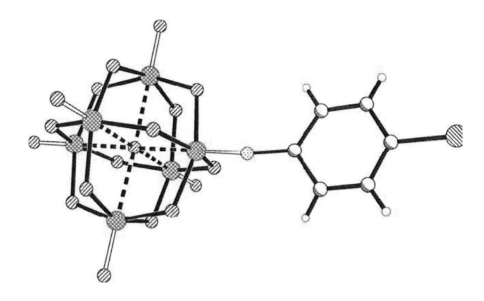

Figure 13. Ball-and-stick showing of the *p*-Bromophenylimido hexamolybdate 19.

It is worth pointing out that, the proton is crucial to the above route for the formation of monofunctionalized derivatives. It was observed that, in the absence of hydrochloride salts– which offer the protons, activated aromatic amines containing electron-donating substituents such as alkyl, alkoxy and alkylamino groups, react with α-$[Mo_8O_{26}]^{4-}$ to yield only the bifunctionalized arylimido derivatives of hexamolybdate. On the other hand, under the same conditions, no dehydrating reaction occurs between α-$[Mo_8O_{26}]^{4-}$ and inert aromatic amines with electron-withdrawing substituents such as chloro, fluoro and nitro groups, and the amines are oxidized instead. Furthermore, increasing the amount of hydrochloride salts can speed up the reaction, and correspondingly, it can be carried out even at room temperature. While more work needs to be done to shed light on the detailed reaction mechanism, the imaginable role of the proton is to complex with DCC and hence to strengthen the electrophilic ability of DCC to attack the oxo group of Mo-O group. Additionally, it also promotes the conversion of the α-octamolybdate into the hexamolybdate through a degradation and re-assembly process since in an acidic organic solvent, the hexamolybdate ions and their derivatives are much more stable than the α-octamolybdate ions.

There are a lot of spectroscopic techniques having been used to characterize the organoimido derivatives. In the low wavenumber region of the IR spectra ($\tilde{\nu}$ <1000 cm^{-1}), the monofunctionalized organoimido derivatives display similar characteristic patterns of the parent hexamolybdate anion: there are two very strong bands of the Mo-O$_t$ stretching and Mo-O$_b$-Mo asymmetric stretching vibrations at *ca* 955 ($\pm$5) and 795 ($\pm$5) cm^{-1}, respectively. Different from the parent cluster, there is also a strong shoulder peak appeared near 975 ($\pm$5) cm^{-1}, around the Mo-O stretching vibrations. This band may derive from the Mo-N stretching vibration [92] and is a diagnostic character for mono-organoimido substitution.

Compared to the ^{1}H NMR spectra of corresponding free amines, the protons attached to the α-carbon atoms or benzene rings in the monofunctionalized alkylimido and arylimido derivatives, respectively, all exhibit significantly downfield chemical shifts, indicating that the shielding nature of $[Mo_5O_{18}(Mo\equiv N-)]^{2-}$ is much weaker than that of the amino group NH_2- and in agreement with the strong electron acceptability of POM clusters.

While the UV/Vis spectra of monofunctionalized alkylimido derivatives resemble that of the hexamolybdate parent intimately in the low energy region, those of monofunctionalized arylimido derivatives show considerable changes very distinctly due to their different molecular and electronic structures. The lowest energy electronic transition at 325nm in $[Mo_6O_{19}]^{2-}$, which was assigned to a L→M charge-transfer transition from the terminal oxygen non-bonding π-type HOMO to the molybdenum π-type LUMO [92], is bathochromically shifted by more than 20 nm in the monofunctionalized arylimido derivatives and becomes significantly more intense. The observed bathochromic shifts imply that the Mo-N π-bond is formed and delocalized over the organic conjugated π systems in such arylimido derivatives. In other words, there are strong electronic interactions between the hexamolybdate cluster and the conjugated organic ligands. In addition, the extent of bathochromic shifts in the reported mono-arylimido derivatives is also consistent with the length of conjugated segments and the trend in the π-donor ability of arylimido ligands.

Electrochemical data and ^{95}Mo, ^{17}O, and ^{14}N NMR data have also been recorded for a range of mono-imido derivatives of the hexamolybdate [63, 71]. The following trends have been observed [53]: (i) All species display a one-electron reduction process, the potential of which undergoes a cathodic shift in the order O < NAr < NR and, for arylimido derivatives, in the order of increasing electron-donating properties of the substituent. (ii) The $[^{95}Mo\equiv NR]$ resonance is shifted upfield in the order O < NAr < NR, and in the series of arylimido complexes, in the order of increasing electron-attracting properties of the substituent. (iii) The ^{17}O NMR spectra each are globally shielded with respect to that of $[Mo_6O_{19}]^{2-}$. These data are consistent with the trend in π-donor ability of the ligands, O < NAr < NR. For arylimido derivatives, molybdenum and nitrogen shieldings increase with increase in the $(\sigma + \pi)$ acceptor ability of the substituent.

The X-ray crystal structures of all the mono-organoimido derivatives have some common features: They resemble that of their hexamolybdate parent, but one terminal oxo group has been replaced by a organoimido substituent. The short Mo-N bond lengths range from 1.68-1.75 Å and nearly linear linkages of Mo-N-C are consistent with a substantial degree of triple bond character, although the angles sometimes deviate from linearity more than expected due to the force of the crystal packing [75]. The central oxygen atoms become close to the imido-bearing molybdenum atoms since the *trans* influence of an imido group is weaker compared to a terminal oxo group [76]. While the bond length involving the bridging-oxygen atoms shows considerable changes, there are no appreciable variations in the terminal bond lengths, raging from 1.66 to 1.72 Å. The pattern of short/long trans bond alternation in the three mutually orthogonal Mo_4O_4 belts observed in the original parent cluster still remains in the Mo_4O_4 belts perpendicular to the Mo-N bonds. In addition, the supramolecular "pairing self-assembly" of the cluster anions through π-π stacking between parallel phenyl rings on two neighboring clusters has also been observed in some mono-arylimido derivatives [75-77], which results the formation of dimers in their crystalline states (figure 14).

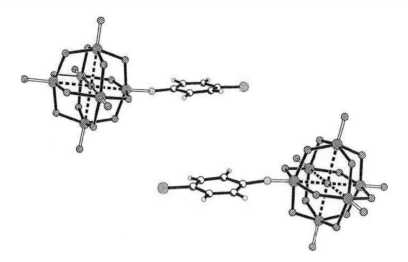

Figure 14. Dimer formation of the cluster anion in the solid of compound 18.

2.1.2. Disubstituted Derivatives

A characteristic feature of the derivatization of the hexamolybdate with organoimido ligands is the capacity for polyfunctionalization. Of which the difunctionalization is the easiest and most common. A. Proust et al noticed that, the *cis*-diimido derivative formed even when only one equivalent of $Ph_3P=NPh$ was exploited to react with $[Mo_6O_{19}]^{2-}$ [53]. However, using two equivalent of $Ph_3P=NPh$, it still gave a mixture of un-reacted hexamolybdate anions and, mono- and *cis*-diimido derivatives. The isolation of the successive derivatives is also difficult because of they cocrystallize usually in the crystalline state [63].

The first example of well characterized di-organoimido derivative was reported by E. A. Maatta and co-workers [65, 71]. They found that the reaction of $(n\text{-}Bu_4N)_2[Mo_6O_{19}]$ (6) with two equivalent of 2,6-diisopropylphenyl isocyanate in refluxing pyridine for 8 days occurs to form the *cis*-disubstituted imido hexamolybdate, $(n\text{-}Bu_4N)_2[cis\text{-}Mo_6O_{17}(NAr)_2]$, Ar = 2,6-diisopropylphenyl (20), in 67% yield. It is obvious that this is a poor route in efficiency. Moreover, they also mentioned that the crude products contain an admixture of higher- and lower-substituted hexamolybdates, which makes the separation of large amounts of the desired difunctionalized product tedious and impractical, if not impossible.

In the past years, we also checked the possibility of our DCC-dehydrating protocol in the preparation of di-organoimido derivatives of the hexamolybdate. In most case, attempts to use the same route as for the mono-substituted derivatives but with stoichiometric ratio of two equivalent aromatic primary amines failed to obtain the disubstituted derivatives in high yields. Instead, a mixture of various polysubstituted hexamolybdates are generated since there are six equally reactive sites on a hexamolybdate ion, and thus repeated recrystallization is needed for purification of the anticipant product. However, an attempt to functionalize the α-isomer of octamolybdate ion, $\alpha\text{-}[Mo_8O_{26}]^{4-}$, with this reaction routine resulted in the selective synthesis of difunctionalized arylimido derivatives of hexamolybdate with good yield and high purity [79].

This novel route is shown in Eq(6), which allows the efficient preparation of a number of di-imido hexamolybdates more conveniently with easy and simple bench operations [79-81]. For example, the refluxing reaction of $(n\text{-}Bu_4N)_2\cdot\alpha\text{-}[Mo_8O_{26}]$ with two equivalent of DCC and two equivalent of 4-iodo-2,6-diisopropylaniline or 4-ethynyl-2,6-dimethylaniline in dry acetonitrile for 6 to 12 hours affords the corresponding di-organoimido hexamolybdates, (n-Bu$_4$N)$_2$[cis-Mo$_6$O$_{17}$(NAr)$_2$], Ar = 4-iodo-2,6-diisopropyl-C$_6$H$_2$ (21) [79], 4-ethynyl-2,6-diisopropyl-C$_6$H$_2$ (22) [81], respectively, in good yields of 60-80%. The X-ray structures of compounds 21 and 22 are illustrated in figures 15 and 16, respectively. They both have the cis disubstituted structures with two activate remote functional groups at the end, which make them very important building bocks in fabricating novel organic-inorganic hybrids.

$$[Mo_8O_{26}]^{4-} + 2ArNH_2 \xrightarrow[\text{Refluxing, 6-12h}]{\text{2DCC, CH}_3\text{CN}} [cis\text{-}Mo_6O_{17}(NAr)_2]^{2-}$$

$$(6)$$

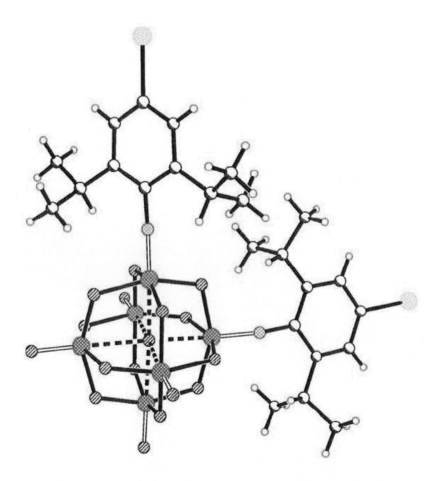

Figure 15. Ball-and-stick structure of the cis-disubstituted hexamolybdate imido derivative 21.

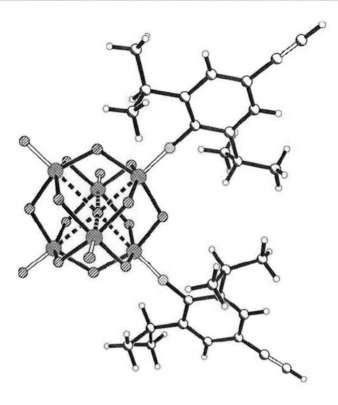

Figure 16. Ball-and-stick structure of the cis-didisubstituted hexamolybdate imido derivative 22.

In principle, there are, *cis* and *trans*, two position isomers in the terminal substituted di-imido hexamolybdates. If the steric and electronic effects of the imido groups are ignored, both of them can be formed in the same reaction, theoretically at a *cis* and *trans* ratio of 4:1, since there are four *cis* sites and only one *trans* terminal oxo reaction sites in a mono-imido derivative. While the steric repulsion is considered, which, in fact, is very common in organic substitution reactions, the *trans*-isomer will be more than 25% and becomes the principal product. But contrary to expectations, except for the *cis*-isomers, the *trans*-disubstituted hexamolybdates have scarcely been detected in the above reaction routes. It is seems that the disubstituted hexamolybdate with aryliimdo prefers to form a *cis*-isomeric derivative. Such a result is amazing, which suggests that the steric effect is less important here and on the other hand, the present of an extant imido substituent exerts an activating effect at proximal molybdyl groups. Indeed, a recent density functional theory study reveals that the *cis*-disubstituted derives are more stable in energy than the corresponding *trans*-isomers [93].

While the above routes give only the *cis*-disubstituted derives, R. J. Errington and co-workers discovered that, using their aromatic primary amines route [Eq(3)], a *trans*-disubstituted hexamolybdate, $(n\text{-Bu}_4\text{N})_2[trans\text{-Mo}_6\text{O}_{17}(\text{NC}_6\text{H}_4\text{NH}_2)_2]$ (23), the unique example reported in the literature so far [66], was formed predominantly in the reaction between $(n\text{-Bu}_4\text{N})_2[\text{Mo}_6\text{O}_{19}]$ (6) and two equivalent of 1,4-phenylenediamine. Moreover, ^{1}H NMR also confirmed the formation of the corresponding *cis*-isomer in the resulting products. Their work implies that the *trans*- disubstituted hexamolybdate is a kinetic stable product under certain supported conditions and can be isolated indeed. It also intensively suggests that there is a obvious different mechanism of the directly functionalization of the hexmolybdate with aromatic primary amines from those with phosphinimines or isocyanates and the aromatic primary amines approach is potential in the *trans*-disubstituted hexamolybdate derivatives.

Keeping the above idea in mind, we have had a renewed check very recently on our DCC-dehydrating protocol for the direct functionalization of the hexamolybdate with stoichiometric ratios of reactants [82- 83]. Indeed, for some activated aromatic primary amines containing electron-donating substitents, such as 2,6-dimethylaniline and *ortho*-anisidine, in the presence of one equivalent of DCC, the reaction of them and 0.5 equivalent of $(n\text{-}Bu_4N)_2[Mo_6O_{19}]$ (6) in heated acetonitrile can afford both the *cis*- and *trans*-disubstituted imido complexes in reasonable yields by a careful control of the refluxing time [Eq(7)].

$$[Mo_6O_{19}]^{2-} + 2ArNH_2 \xrightarrow[\text{Refluxing, 6-24h}]{\text{2DCC, CH}_3\text{CN}} [Mo_6O_{17}(NAr)_2]^{2-}$$

(7)

In general, a prolonged refluxing time promotes the formation of the *cis*-isomer, for example, the compound, $(n\text{-}Bu_4N)_2[cis\text{-}Mo_6O_{17}(NAr)_2]$, Ar = 2-methoxy-$C_6H_4$, (24) [82], was obtained from 24 hours refluxing reaction; On the other hand, the *trans*-isomer can be isolated from a relative short-time reaction system, for example, the compound, $(n\text{-}Bu_4N)_2[trans\text{-}Mo_6O_{17}(NAr)_2]$, Ar = 2,6-dimethyl-$C_6H_3$, (25) [83], was acquired from 6 hours refluxing reaction. In figures 17 and 18, the X-ray structures of compound 24 and 25 are shown, respectively. Compared to the parent hexamolybdate, compound 24 has a smaller optical band gap of 1.75 eV due to the incorporation of two strong electron-donating anisidido groups, which suggests that compound 24 is a potential molecular semiconductor and the optical band gap of POM-based semiconducting materials could be tuned exactly and controllably via chemical modification using suitable organic ligands. The chemical and physical properties of compound 25 are still underway in our laboratory.

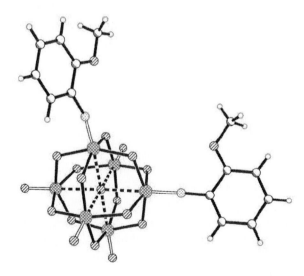

Figure 17. Ball-and-stick structure of the cis-didisubstituted hexamolybdate imido derivative 24.

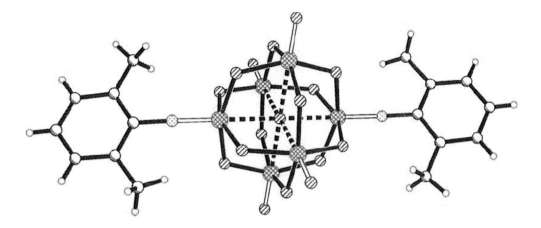

Figure 18. Ball-and-stick structure of the trans-didisubstituted imido complex 25.

Similar to their monosubstituted analogues, all the reported diimido hexamolybdates have of their own common features and some general trends in the structures and spectroscopies: (i) The bond lengths of [Mo≡NAr] triple bond are within the range of 1.71 to 1.75 Å, together with nearly linear linkages of Mo-N-C angles close to 180°. For the *cis*-isomers, one of the angles sometimes deviate from linearity more than expected due to the aggregation of the cluster anions through π-π stacking and the force of the crystal packing [80]. (ii) The central oxygen atoms inside the hexamolybdate cage are drawn closer to the imido-bearing Mo atoms, while the four bridge oxygen atoms are pushed away from the imido-bearing Mo atoms. (iii) The short-long bond alternation in the in the Mo_4O_4 belts of the parent hexamolybdate disappears in the *cis*-isomers. However, in the trans-isomers, this pattern is still kept in the belt perpendicular to the long axis of the cluster along the two Mo-N bonds. (iv) The patterns of supramolecular self-assembly due to π-π stacking or/and C-H...O hydrogen bonding are very popular. In particular, the I-O interaction has also been observed in compound 21 (figures 19). (v) In their IR spectra, the Mo-N stretching vibrations at 968 (±2) cm^{-1}, appear as a strong shoulder peak of the Mo-O stretching vibrations at 945 (±2) cm^{-1}. Compared to the corresponding mono-imido derivatives, there are obviously blue-shifts of *ca* 5 and 10 cm^{-1} for the Mo-N and Mo-O stretching vibrations, respectively. Correspondingly, the $Mo-O_b-Mo$ asymmetric stretching vibrations at *ca* 795 (±5) cm^{-1} of the mono-imido and parent hexamolybdates are significantly blue-shifted more than 20 cm^{-1} and appear as a broad and strong band at *ca* 775 cm^{-1}. (vi) Compared to the corresponding mono-imido derivatives, the lowest energy electronic transition of L→M charge-transfer is further red-shifted by *ca* 5 nm due to the incorporation of the other arylimido ligand. In the meanwhile, ^{1}H NMR of the aromatic protons shows visible upfield shifts, and the reduction potentials are also observed to shift cathodically due to the increase of electron density in the cluster anions accompanying the incorporation of the second arylimido ligand.

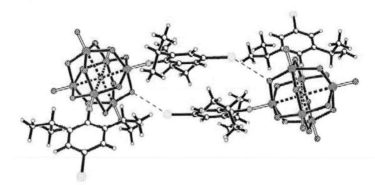

Figure 19. A view of the I-O interaction in a dimer of compound 21 in the crystalline state.

2.1.3. Polysubstituted Derivatives

The polysubstituted hexamolybdates, where three to six terminal oxo groups replaced by 2,6-diisopropylphenylimido ligands, have also been reported by E. A. Maatta and co-workers for nearly a decade [65, 71, 84]. They discovered that, by carefully controlling the molar ratios of 2,6-diisopropylaniline isocyanate over the hexamolybdate cluster , multiple-functionalized hexamolybdates, $[Mo_6O_{(19-x)}(NAr)_x]^{2-}$, Ar = 2,6-diisopropylphenyl, x = 3 (26) [71], 4 (27) [65, 71], 5 (28) [65, 71], 6 (29) [84] could be synthesized in refluxing pyridine after a prolonged reaction time of 13, 12, 11 and 21 days for 26, 27, 28, and 29, respectively. Of which the X-ray molecular structures of compounds 26 and 29 are exemplified in figures 20 and 21, respectively. While all these polysubstituted derivatives have been isolated and characterized, the nonselective nature of the reaction, which always results in an admixture of various multiple functionalized products, makes their separation process rather difficult. Apparently, the rational synthesis of polysubstituted hexamolybdates is a formidable challenge for a chemist.

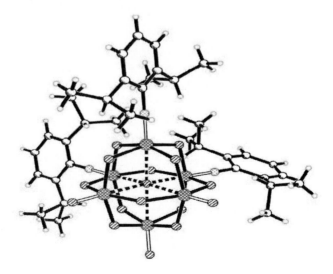

Figure 20. X-ray molecular structure of the trisubstituted hexamolybdate 26.

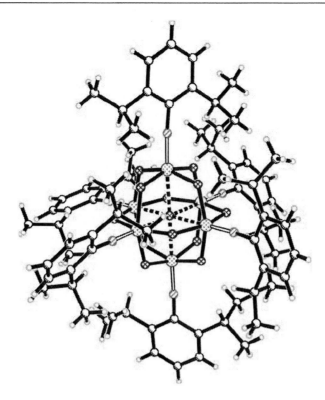

Figure 21. X-ray molecular structure of the hexasubstituted hexamolybdate 29.

We have also interested in the preparation of the polysubstituted hexamolybdates in the context of our DCC-dehydrating protocol using aromatic primary amines as the imido-releasing agents. In a recent attempt to synthesizing the *fac*-tri-2,6-dimethylanilido hexamolybdate [Eq(8)], an analogue of compound 26, contrary to our expectation, the as-resulted tri-imido derivative, $(n\text{-}Bu_4N)_2[cis\text{-}Mo_6O_{16} \mu_2\text{-}(NAr)(NAr)_2]$, Ar = 2,6-dimethyl-$C_6H_3$, (30) is not exclusively terminal-oxo replaced complex, and instead, one of bridging-oxo groups is substituted.

$$[Mo_6O_{19}]^{2-} + 3ArNH_2 \xrightarrow[\textbf{Refluxing, 24h}]{\textbf{3DCC, CH}_3\textbf{CN}} [Mo_6O_{16}(NAr)_3]^{2-}$$

(8)

As is illustrated in figure 22, the compound 30 has a feature that the μ_2-bridging oxygen atom sharing by the two imido-bearing Mo atoms in a *cis*-diimido hexamolybdate is substituted with a μ_2-bridging organoimido ligand. It suggests that such oxygen atoms in the *cis*-diimido hexamolybdates have been doubly activated by the neighboring imido groups and become more negative (nucleophilic) than other oxygen atoms, resulting in the easy electrophilic attack by DCC. Compound 30 stands for the first example of a bridging oxygen atom in POMs replaced by the bridging imido groups. It breaks the myth that only the terminal oxo groups can be directly replaced with imido ligands before and brings us a wonderful prospect in the chemistry of organoimido derivatives of POMs.

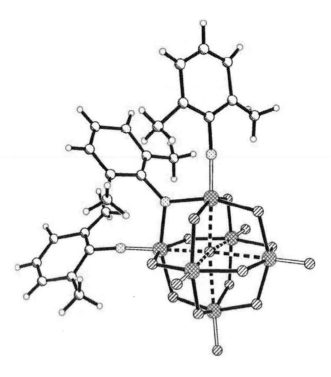

Figure 22. Ball-and-stick model of the trisubstituted hexamolybdate 30.

2.2. Organoimido Derivatives of other Lindqvist POMs

While there is fruitful in the synthetic chemistry of hexamolybdate, related studies on other POMs keep scarcely well developed. So far, there only has been obtained quite limited success in extending the above approaches to the other Lindqvist POMs.

Attempts to directly functionalize the hexatungstate anion, $[W_6O_{19}]^{2-}$, another important Lidquvist homopolyoxometalate besides the hexamolybdate, have failed to utilize all the above-mentioned approaches, since $[W_6O_{19}]^{2-}$ does not react at all with phosphinimines, isocyanates or aromatic primary amines under the present developed conditions. It seems that compared to the molybdyl groups in the hexamolybdate, the tungstyl groups in the hexatungstate is rather unreactive. The only one example of imidosubstituted hexatungstates, $(n\text{-}Bu_4N)_2[W_6O_{18}(NAr)]$, Ar = 2,6-diisopropyl-$C_6H_3$, (31) was obtained about ten years ago by E. A. Maatta et al via the reaction of $(n\text{-}Bu_4N)_2[WO_4]$ with 2,6-diisopropylphenyl isocyante in either 1,2-dichloroethane or pridine, but the yield is very low (*ca* 10% based on $n\text{-}Bu_4N)_2[WO_4]$) [94]. They thought that the plausible mechanism involves the initial formation of the imdo-tungstate complex, $[WO_3(NAr)]$, which then aggregates with additional $[WO_4]^{2-}$, in a self-assembly manner, to yield the target product. In addition, it is probably the proton in acidic contaminant, not water mentioned by Maatta *et al*, that involves in the assembly of the hexanuclear cluster, just like the above-mentioned acid-assisting route developed by us for the mono-imido hexamolybdates [75]. The deliberate addition of a suitable acid will perhaps bring us an improvement of this route.

Although the hexatungstate cluster is less labile, replacement of tungstyl groups with reactive molybdyl groups will exert an activate effect on the cluster. Surely, we discovered that the pentatungstenmolybdate ion, $[MoW_5O_{19}]^{2-}$, a Lidquvist heteropolyoxometalate, can be directly functionalized with aromatic primary amines using the DCC-dehydrating protocol [95].

$$[MoW_5O_{19}]^{2-} + ArNH_2 \xrightarrow[\text{Refluxing, 12h}]{\text{1.5DCC, CH}_3\text{CN}} [W_5O_{18}(MoNAr)]^{2-}$$

(9)

As shown in [Eq(9)], on addition of one and a half equivalents of DCC, the reaction of one equivalent of 2,6-dimethylaniline or 4-iodo-2,6-dimethylaniline with one equivlant of (n-Bu$_4$N)$_2$[MoW$_5$O$_{19}$] does occur in hot acetonitrile under nitrogen, and in *ca* 60% yields affords the mono-imido derivatives [95], (n-Bu$_4$N)$_2$[W$_5$O$_{18}$(MoNAr)], Ar = 2,6-dimethyl-C$_6$H$_3$, (32), 4-iodo-2,6-dimethyl-C$_6$H$_2$, (33), respectively. The X-ray crystal structure of compound 33 is given in figure 23. Indeed as expected, the terminal oxygen atom bonded to the molybdenum atom is selectively replaced by an imido substituent. This result opens a general road for the synthesis of organo imido derivatives of less reactive POM clusters, namely, by replacement one of the inert terminal metal-oxo groups with a functionalizable Mo-O group.

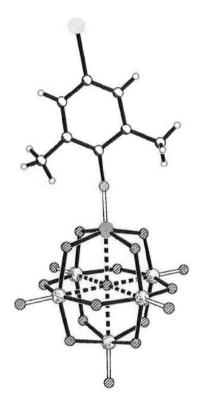

Figure 23. Ball-and-stick model of the imido pentatungstenmolybdate 33.

2.3. Organoimido Derivatives of Keggin POMs

Keggin-type POMs have a general formula of $[XMO_{40}]^{n-}$ and feature a central tetrahedral oxyanion $[XO_4]^{m-}$ encapsulated in a cage made up of twelve MO_6 octahedra by sharing corners, which represent the most popular and largest subclasses of POMs with a variety of transformations in structure and composition of the incorporated hetero-atom in the interior or metal atoms on the surface [1]. While the surface oxygen atoms of the α-phosphomolybdate ion, $[\alpha\text{-}PMo_{12}O_{34}]^{3-}$, and α-phosphotungstate ion, $[\alpha\text{-}PW_{12}O_{34}]^{3-}$, are readily alkylated with alkylation agents such as methyl sulfate and trimetyloxonium tetrafluoborate [54,55], they are unable to be replaced directly by any an imdio delivery reagent so far. For example, reaction of $[\alpha\text{-}PMo_{12}O_{34}]^{3-}$ with p-methyphenyl isocyanate in hot pyridine failed to afford the expected imidosubstituted α-phosphotungstate, but a degradated decanuclear imido complex instead [96]. We also tried the reaction of several Keggin-structure POMs containing molybdyl groups, including the α-phosphomolybdate, with aromatic primary amines in the presence of DCC. However, in most case, aromatic amines were oxidized instead of acting as imido delivery reagents due to the high oxidation nature of the Keggin POMs of interest.

The only successful case in the synthesis of Keggin-type imido derivatives was just reported last year by A. Proust and coworkers [97]. They obtained an imido complex, $[Bu_4N]_4$ $[PW_{11}O_{39}(ReNC_6H_5)]$ (34), by the substitution reaction of the monovacant Keggin compound $[Bu_4N]_4[H_3PW_{11}O_{39}]$, a penta-dentate ligand, with the mononuclear complex $[Re(NPh)Cl_3(PPh_3)_2]$ containing a metal-imido functional group in acetonitrile [Eq(10)]. Compound 34 have clearly demonstrated by various techniques including ^1H and ^{14}N NMR, electrochemistry, and especially by ESI mass spectrometry. However, its X-ray structure has not yet been completed.

$$[H_3PW_{11}O_{39}]^{4-} + [Re(NPh)Cl_3(PPh_3)_2] \xrightarrow[\text{Refluxing, 1h}]{\text{Et}_3\text{N, CH}_3\text{CN}} [PW_{11}O_{39}(ReNC_6H_5)]^{4-} \quad (10)$$

In all, the functionalization of Keggin-type POMs with imido groups faces still formidable challenges and much more attentions should be paid into this interesting field in the future since the potential of Keggin POM-based molecular materials is expected to be greater than that of Lindqvist derivatives.

3. REACTIVE CHEMISTRY OF ORGANOIMIDO DERIVATIVES OF POMS

While organoimido derivatives of POMs have been discovered for more than one decade [58] and a variety of synthetic approaches have been developed as mentioned above, the reactive chemistry of them has just been investigated very recently and only the organoimido derivatives of hexamolybdate have been involved [72, 73, 81, 98-100]. Both the inorganic cluster skeleton and the incorporated organic ligand in an organoimido POM can act as the active sites for their own enjoyed reactions. The Mo-N bond is also a possible reaction site, the correlative reactions such as hydrolysis, however, are trivial due to break of the Mo-N bond and leaving of the organic ligand from the cluster in most cases. So far, the reaction chemistry of organoimido derivatives of POMs has focused mainly on the incorporated organic ligands because it provides great potentials in controllable synthesis of organic-inorganic hybrids exploiting well-developed common organic synthesis techniques. In particular, for the imidosubstituted hexamolybdates, several important inorganic and organic reactions, which do not cleave the Mo-N bond, have been set up. These includes coordination [73], addition polymerization [72] and carbon-carbon coupling reactions [98], of which the carbon-carbon coupling has been proved to be a powerful tool for our molecule-module-assembly strategy in fabricating POM-based organic-inorganic hybrid materils [56].

3.1. Coordination

Compared to the corresponding parent cluster, the surface oxygen atoms on the cluster skeleton of an organically derivatived POM should be more reactive to coordinate with a metal ion due to the activation effect of incorporated organic ligands [53], nevertheless, such reactions have not been realized in the imidosubstituted hexamolybdates. On the other hand, if the incorporated imido groups bearing ligatable substituents such as hydroxyl, carboxyl and pryidyl, such derivatives can in principle behave as POM-functionalized organic ligands. According to this idea, we synsesized the compound 15 with a terpyridyl group (see figure 10) [73]. The coordination reaction of this hexamolybdate-functionalized ligand 15 with some bivalent transitional metal ions such as Zn^{2+} and Ru^{2+} has been confirmed by preliminary *in situ* UV studies. For example, on adding an excess of Ru(tPy)Cl$_3$ to a solution of functionalized ligand 15 in DMSO, the intense absorption at 392 nm of ligand 15 diminishes significantly, in the meanwhile, a new absorption band at 504 nm appears gradually. The new absorption is a characteristic M→L charge-transfer (MLCT) transition of the Ru-bisterpyridine complexes, which implies definitely that the new hexamolybdate-functionalized bisterpyridine-Ru complex 35 has formed by the complexation reaction shown in [Scheme 1]. Isolation of the complex 35 and its X-ray single crystal structure determination are currently in progress.

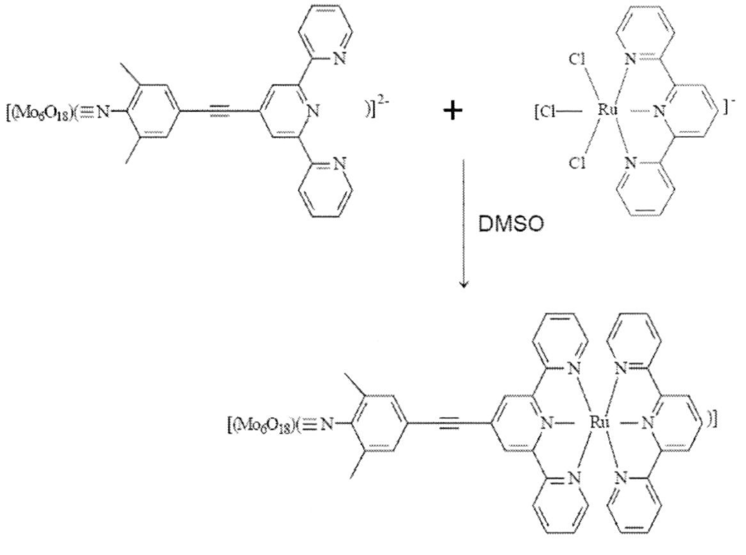

Scheme 1. Coordination reaction of compound 15 with Ru(tPy)Cl$_3$.

3.2. Addition Polymerization

The addition polymerization of organoimido derivatives of POMs bearing an unsaturated functional group such as the C=C double bond was just reported at the beginning of this new centenary [72]. In this report, E. A Maatta and co-workers discovered that, the *p*-styrenylimido hexamolybdate **9** and 4-methylstyrene could undergo conventional free radical-induced copolymerization initiated by AIBN [2,2'-azobis(2-methylpropionitrile)] in 1,2-dichloroethane and afforded a copolymer 36 with organoimido-POMs as pendants [Scheme 2]. The formation of copolymer 36 was confirmed by its IR spectrum, where the band of C=C stretching disappeared, but the characteristic doublet pattern of the terminal Mo-O and Mo-N vibrations in an monoimido hexamolybdate were still kept, compared to the precursor **9**. By comparison of the relative intensities of ^{1}HNMR due to the aryl groups, the α-CH$_2$ groups of the [*n*-NBu$_4$]$^+$ cations, and the styrenic CH$_3$ group, copolymer 36 was suggested to have a composition of nearly three 4-methyl styrene units per unit of **9**. This work demonstrates that the styrylimido ligand has similar chemical reactivity to that of the corresponding free stryrene and can be employed as a reactive site capable of delivering its metal complex into a conventional polymeric environment. Such a methodology should have broadly applicable potentials in the synthesis of novel inorganic/organic hybrid polymeric materials which incorporate covalently attached organoimido-POM complexes.

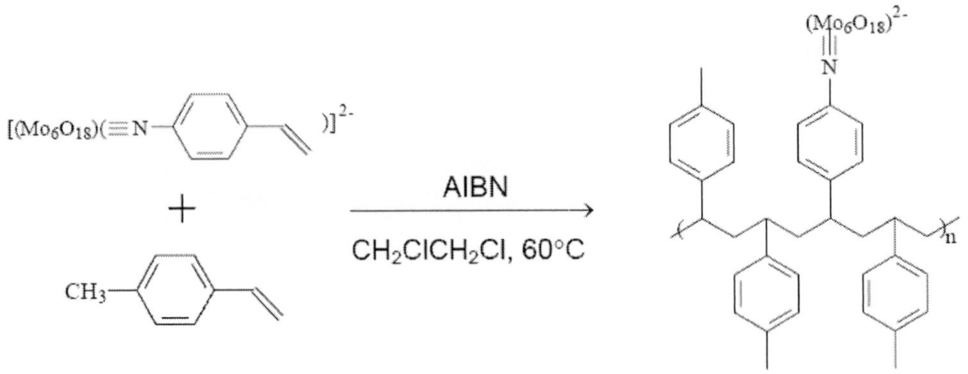

Scheme 2. Free radical-induced copolymerization Between compound **9** and 4-methylstyrene.

3.3. Carbon-Carbon Coupling and Its Application in POM-Based Organic-Inorganic Hybrids

It is well known that aryl halides and terminal acetylenes can undergo Pd catalyzed carbon-carbon coupling reactions [101-103]. The arylimido derivatives of POMs with a remote iodo or ethynyl groups, such as compounds 13 and 14, are in fact, POM-functionalized aryl halides and terminal acetylenes. If the iodo or ethynyl groups on functionalized POM clusters can undergo common Pd-catalyzed coupling reactions, and the POM clusters survive and the imido linkage does not break under the coupling reaction conditions, such organically-functionalized clusters would be invaluable building blocks for the rational synthesis of novel molecular POM–organic hybrids.

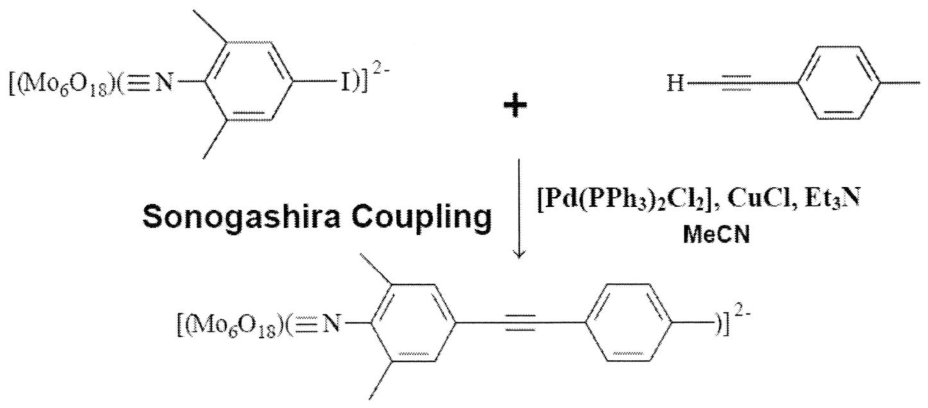

$[(Mo_6O_{18})(\equiv N-\!\!\!-\!\!\!-I)]^{2-}$ **+** $H\!\!-\!\!\equiv\!\!-$

Sonogashira Coupling | $[Pd(PPh_3)_2Cl_2]$, **CuCl, Et₃N**
MeCN

$[(Mo_6O_{18})(\equiv N-\!\!\!-\!\!\!-\!\!\equiv\!\!-)]^{2-}$

Scheme 3. Sonogashira coupling of 13 with 4-methyl-1-ethynylbenzene in acetonitrile at room temperature.

The Sonogashira coupling reaction of 13 with 4-methyl-1-ethynylbenzene [Scheme 3] or 3,5-di-tertbutyl-1-ethynylbenzene has been carried out in acetonitrile at room temperature under the protection of N₂ gas [98]. Compared to classic coupling reactions in organic synthesis [101-103], this reaction is extremely fast and finishes within only 20 minutes as monitored by thin-layer chromatography. As confirmed by the single X-ray diffraction study (figure 24), the hexamolybdate cluster survived indeed in the coupling product 37 under the reaction conditions, and obviously it activates the coupling reaction. The enhanced reactivity of the iodo function in 13 is presumably due to the electron-withdrawing nature of the hexamolybdate cluster. Prolonged reaction time and excess base were found to be harmful to the desired coupling products 37 and 38, as they slowly decomposed to the parent hexamolybdate cluster under the reaction conditions. Nevertheless, pure products can be usually obtained in more than 80% excellent yields.

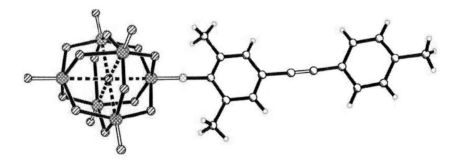

Figure 24. Ball-and-stick structure of the coupling product 37.

Hybrid Molecular Nanodumbbells

With our success in which iodo- or ethynylphenylimido-functionalized POM clusters can undergo Pd-catalyzed carbon-carbon coupling reactions, a great number of POM-based hybrids with various linking geometries, including covalently linked POM networks, can be rationally designed and easily prepared. For example, hybrid molecular nanodumbbells 39 and 40 with two hexamolybdate clusters linked by a rigid conjugated organic rod with different lengths in nanosize, shown in Scheme 4, have been synthesized by the Sonogashira coupling reaction between 13 and 14 in good yields [99].

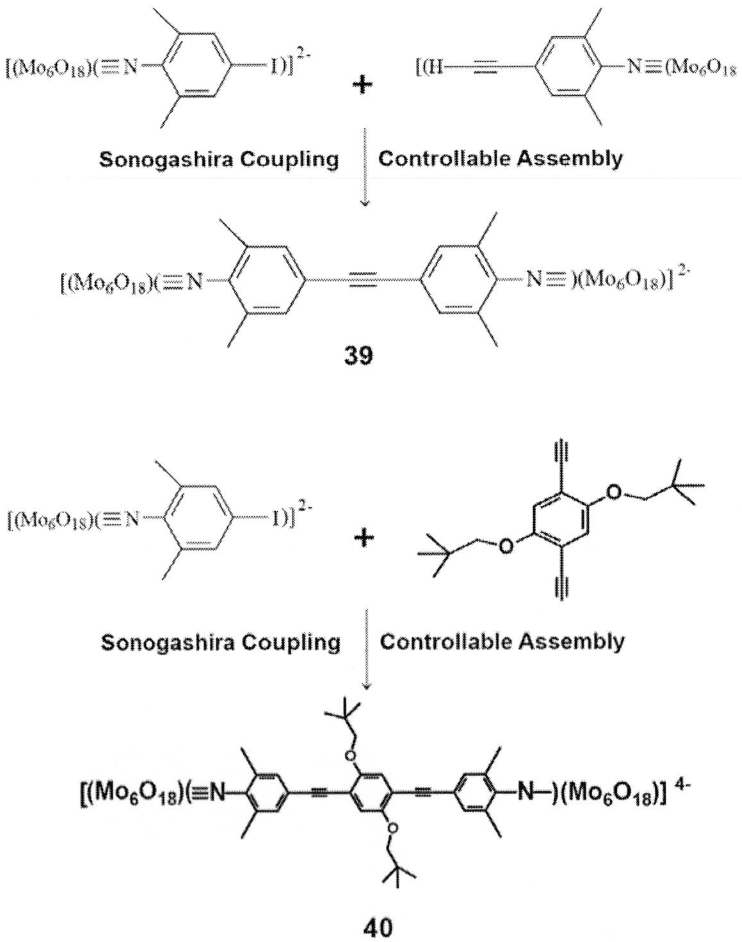

Scheme 4. Controllable assembly of hybrid nanodumbbells.

The X-ray crystal structure of nanodumbbell 39 has been determined (figure 25). Cyclic-voltammetry studies reveal a reversible one-electron reduction wave at -891 mV (vs Ag/Ag$^+$) for both nanodumbbells 39 and 40, which is very close to the reduction potential of other momosunstituted imido derivatives such as 13 (-909 mV), 14 (-893 mV), 37 (-893 mV), and 38 (-895 mV) [56]. UV/Vis absorption spectroscopy studies show that the lowest energy electronic transition is significantly bathochromically shifted from compounds 13 (362 nm) and 14 (368 nm), to hybrid dumbbells 39 (426 nm) and 40 (438 nm), thus indicating that the Mo≡N π bond is delocalized within the organic conjugated π system. Compared to the original oligo(phenylene ethynylene) ligands, which are strongly fluorescent molecules [104], both of two hybrid dumbbells show no fluorescence under excitation from 200 nm to 500 nm, which implies clearly that the conjugatedly-attached hexamolybdate cluster acts as an efficient fluorescence quencher in the dumbbells. These results suggest that strong electronic interactions through the conjugated organic bridge exist between the two clusters in a nanodumbbell.

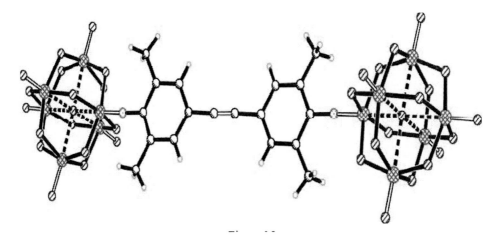

Figure 25. Ball-and-stick structure of the hybrid nanodumbbell 39.

Molecular Charge Transfer Hybrids

Charge-transfer hybrids based on organic donors and inorganic acceptors have being focused by chemists for several decades [100, 107, 108], due to their potential applications in solar energy conversion and storage such as photovoltaic cells, conducting materials including novel molecular conductors and superconductors, and nonlinear optical devices. POMs are one of the most important electron acceptors [50, 105], which have been widely exploited in preparing charge-transfer salts with organic cation donors, or charge-transfer supramolecular aggregates with neutral organic π donors through cocrystallization [106]. Efforts have recently been made to develop intramolecular charge transfer hybrids based on POMs[100]. Compound 11 is one of such examples, in which a ferrocenyl donor is bound covalently to a hexamolybdate acceptor and a charge-transfer transition from the Fe^{2+} center to the cluster is observed at 536nm [70]. However, this hybrid was obtained in only 19% yield through a direct functionalization reaction on the hexamolybdate cluster. Additionally, the short separation distance between the ferrocenyl donor and the cluster acceptor also renders the charge recombination facile as a coherent superexchange mechanism dominates [109]. These problems, however, are easy to be overcome with iodophenylimido-functionalized POM cluster building blocks through the Pd-catalyzed carbon-carbon coupling reactions.

For example, two novel charge-transfer hybrids 41 and 42, shown in Scheme 5, in which one or two ferrocenyl groups are bonded indirectly to a hexamolybdate cluster through an extended π-conjugated bridge, have been prepared in over 65% good yields from the Sonogashira coupling reactions of 13 and 21 with ethynylferrocene [100], respectively. The X-ray molecular structure of 41 was shown in figure 26.The electronic spectra of these hybrids show a broad absorption tail extending beyond 550 nm, indicating the existence of charge-transfer transition from the ferrocenyl donor to the cluster acceptor. The observation of the clear charge-transfer transition indicates the contribution of charge-transfer resonance to the ground state in both 41 and 42 even though the donor-acceptor separation distance of 11.29 Å is rather long, signaling a through-bond charge-transfer nature made possible by the organic π-conjugated bridge. Cyclic voltammetry studies reveal a single-electron oxidation wave at 0.328 V and a single electron reduction wave at -0.831 V for 41 with one ferrocenyl unit. For the hybrid with two ferrocenyl units compound 42, a lower single electron reduction potential at -1.055 V and a two-electron oxidation wave at 0.331 V are observed. The reduction process in 41 and 42 is associated with the cluster skeleton. The observed cathodical shift of reduction potential from 41 to 42 is in agreement with the increasing number of the electron-donating imido ligand in 42. The oxidation wave in these two hybrids is attributed to the ferrocenyl unit. The almost same oxidation potential of them indicates negligible electronic interactions between the two ferrocenyls in 42.

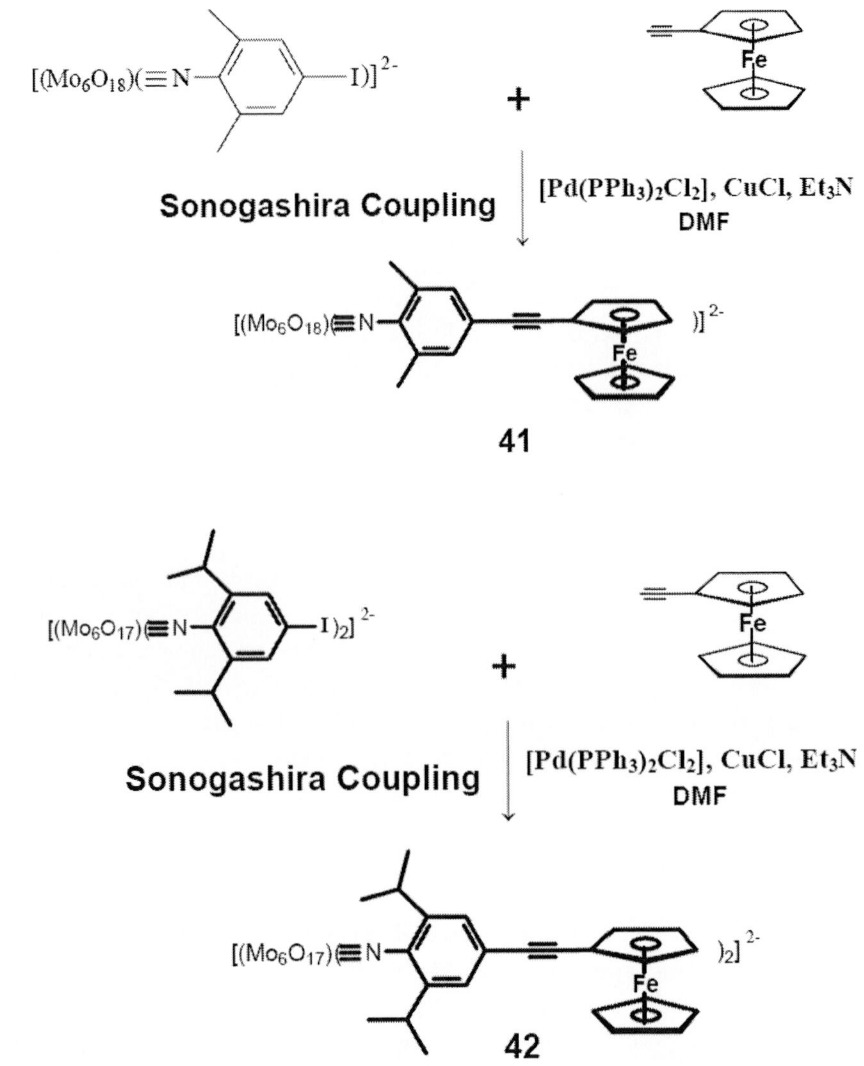

Scheme 5. Synthesis of two novel charge-transfer hybrids based on hexamolybdate cluster.

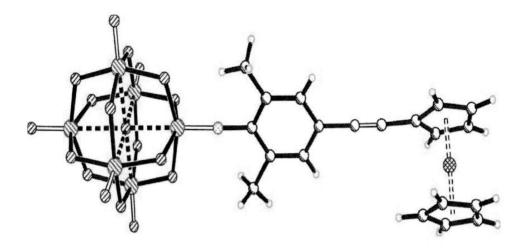

Figure 26. Ball-and-stick structure of the charge-transfer hybrid 41.

Hybrid Ploymers

Hybrid polymers containing POMs have been intensively studied for seeking new conducting materials. Conjugated polymeric hybrids containing covalently linked POM clusters, either in the main chain or in the side chains, at present, can be synthesized conveniently with the Pd-catalyzed coupling reactions. Starting from difunctionalized cluster 21 or 22, main-chain POM-containing polymers have been prepared as shown in Scheme 6. These polymers have an average molecular weight of about 160,000. Their electronic spectra are nearly identical to that of hybrid dumbbell 39, indicating that the organic π electrons delocalize only within the two imidobearing Mo atoms and the electronic interaction between the two orthogonally jointed organic π systems is limited. Cyclicvoltammetry studies reveal a reversible single-electron reduction wave at -1.2 V (vs Ag/Ag+) which is consistent with bis(arylimido)-substituted hexamolybdates. These polymers also show intense absorption in the visible range but with little fluorescence emissions, suggesting efficient fluorescence quenching of the embedded POM cluster on the organic phenylene acetylene units. Peng et al has studied their applications in photovoltaic cells and a power conversion efficiency of 0.15% was obtained, which is significantly higher than photovoltaic cells fabricated with other conjugated polymers in the same device configuration.

Recently, the hexamolybdate cluster has been incorporated into poly(phenylene ethynylene) as conjugated side-chain pendants through the Pd-catalyzed coupling reactions in Peng's groups too. These polymeric hybrids have much more potential for applications in photovoltaic cells, owing to efficient fluorescence quenching in such materials to make free charge carriers reside in different structural units [110].

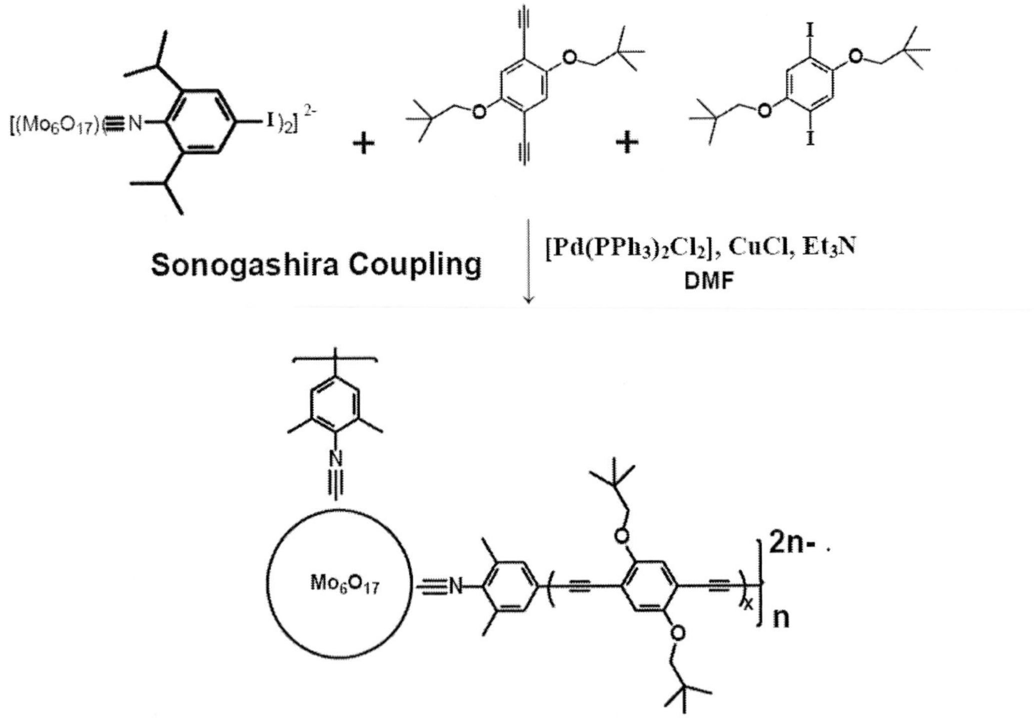

Scheme 6. Synthesis of polymeric hybrids based on hexamolybdate cluster.

4. CONCLUSION AND OUTLOOK

The chemistry of organoimido derivatives of polyoxometalates, especially their reactive chemistry, is still in its infancy. Although classically organic synthetic techniques have given a fascinating promise in this field, to date, only a limited number of methods have been developed for the preparation of covalently bonded organoimido derivatives of POMs and their reactions have just stepped mainly on the Pd-catalyzed coupling reactions. Nonetheless, this new chemistry indeed opens a wide road to make novel molecular hybrids and polymeric hybrids based on POMs in a much more controlled manner. With the development of this chemistry, a large number of covalently bonded POM–organic hybrids, particularly those involving organic π-conjugated systems, including molecular, dendritic, and polymeric systems are sure to be prepared in the near future. In addition, efforts on exploring the electrical, optical and photophysical properties of this new class of hybrids are expected.

ACKNOWLEDGEMENTS

Y. G. Wei sincerely thanks Professor Zhonghua Peng for professional support to start the project when he was at UMKC. He also gratefully thanks Professors Hongyou Guo and Sijia Xue for helps in continuing this research in china. The contributions of undergraduate and graduate students: Ning Ge, Na Wang, Qiang Li, Yun Xia, Zicheng Xiao, Yi Zhu and Shaoyong Ke, are also gratefully acknowledged. This work is sponsored by NFSC No.

20201001, 20373001, 20671054 and 20871073, SRF for ROCS of SEM, MOST-TG2000077503, and the Natural Science Funding of Hubei Province No.2008CDB020.

REFERENCES

[1] Pope, M. T. *Heteropoly and Isopoly Oxometalates*. Berlin: Springer; 1983.

[2] Berzelius, J. J. Beitrag zur näheren Kenntniss des Molybdäns. Poggend. *Ann. Phys. Chem.* 1826, 6, 369–392.

[3] Pope, M. T.; Müller, A. Editors. Polyoxometalates: *From Platonic Solids to Anti-Retroviral Activity*. Dordrecht: Kluwer; 1994.

[4] Hill, C. L. Guest Editor. A Special Thematic Issue on Polyoxometalates. *Chem. Rev.* 1998, 98, 1-390.

[5] Pope, M. T.; Müller, A. Editors. Polyoxometalate Chemistry: *From Topology via Self-Assembly to Application.* Dordrecht: Kluwer; 2001.

[6] Zhang, S. W.; Wei, Y. G.; Yu, Q.; Shao, M. C.; Tang, Y. Q. Toward Quantum Line-Self-Assembly Process of Nanomolecular Building Blocks Leading to a Novel One-Dimensional Nanomolecular Polymer: Single-Crystal X-ray Structure of $\{[H_3O]^+_{12}\{(H_2O)MoO_{2.5}[Mo_{36}\ O_{108}(NO)_4(H_2O)_{16}]O_{2.5}Mo(H_2O)\}^{12-}\}$. *J. Am. Chem. Soc.* 1997, 119, 6440-6441.

[7] Jiang, C. C.; Wei, Y. G.; Liu, Q.; Zhang, S. W.; Shao, M. C.; Tang, Y. Q. Self-assembly of a Novel Nanoscale Giant Cluster: $[Mo_{176}O_{496}(OH)_{32}(H_2O)_{80}]$. *Chem. Commun.* 1998, 1937-1938.

[8] Liu, G.; Wei, Y. G.; Yu, Q.; Liu, Q.; Zhang, S. W. Polyoxometalate Chain-like Polymer: the Synthesis and Crystal Structure of a 1D Compound, $[Mo_{36}O_{108}(NO)_4(MoO)_2La_2(H_2O)_{28}]_n \cdot 56nH_2O$. *Inorg. Chem. Commun.* 1999, 2, 434-437.

[9] Wei, Y. G..; Zhang, C.; Zhang, S. W.; Shao, M. C. A Spherical Nanascale Polyoxomolybdate Encapsulating Thirty Coordinated Butyrate: Synthesis and Structure Characterization of $(NH_4)_{42}[Mo_{132}O_{372}(CH_3CH_2CH_2COO)_{30}(H_2O)_{72}]$. Unpublished result.

[10] Müller, A.; Krickemeyer, E.; Meyer, J.; Bögge, H.; Peters, F.; Plass, W.; Diemann, E.; Dillinger, S.; Nonnenbruch, F.; Randerath, M.; Menke, C. $[Mo_{154}(NO)_{14}O_{420}(OH)_{28}(H_2O)_{70}]^{(25 \pm 5)-}$: A Water-Soluble Big Wheel with More than 700 Atoms and a Relative Molecular Mass of About 24000. *Angew. Chem. Int. Ed.* 1995, 34, 2122-2124.

[11] Müller, A.; Krickemeyer, E.; Bögge, H.; Schmidtmann, M.; Peters, F.; Menke, C.; Meyer, J. An Unusual Polyoxomolybdate: Giant Wheels Linked to Chains. *Angew. Chem. Int. Ed.* 1997, 36, 484-486.

[12] Müller, A.; Krickemeyer, E.; Bögge, H.; Schmidtmann, M.; Beugholt, C.; Kögerler, P.; Lu, C. Z. Formation of a Ring-Shaped Reduced "Metal Oxide" with the Simple Composition $[(MoO_3)_{176}(H_2O)_{80}H_{32}]$. *Angew. Chem. Int. Ed.* 1998, 37, 1220-1223.

[13] Müller, A.; Shah, S. Q. N.; Bögge, H.; Schmidtmann, M. Molecular Growth from a Mo-176 to a Mo-248 Cluster. *Nature.* 1999, 397, 48-50.

[14] Müller, A.; Serain C. Soluble Molybdenum Blues-"des Pudels Kern". *Acc. Chem. Res.* 2000, 33, 2-10.

[15] Müller, A.; Krickemeyer, E.; Bögge, H.; Schmidtmann, M.; Peters, F. Organizational Forms of Matter: An Inorganic Super Fullerene and Keplerate Based on Molybdenum Oxide. *Angew. Chem. Int. Ed.* 1998, 37, 3359-3363.

[16] Müller, A.; Sarkar, S.; Shah, S. Q. N.; Bögge, H.; Schmidtmann, M.; Sarkar, S.; Kögerler, P.; Hauptfleisch, B.; Trautwein, A. X.; Schünemann, V. Archimedean Synthesis and Magic Numbers: "Sizing" Giant Molybdenum-Oxide-Based Molecular Spheres of the Keplerate Type. *Angew. Chem. Int. Ed.* 1999, 38, 3238-3241.

[17] Müller, A.; Polarz, S.; Das, S. K.; Krickemeyer, E.; Bögge, H.; Schmidtmann, M.; Hauptfleisch, B. "Open and Shut" for Guests in Molybdenum-Oxide-Based Giant Spheres, Baskets, and Rings Containing the Pentagon as a Common Structural Element. *Angew. Chem. Int. Ed.* 1999, 38, 3241-3245.

[18] Müller, A.; Koop, M.; Bögge, H.; Schmidtmann, M.; Peters, F.; Kogerler, P. Building blocks as disposition in solution: $[\{Mo^{VI}O_3(H_2O)\}_{10}\{V^{IV}O(H_2O)\}_{20}\{(Mo^{VI}/Mo^{VI}{}_5O_{21})(H_2O)_3\}_{10}\ (\{Mo^{VI}O_2(H_2O)_2\}_{5/2})_2(\{NaSO_4\}_5)_2]^{20-}$, a giant spherical cluster with unusual structural features of interest for supramolecular and magneto chemistry. *Chem. Commun.* 1999, 1885-1886.

[19] Müller, A.; Shah, S. Q. N.; Bögge, H.; Schmidtmann, M.; Kogerler, P.; Hauptfleisch, B.; Leiding, S.; Wittler, K. Thirty Electrons "Trapped" in a Spherical Matrix: A Molybdenum Oxide-Based Nanostructured Keplerate Reduced by 36 Electrons. *Angew. Chem. Int. Ed.* 2000, 39, 1614-1616.

[20] Müller, A.; Das, S. K.; Bögge, H.; Schmidtmann, M.; Botar, A.; Patrut, A. Generation of Cluster Capsules (I_h) from Decomposition Products of a Smaller Cluster (Keggin-T_d) while Surviving Ones Get Encapsulated: Species with Core-shell Topology formed by a Fundamental Symmetry-driven Reaction. *Chem. Commun.* 2001, 657-658.

[21] Müller, A.; Kogerler P.; Dress A. W. M. Giant Metal-oxide-based Spheres and Their Topology: from Pentagonal Building Blocks to Keplerates and Unusual Spin Systems. *Coord. Chem. Rev.* 2001, 222, 193-218.

[22] Müller, A.; Beckmann, E.; Bögge, H.; Schmidtmann, M.; Dress A. Inorganic Chemistry Goes Protein Size: A Mo_{368} Nano-Hedgehog Initiating Nanochemistry by Symmetry Breaking. *Angew. Chem. Int. Ed.* 2002, 41, 1162-1167.

[23] Müller, A.; Roy S. En Route from the Mystery of Molybdenum Blue via Related Manipulatable Building Blocks to Aspects of Materials Science. *Coord. Chem. Rev.* 2003, 245, 153-166.

[24] Müller, A.; Kuhlmann C.; Bögge, H.; Schmidtmann M.; Baumann M.; Krickemeyer E. $Mo^V{}_2O_4{}^{2+}$ directs the formation and subsequent linking of potential building blocks under different boundary conditions: A related set of novel cyclic polyoxomolybdates. *Eur. J. Inorg. Chem.* 2001, (9), 2271-2277.

[25] Yamase T.; Prokop P. V. Photochemical Formation of Tire-shaped Molybdenum Blues: Topology of a Defect Anion, $[Mo_{142}O_{432}H_{28}(H_2O)_{58}]^{12-}$. *Angew. Chem. Int. Ed.* 2002, 41, 466-469.

[26] Yang, W. B.; Lu C. Z.; Lin X.; Zhuang H. H. A Novel Acetated 54-member Crown-shaped Polyoxomolybdate with Unprecedented Structural Features: $Na_{26}[\{Na(H_2O)_2\}_6(\{\mu_3\text{-}OH\}_4Mo^V{}_{20}Mo^{VI}{}_{34}O_{164}(\mu_2\text{-}CH_3COO)_4)]\cdot120H_2O$. *Chem. Commun.* 2000, 1623-1624.

[27] Yang, W. B.; Lu C. Z.; Lin X.; Zhuang H. H. Novel Acetate Polyoxomolybdate "Host" Accommodating a Zigzag-Chainlike "Guest" of Five Edge-Shared Sodium Cations: $Na_{21}\{[Na_5(H_2O)_{14}]\subset[Mo_{46}O_{134}(OH)_{10}(\mu-CH_3COO)_4]\}\cdot CH_3COONa\cdot 95H_2O$. *Inorg. Chem.* 2002, 41, 452-454.

[28] Wassermann K.; Dickman M. H.; Pope M. T. Self-Assembly of Supramolecular Polyoxometalates: The Compact, Water-Soluble Heteropolytungstate Anion $[As_{12}^{III}Ce_{16}^{III}(H_2O)_{36} W_{148}O_{524}]^{76-}$. *Angew. Chem. Int. Ed.* 1997, 36, 1445-1448.

[29] Liu T. B. Supramolecular Structures of Polyoxomolybdate-Based Giant Molecules in Aqueous Solution. *J. Am. Chem. Soc.* 2002, *124,* 10942-10943.

[30] Liu T. B.; Diemann E.; Li H. L.; Dress A. W. M., Müller, A. Self-assembly in Aqueous Solution of Wheel-shaped Mo_{154} Oxide Clusters into Vesicles. *Nature.* 2003, 426, 59-62.

[31] Liu T. B. An Unusually Slow Self-Assembly of Inorganic Ions in Dilute Aqueous Solution. *J. Am. Chem. Soc.* 2003, 125, 312-313.

[32] Liu, G.; Cai, Y. G.; Liu, T. B. Automatic and Subsequent Dissolution and Precipitation Process in Inorganic Macroionic Solutions. *J. Am. Chem. Soc.* 2004, 126, 16690-16691.

[33] Bredas, J. L. Silbey, R. Editors. *Conjugated Polymers.* Dordrecht/ The Netherlands: Kluwer; 1991.

[34] Nalwa, H. S. Editor. Handbook of Organic Conductive Molecules and Polymers I – IV. Chichester/ England: Wiley; 1997.

[35] Pope, M.; Swenberg, C. E. Editors. *Electronic Processes in Organic Crystals and Polymers.* 2nd edition. Oxford: Oxford University Press; 1999.

[36] Heeger, A. J. Semiconducting and Metallic Polymers: The Fourth Generation of Polymeric Materials. *J. Phys. Chem. B.* 2001, *105*, 8475-8491.

[37] Kraft, A.; Grimsdale, A. C.; Holmes A. B. Electroluminescent Conjugated Polymers - Seeing Polymers in a New Light. *Angew. Chem. Int. Ed.* 1998, 37, 402-428.

[38] Kelly, S. M. *Flat Panel Displays: Advanced Organic Materials.* Cambridge: Royal Society of Chemistry; 2002.

[39] Katz, H. E.; Bao, Z. The Physical Chemistry of Organic Field-Effect Transistors. *J. Phys .Chem. B.* 2000, 104, 671-678.

[40] Katz, H. E.; Bao, Z.; Gilat, S. L. Synthetic Chemistry for Ultrapure, Processable, and High-Mobility Organic Transistor Semiconductors. *Acc. Chem. Res.* 2001, 34, 359-369.

[41] Katz, H. E. Recent Advances in Semiconductor Performance and Printing Processes for Organic Transistor-Based Electronics. *Chem. Mater.* 2004, *16*, 4748-4756.

[42] Newman, C. R.; Frisbie, C. D.; da Silva Filho, D. A.; Bredas, J.-L.; Ewbank, P. C.; Mann, K. R. Introduction to Organic Thin Film Transistors and Design of n-Channel Organic Semiconductors. *Chem. Mater.* 2004, *16*, 4436-4451.

[43] Hide, F.; Diaz-Garcia, M. A.; Schwartz, B. J.; Heeger, A. J. New Developments in the Photonic Applications of Conjugated Polymers. *Acc. Chem. Res.* 1997, 30, 430-436.

[44] McGehee, M. D.; Heeger, A. J. Semiconducting (Conjugated) Polymers as Materials for Solid-State Lasers. *Adv. Mater.* 2000, 12, 1655-1668.

[45] Dagani, R. Nobel Prize in chemistry. *Chem. Eng. News.* 2000, 78(42), 4-4.

[46] Shirakawa H. The Discovery of Polyacetylene Film: The Dawning of an Era of Conducting Polymers (Nobel Lecture). *Angew. Chem. Int. Ed.* 2001, 40, 2574-2580.

[47] MacDiarmid, A. G. "Synthetic Metals": A Novel Role for Organic Polymers (Nobel Lecture). *Angew. Chem. Int. Ed.* 2001, 40, 2581-2590.

[48] Heeger, A. J. Semiconducting and Metallic Polymers: The Fourth Generation of Polymeric Materials (Nobel Lecture). *Angew. Chem. Int. Ed.* 2001, 40, 2591-2611.

[49] Pope, M. T.; Müller, A. Polyoxometalate Chemistry: An Old Field with New Dimensions in Several Disciplines. *Angew. Chem. Int. Ed.* 1991, 30, 34-48.

[50] Yamase T. Photo- and Electrochromism of Polyoxometalates and Related Materials. *Chem. Rev.* 1998, 98, 307-326.

[51] Sanchez, C.; Soler-Illia, G. J. A. A.; Ribot, F.; Lalot, T.; Mayer, C. R.; Cabuil, V. Designed Hybrid Organic-Inorganic Nanocomposites from Functional Nanobuilding Blocks. *Chem. Mater.* 2001, *13*, 3061-3083.

[52] Katsoulis D. E. A Survey of Applications of Polyoxometalates. *Chem. Rev.* 1998, 98, 359-388.

[53] Gouzerh, P. Proust, A. Main-Group Element, Organic, and Organometallic Derivatives of Polyoxometalates. *Chem. Rev.* 1998, 98, 77-112.

[54] Knoth, W. H. Harlow, R. L. Derivatives of heteropolyanions. 3. O-alkylation of $Mo_{12}PO_{40}^{3-}$ and $W_{12}PO_{40}^{3-}$. *J. Am. Chem. Soc.* 1981, *103*, 4265-4266.

[55] Knoth, W. H. Farlee, R. D. Esters of phosphotungstic acid. Anhydrous phosphotungstic acid. *Inorg. Chem.* 1984, 23, 4765-4766.

[56] Peng, Z. H. Rational Synthesis of Covalently Bonded Organic-Inorganic Hybrids. *Angew. Chem. Int. Ed.* 2004, 43, 930-935.

[57] Kang, H.; Zubieta, J. Co-ordination complexes of polyoxomolybdates with a hexanuclear core: synthesis and structural characterization of $(NBu^{n}_{4})_{2}[Mo_{6}O_{18}(NNMePh)]$, *Chemical Communications*, 1988, (17), 1192-1193.

[58] Du, Y.; Rheingold, A. L.; Maatta, E. A. A Polyoxometalate Incorporating an Organoimido Ligand: Preparation and Structure of $[Mo_{5}O_{18}(MoNC_{6}H_{4}CH_{3})]^{2-}$ *J. Am. Chem. Soc.* 1992, *114*, 345-346.

[59] Hur, N. H.; Klemperer, W. G.; Wang, R.-C. Tetrabutylammonium Hexamolybdate(VI) *Inorg. Synth.* 1990, 27, 77-78.

[60] Allcock, H. R.; Bissell, E. C.; Shawl, E. T. Crystal and molecular structure of a new hexamolybdate-cyclophosphazene complex. *Inorg. Chem.* 1973, 12, 2963-2967.

[61] Hsieh, T.- C.; Zubieta, J. A. Synthesis and Characterization of Oxomolybdate Clusters Containing Coordinatively Bound Organo-Diazenido Units: The Crystal and Molecular Structure of the Hexanuclear Diazenido-Oxomolybdate, $(NBu^{n}_{4})_{3}[Mo_{6}O_{18}(N_{2}C_{6}H_{5})]$ *Polyhedron.* 1986, 5, 1655-1657.

[62] Kwen, H.; Young, V. G., Jr.; Maatta, E. A. A Diazoalkane Derivative of a Polyoxometalate: Preparation and Structure of $[Mo_{6}O_{18}(NNC(C_{6}H_{4}OCH_{3})CH_{3})]^{2-}$ *Angew. Chem. Int. Ed.* 1999, 38, 1145-1146.

[63] Proust, A.; Thouvenot, R.; Chaussade, M.; Robert, F.; Gouzerh, P. Phenylimido Derivatives of $[Mo_{6}O_{19}]^{2-}$: Syntheses, X-ray Structures, Vibrational, Electrochemical, [95]Mo and ^{14}N NMR Studies. *Inorg. Chim. Acta* 1994, 224, 81-95.

[64] Errington, R. J.; Lax, C.; Richards, D. G.; Clegg, W.; Fraser, K. A. New Aspects of Non-Aqueous Polyoxometalate Chemistry: 3. Suface Oxide Reactivity. In: Pope, M. T.; Müller, A. Editors. *Polyoxometalates: From Platonic Solids to Anti-Retroviral Activity.* Dordrecht: Kluwer; 1994; p 112-113.

[65] Strong, J. B.; Ostrander, R.; Rheingold, A. L.; Maatta, E. A. Ensheathing a Polyoxometalate: Convenient Systematic Introduction of Organoimido Ligands at Terminal Oxo Sites in $[Mo_{6}O_{19}]^{2-}$. *J. Am. Chem. Soc.* 1994, 116, 3601-3602.

[66] Clegg, W.; Errington, R. J.; Fraser, K. A.; Holmes, S. A.; Schäfer, A. Functionalisation of $[Mo_6O_{19}]^{2-}$ with Aromatic Amine: Synthesis and Structure of a hexamolybdate Building Block with Linear Difunctionality. *Chem. Commun.* 1995, 455-456.

[67] Roesner, R. A.; McGrath, S. C.; Brockman, J. T.; Moll, J. D.; West, D. X.; Swearingen, J. K.; Castineiras, A. Mono- and di-functional aromatic amines with *p*-alkoxy substituents as novel arylimido ligands for the hexamolybdate ion. *Inorg. Chim. Acta.* 2003, 342, 37-47.

[68] Wei, Y. G.; Xu, B. B.; Barnes, C. L.; Peng, Z. H. An Efficient and Convenient Reaction Protocol to Organoimido Derivatives of Polyoxometalates. *J. Am. Chem. Soc.* 2001, *123*, 4083-4084.

[69] Stark, J. L.; Rheingold, A. L.; Maatta, E. A. Polyoxometalate clusters as building blocks: preparation and structure of bis(hexamolybdate) complexes covalently bridged by organodiimido ligands. *Chem. Commun.* 1995, 1165-1166.

[70] Stark, J. L.; Young, V. G., Jr.; Maatta, E. A Functionalized Polyoxometalate Bearing a Ferrocenylimido Ligand: Preparation and Structure of $[(FcN)Mo_6O_{18}]^{2-}$. A. *Angew. Chem. Int. Ed.* 1995, 34, 2547-2548.

[71] Strong, J. B.; Yap, G. P. A.; Ostrander, R.; Liable-Sands, L. M.; Rheingold, A. L.; Thouvenot, R.; Gouzerh, P.; Maatta, E. A. A New Class of Functionalized Polyoxometalates: Synthetic, Structural, Spectroscopic, and Electrochemical Studies of Organoimido Derivatives of $[Mo_6O_{19}]^{2-}$. *J. Am. Chem. Soc.* 2000, *122*, 639-649.

[72] Moore, A. R.; Kwen, H.; Beatty, A. M.; Maatta, E. A. Organoimido-polyoxometalates as polymer pendants. *Chem. Commun.* 2000, 1793-1794.

[73] Xu, B. B.; Peng, Z. H.; Wei, Y. G.; Powell, D. R. Polyoxometalates covalently bonded with terpyridine ligands. *Chem. Commun.* 2003, 2562-2563.

[74] Zhu, Y.; Xiao, Z. C.; Ge, N.; Wang, N.; Wei, Y. G.; Wang, Y. Naphthyl amines as novel organoimido ligands for design of POM-based organic-inorganic hybrids: Synthesis, structural characterization, and supramolecular assembly of $(Bu_4N)_2[Mo_6O_{18}(NC_{10}H_7-1)]$. *Cryst Growth. & Des.* 2006, *6*, 1620-1625.

[75] Wu, P. F.; Li, Q.; Ge, N.; Wei, Y. G.; Wang, Y.; Wang, P.; Guo, H. Y. An Easy Routine to Monofunctionalized Organoimido Derivatives of the Lindqvist Hexamolybdate. *Eur. J. Inorg. Chem.* 2004, 2819-2822.

[76] Li, Q.; Wu, P. F.; Wei, Y. G.; Xia, Y.; Wang, Y.; Guo, H. Y. Organic-Inorganic Hybrids: Preparation and Structural Characterization of $(Bu_4N)_2[Mo_6O_{17}(NAr)_2]$ and $(Bu_4N)_2[Mo_6O_{18}(NAr)]$ (Ar = o-$CH_3C_6H_4$). *Z. Anorg. Allg. Chem.* 2005, *631*, 773-779.

[77] Li, Q.; Wu, P. F.; Xia, Y.; Wei, Y. G.; Guo, H. Y. Synthesis, spectroscopic studies and crystal structure of a polyoxoanion cluster incorporating para-bromophenylimido ligand, $(Bu_4N)_2[Mo_6O_{18}(NC_6H_4Br-p)]$. *J. Organomet. Chem.* 2005, *691*, 1223-1228.

[78] Xue, S. J.; Ke, S. Y.; Yan, L.; Cai, Z. J.; Wei, Y. G. A trifluoromethyl substituted organoimido derivative of the hexametalate cluster: Synthesis, crystal structure and bioactivity of $[Mo_6O_{17}(NAr)_2]^{2-}$ (Ar = o-$CF_3C_6H_4$). *J. Inorg. Biochem.* 2005, 99, 2276-2281.

[79] Xu, L.; Lu, M.; Xu, B. B.; Wei, Y. G.; Peng, Z. H.; Powell, D. R. Towards Main-Chain-Polyoxometalate-Containing Hybrid Polymers: A Highly Efficient Approach to Bifunctionalized Organoimido Derivatives of Hexamolybdates. *Angew. Chem. Int. Ed.* 2002, 41, 4129-4132.

[80] Li, Q.; Wu, P. F.; Wei, Y. G.; Wang, Y.; Wang, P.; Guo, H. Y. Synthesis, structure and supramolecular assembly in the crystalline state of a bifunctionalized arylimido derivative of hexamolybdate. *Inorg.Chem. Commun.* 2004, 7, 524-527.

[81] Lu, M.; Xie, B.; Kang, J.; Chen, F.-C.; Yang, Y.; Peng, Z. H. Synthesis of Main-Chain Polyoxometalate-Containing Hybrid Polymers and Their Applications in Photovoltaic Cells. *Chem. Mater.* 2005, *17*, 402-408.

[82] Xia, Y.; Wu, P. F.; Wei, Y. G.; Wang, Y.; Guo, H. Y. Synthesis, crystal structure, and optical properties of a polyoxometalate-based inorganic-organic hybrid solid, (n-Bu$_4$N)$_2$[Mo$_6$O$_{17}$($\equiv$NAr)$_2$] (Ar = o-CH$_3$OC$_6$H$_4$). *Cryst Growth. & Des.* 2006, *6*, 253-257.

[83] Xia, Y.; Wei, Y. G.; Wang, Y.; Guo, H. Y. A kinetically controlled trans bifunctionalized organoimido derivative of the Lindqvist-type hexamolybdate: Synthesis, spectroscopic characterization, and crystal structure of (n-Bu$_4$N)$_2$[Mo$_6$O$_{17}$($\equiv$NAr)$_2$] (Ar = 2,6-(CH$_3$)$_2$C$_6$H$_4$). *Inorg. Chem.* 2001, *44*, 9823-9828.

[84] Strong, J. B.; Haggerty, B. S.; Rheingold, A. L.; Maatta, E. A. A Superoctahedral Complex Derived from a Polyoxometalate: the Hexakis(arylimido)hexamolybdate Anion [Mo$_6$(NAr)$_6$O$_{13}$H]$^-$. *Chem. Commun.* 1997, 1137-1138.

[85] Xia, Y.; Wu, P. F.; Wei, Y. G.; Guo, H. Y. The Unprecedent Bridging Oxo Replacement in a Trifunctionalized Organoimido Derivative of the Hexmolybate, [Mo$_6$O$_{16}$($\equiv$NAr)$_2$(μ_2-NAr)]$^{2-}$ (Ar = 2,6-(CH$_3$)$_2$C$_6$H$_4$). Submitted to publication.

[86] Johnson, A. W. *Ylides and Imines of Phosphorus*. Wiley: New York; 1993.

[87] Saunders, J. H.; Slocombe, R. J. The Chemistry of the Organic Isocyanates. *Chem. Rev.* 1948, *43*, 203-218.

[88] Patrick G. Section J– Aldehydes and Ketones: J6. Nucleophilic Addition– Nitrogen Nucleophiles. *Instant Notes in Organic Chemistry*. Reprint Edition; BIOS Scientific Publishers Limited: Beijing; 2002; p 183-184.

[89] Williams, A.; Ibrahim, I. T. Carbodiimide Chemistry: Recent Advances. *Chem. Rev.* 1981, *81*, 589-636.

[90] Kurzer, F.; Douraghi-Zadeh, K. Advances in the Chemistry of Carbodiimides. *Chem. Rev.* 1967, *67*, 107-152.

[91] Khorana, H. G. the Chemistry of Carbodiimides. *Chem. Rev.* 1953, *53*, 145-166.

[92] Nugent, W.; Mayer, J. E. *Metal-Ligand Multiple Bonds*. Wiley: New York; 1988; p122-125.

[93] Yan, L.-K.; Su, Z.-M.; Guan, W.; Zhang, M.; Chen, G.-H.; Xu, L.; Wang, E.-B. Why Does Disubstituted Hexamolybdate with Arylimido Prefer to Form an Orthogonal Derivative? Analysis of Stability, Bonding Character, and Electronic Properties on Molybdate Derivatives by Density Functional Theory (DFT) Study. *J. Phys. Chem. B.* 2004, *108*, 17337-17343.

[94] Mohs, T. R.; Yap, G. P. A.; Rheingold, A. L.; Maatta, E. A. An Organoimido Derivative of the Hexatungstate Cluster: Preparation and Structure of [W$_6$O$_{18}$(NAr)]$^{2-}$ (Ar = 2,6-(i-Pr)$_2$C$_6$H$_3$). *Inorg. Chem.* 1995, *34*, 9-10.

[95] Wei, Y. G.; Lu, M.; Cheung, C. F.- C.; Barnes, C. L.; Peng, Z. H. Functionalization of [MoW$_5$O$_{19}$]$^{2-}$ with Aromatic Amines: Synthesis of the First Arylimido Derivatives of Mixed-Metal Polyoxometalates. *Inorg. Chem.* 2001, *40*, 5489-5490.

[96] Proust, A.; Taunier, S.; Artero, V.; Robert, F.; Thouvenot, R.; Gouzerh, P. The Unexpected Reactivity of *p*-Tolylisocyanate towards the Keggin Anion α-[PMo$_{12}$O$_{40}$]$^{3-}$. *Chem. Commun.* 1996, 2195 – 2196.

[97] Dablemont, C.; Proust, A.; Thouvenot, R.; Afonso, C.; Fournier, F.; Tabet, J.-C. Functionalization of Polyoxometalates: From Lindqvist to Keggin Derivatives. 1. Synthesis, Solution Studies, and Spectroscopic and ESI Mass Spectrometry Characterization of the Rhenium Phenylimido Tungstophosphate $[PW_{11}O_{39}\{ReNC_6H_5\}]^{4-}$. *Inorg. Chem.* 2004, *43*, 3514-3520.

[98] Xu, B. B.; Wei, Y. G.; Barnes, C. L.; Peng, Z. H. Hybrid Molecular Materials Based on Covalently Linked Inorganic Polyoxometalates and Organic Conjugated Systems. *Angew. Chem. Int. Ed.* 2001, 40, 2290-2292.

[99] Lu, M.; Wei, Y. G.; Xu, B. B.; Cheung, C. F.- C.; Peng, Z. H.; Powell, D. R. Hybrid Molecular Dumbbells: Bridging Polyoxometalate Clusters with an Organic π-Conjugated Rod. *Angew. Chem. Int. Ed.* 2002, 41, 1566-1568.

[100] Kang, J.; Nelson, J. A.; Lu, M.; Xie, B.; Peng, Z. H.; Powell, D. R. Charge-Transfer Hybrids Containing Covalently Bonded Polyoxometalates and Ferrocenyl Units. *Inorg. Chem.* 2004, *43*, 6408-6413.

[101] Sonogashira, K.; Tohda, Y.; Hagihara, N. Convenient synthesis of acetylenes. Catalytic substitutions of acetylenic hydrogen with bromo alkenes, iodo arenes, and bromopyridines. *Tetrahedron Lett.* 1975, 50, 4467-4470.

[102] deSousa, P. T. The Palladium-catalysed Cross Coupling Reaction of Acetylenic Compounds with Aryl Halides and Related Compounds. *Quimica Nova.* 1996, 19, 377-382.

[103] Wang, Y. F.; Deng, W.; Liu, L.; Guo, Q. X. Recent Progress in Sonogashira Reaction. *Chin. J. Org. Chem.* 2005, 25, 8-24.

[104] Ziener, U.; Godt, A. Synthesis and Characterization of Monodisperse Oligo(phenyleneethynylene)s. *J. Org. Chem.* 1997, 62, 6137-6143.

[105] Sadakane, M.; Steckhan, E. Electrochemical Properties of Polyoxometalates as Electrocatalysts. *Chem. Rev.* 1998, 98, 219-238.

[106] Coronado, E.; Gómez-García, C. J. Polyoxometalate-Based Molecular Materials. *Chem. Rev.* 1998, 98, 273-296.

[107] Gust, D.; Moore, T. A.; Moore, A. L. Molecular Mimicry of Photosynthetic Energy and Electron Transfer. *Acc. Chem. Res.* 1993, 26, 198-205.

[108] Wasielewski, M. R. Photoinduced Electron Transfer in Supramolecular Systems for Artificial Photosynthesis. *Chem. Rev.* 1992, 92, 435-461.

[109] Weiss, E. A.; Ahrens, M. J.; Sinks, L. E.; Gusev, A. V.; Ratner, M. A.; Wasielewski, M. R. Making a Molecular Wire: Charge and Spin Transport through *para*-Phenylene Oligomers. *J. Am. Chem. Soc.* 2004, *126*, 5577-5584.

[110] Xu, B. B.; Lu, M.; Kang, J.; Wang, D. G.; Brown, J.; Peng, Z. H. Synthesis and Optical Properties of Conjugated Polymers Containing Polyoxometalate Clusters as Side-Chain Pendants. *Chem. Mater.* 2005, *17*, 2841-2851.

INDEX

B

D

G

H

P

T

U

V

W